st ate belo

CIRCUIT SIMULATION

CIRCUIT SIMULATION

Farid N. Najm

A JOHN WILEY & SONS, INC., PUBLICATION

For general information on our other products and services or for technical support, please contact our Customer Care Department within the United States at (800) 762-2974, outside the United States at (317) 572-3993 or fax (317) 572-4002.

Wiley also publishes its books in a variety of electronic formats. Some content that appears in print may not be available in electronic formats. For more information about Wiley products, visit our web site at www.wiley.com.

Library of Congress Cataloging-in-Publication Data:

Najm, Farid N.
   Circuit simulation / Farid N. Najm.
      p. cm.
   Includes bibliographical references and index.
   ISBN 978-0-470-53871-5
  1.   Electronic circuits–Computer simulation. 2.   Electronic circuits–Mathematical models.
3.   Integrated circuits–Computer simulation. I. Title.
   TK7867.N33 2010
   621.381501'13–dc22

                                                                      2009022673

Printed in the United States of America.

10 9 8 7 6 5 4 3 2 1

To my wife, *Diana,*
and to our two daughters,
Lily Marie and *Tanya Kristen.*

■■■■ CONTENTS

This text describes in detail the numerical techniques and algorithms that are part of modern circuit simulators, with a focus on the most commonly used simulation modes: DC Analysis and Transient Analysis. After a general introduction in chapter 1, network equation formulation is covered in chapter 2, with emphasis on modified nodal analysis (MNA). The coverage also includes the network cycle space and bond space, element stamps, and the question of unique solvability of the system. Solving linear resistive circuits is the focus of chapter 3, which gives a comprehensive treatment of the most relevant aspects of linear system solution techniques. This includes the standard methods of Gaussian elimination (GE) and *LU* factorization, as well as some in-depth treatment of numerical error in floating point systems, pivoting for accuracy, sparse matrix methods, and pivoting for sparsity. Indirect solution methods, such as Gauss-Jacobi (GJ) and Gauss-Seidel (GS) are also covered. As well, some discussion of node tearing and partitioning is given, in recognition of the recent trend of increased usage of parallel software on multi-core computers.

Solving nonlinear resistive circuits is covered in chapter 4, with a focus on Newton's method. A detailed study is given of Newton's method, including its links to the fixed point method and the conditions that govern its convergence. A rigorous treatment is then provided of how this method applies to circuit simulation, leading up to the notion of companion models for nonlinear resistive elements, with coverage of multiterminal elements. As well, a coverage of quasi-Newton methods in simulation is provided, which includes the three commonly used homotopy methods for DC Analysis: source stepping, Gmin stepping, and pseudo-transient. Simulation of dynamic circuits, both linear and nonlinear, is covered in chapter 5. This chapter gives a detailed treatment of methods for solving ordinary differential equations (ODEs), with a focus on those methods that have been found useful for circuit simulation. Issues of accuracy and stability of linear multistep methods are covered in some depth. These methods are then applied to circuit simulation, illustrating how the companion models of dynamic elements are derived. Here too, multiterminal elements are addressed, as well as other advanced topics of time-step control, variable time-step, charge conservation, and the use of charge-based models in simulation.

My aim throughout has been to produce a text that has two key features: 1) sufficient depth and breadth so that it can be used in a graduate course on the topic, and 2) enough detail so as to allow the reader to write his/her own basic circuit simulator. I hope that I have succeeded. Indeed, the book has already been tested for this dual purpose, as I have used it to teach a graduate course on circuit simulation at the University of Toronto. As part of this course, students write a rudimentary circuit simulator, in a sequence of five computer projects, all of which are included in the problem sets in this text. The first project is simply to develop a parser; the second is to develop code that builds the MNA system of equations for any linear resistive circuit, using element stamps; the third requires the implementation of an *LU* factorization capability to solve the MNA system. The fourth project implements a Newton loop around the MNA solver, allowing the simulation of basic nonlinear resistive circuits. The fifth and final project builds a time-domain simulation loop around the Newton loop, using the trapezoidal rule. The result is a basic simulator that can simulate circuits containing MOSFETs, BJTs, and diodes (using the simplest first-order models for these devices), along with the standard linear elements. With problem sets and computer projects at the end of every chapter, this text is suitable as the main textbook for a course on the topic. As well, the text has sufficient depth that I hope it would serve as a reference for practicing design engineers and computer-aided design practitioners.

Throughout the text, detailed coverage is given of the mathematical and numerical techniques that are the basis for the various simulation topics. Such a theoretical background is important, I feel, for a full understanding of the practical simulation techniques. However, this theoretical background is given piecemeal, as the need arises, and is never presented as an end in itself; it is scattered throughout the text and paired up with the various simulation topics. Furthermore, and in order to maintain the focus on the end-goal of practical simulation methods, I have found it necessary to *state all theorems without proof*. Ample references are provided, however, which the interested reader can consult for a deeper study.

Finally, the reader is encouraged to consult the following web site, where I hope to maintain various resources that are relevant to this book, including an up-to-date list of any known errors:

`http://www.eecg.utoronto.ca/~najm/simbook`

ACKNOWLEDGMENTS

I owe much gratitude to Professors Ibrahim N. Hajj and Vasant B. Rao, then at the University of Illinois, who taught me much of what I know about circuit simulation. Ibrahim was also kind enough to provide his extensive set of notes on the topic, which were very helpful as, many years later, I contemplated teaching a course on this topic. It has taken a lot more reading and discussions with many other colleagues to help bring this book project to fruition. I would especially like to thank my friends from the Texas Instruments circuit simulation group, Lawrence A. Arledge and David C. Yeh, for extensive discussions over the last two years. Thanks as well to my friend John F. Croix for answering my many questions on time-step control and transient simulation. Thanks to my graduate students, Khaled R. Heloue, Nahi H. Abdul Ghani, and Sari Onaissi, for their help in proof-reading parts of the manuscript and in helping to develop the computer projects. Last but not least, I owe a big *thank you* to my wife, Diana Tawil Najm, for her patience over the last two years, during which I was virtually "absent" as I repeatedly revised the manuscript. Without her support, this would not have been possible.

Farid N. Najm
January 2010
Toronto, Canada

CHAPTER 1

Introduction

Circuit simulation is a technique for checking and verifying the design of electrical and electronic circuits and systems prior to manufacturing and deployment. It is used across a wide spectrum of applications, ranging from integrated circuits and microelectronics to electrical power distribution networks and power electronics. Circuit simulation is a mature and established art and also remains an important area of research. This text covers the theoretical background for circuit simulation, as well as the numerical techniques that are at the core of modern circuit simulators. Circuit simulation combines a) mathematical modeling of the circuit *elements*, or *devices*, b) formulation of the circuit/network equations, and c) techniques for solution of these equations. We will focus mainly on the formulation and solution of the network equations and will not cover device modeling in any detail.

Compared to simulators that operate on a design description at higher levels of abstraction, such as logic or functional simulators, circuit simulators use a *detailed* (so-called *circuit level* or *transistor level*) description of the circuit and perform a relatively *accurate* simulation. Typically, such a simulation uses physical models of the circuit elements, solves the resulting differential and algebraic equations, and generates time-waveforms of node voltages and element currents. In general, a circuit simulator allows the use of any simulation primitive, provided it can be described with an appropriate *device model*. In practice, while some devices (e.g., resistors, capacitors) are two-terminal elements, others (e.g., transistors) have more than two terminals. However, a multiterminal element is usually modeled as a subcircuit consisting of only two-terminal elements. Thus, a common starting point for studying circuit simulation is to restrict the formulation to two-terminal elements. For integrated circuits, circuit simulators often work with an *extracted* circuit description, which gives better accuracy. Thus circuit capacitance, resistance, and inductance can be included in the analysis, be they prespecified discrete components or parasitic.

Early techniques for circuit simulation using computers were introduced in the 1950s and 1960s, and several limited-scope simulators were developed. The first general-purpose circuit simulator to experience widespread usage was SPICE, developed by L. W. Nagel at the University of California, Berkeley, in the early

Circuit Simulation, by Farid N. Najm
Copyright © 2010 John Wiley & Sons, Inc.

1970s, under the supervision of Professor D. O. Pederson. According to Nagel (1975), SPICE evolved from an earlier simulator called CANCER that, in turn, *"emerged from a series of courses that were instructed by Professor R. A. Rohrer"* at Berkeley. The original CANCER program, described in Nagel and Rohrer (1971), was followed up by SPICE, described in Nagel and Pederson (1973), and then SPICE2, which is described in Nagel's 1975 thesis. The rest, as they say, is history, as SPICE has become the *de facto* standard circuit simulator. At the time of this writing, SPICE3f is the latest version of the program. Further information on SPICE and its history is available in Kundert (1995), in Vladimirescu (1994), and in Pederson (1984).

SPICE was developed specifically with integrated circuits in mind. Indeed, the acronym stands for *Simulation Program with Integrated Circuit Emphasis*. This, coinciding with the birth and growth of the integrated circuits industry in the 1970s, led to widespread adoption of the program. Furthermore, the availability of the SPICE code and documentation from Berkeley, for a nominal fee, spurred the development of similar circuit simulators elsewhere. Today, there are thousands of copies of circuit simulators in use across the industry, and there are many SPICE-like simulators in the market. Some semiconductor companies average over 1 million circuit simulation runs per week!

Circuit simulation issues to be covered in this book include a) device equations, b) equation formulation, and c) solution techniques. In this chapter, we will briefly introduce each of these issues and then present the overall structure of a circuit simulator.

1.1   DEVICE EQUATIONS

In this context, a *device* is any simulation *primitive*, or *element*, described by means of a current-voltage relationship. Thus, a resistor is described by the (Ohm's law) equation $v = Ri$. By convention, the positive (reference) direction for current is determined from the positive (reference) direction for voltage, as shown in Fig. 1.1, so that current flows in the device from the positive reference node to the negative node.

When the element equation contains no terms with powers of 2 or higher, the element is said to be a *linear element*. A network of linear elements is said to be a *linear circuit*. The resistor element equation is algebraic. On the other hand, a capacitor is described by $i = Cdv/dt$, which is a *first-order ordinary differential*

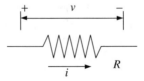

Figure 1.1: A resistor.

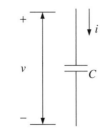

Figure 1.2: A capacitor.

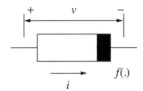

Figure 1.3: Nonlinear resistor.

equation. It is first order because it contains only first-order derivatives and it is ordinary because it contains no partial derivatives. Since there are no powers of 2 or more, this too is a linear equation. Here too, the reference direction for current is based on the reference direction for voltage, as in Fig. 1.2.

Wiring is typically modeled using lumped R, L, or C elements, so that metal interconnect is described by a system of *linear first-order differential equations*. Resistors and capacitors are examples of *two-terminal* linear devices. In general, a two-terminal device may be described by an i-v equation $i = f(v)$, where f can be any function $f : \mathbb{R} \to \mathbb{R}$. When f is a nonlinear function, the device is said to be *nonlinear* and is given the (nonlinear resistor) symbol shown in Fig. 1.3.

A pn-junction diode is an example of a commonly used two-terminal nonlinear device. Transistors, such as BJTs and MOSFETs are three-terminal nonlinear devices (four-terminal, if a detailed model is used that includes the body voltage).

1.2 EQUATION FORMULATION

The behavior of a circuit is captured by a set of equations that are formulated by combining the element equations and Kirchoff's Current and Voltage Laws (KCL and KVL). In general, this results in a set of simultaneous *nonlinear first-order differential equations*. For a purely *resistive*, *linear*, circuit, the equations are simply a system of simultaneous linear algebraic equations.

As an example, consider the linear circuit shown in Fig. 1.4. From KCL, we can write:

$$i = i_1 = i_2 \tag{1.1}$$

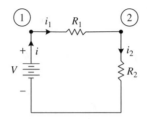

Figure 1.4: A simple linear circuit.

The element equations provide:

$$v_1 = V, \quad i_1 = \frac{v_1 - v_2}{R_1}, \quad i_2 = \frac{v_2}{R_2} \tag{1.2}$$

which, substituted into KCL ($i_1 = i_2$), gives:

$$\frac{1}{R_1}(v_1 - v_2) = \frac{1}{R_2}v_2 \tag{1.3}$$

KVL around the loop then provides:

$$V = R_1 i_1 + R_2 i_2 = (R_1 + R_2)i = (R_1 + R_2)\frac{v_2}{R_2} \tag{1.4}$$

where, in the last step, we have benefited from KCL ($i = i_2$) and the element equation $i_2 = v_2/R_2$, and this then leads to:

$$v_2 = \frac{R_2}{R_1 + R_2}V \tag{1.5}$$

With v_2 in hand then, using (1.3), we get the value of v_1, and the element equation $i_2 = v_2/R_2$ can then be used to solve for $i = i_1 = i_2$.

The above *ad hoc* approach of solving equations by substitution and similar operations does not scale well to large circuits. Instead, we need a *systematic* and *automatic* approach for formulating and solving the circuit equations. For now, we maintain our focus on the simple case of linear resistive circuits. There are two popular approaches for systematic equation formulation, *sparse tableau analysis* (STA) and *modified nodal analysis* (MNA). Sparse tableau analysis, described in Hachtel et al. (1971), involves the following steps:

1. Write KCL as $Ai = 0$, where A is a *reduced incidence matrix* that we will introduce later on, and i is a vector of all branch currents.
2. Write KVL as $u = A^T v$, where u is a vector of all branch voltages and v a vector of all nodal voltages to ground.
3. Write the element equations as $Zi + Yu = s$, where Z and Y are matrices and s is a vector.

The combination of these three sets of linear algebraic equations leads to the sparse tableau system:

$$\begin{bmatrix} A & 0 & 0 \\ 0 & I & -A^T \\ Z & Y & 0 \end{bmatrix} \begin{bmatrix} i \\ u \\ v \end{bmatrix} = \begin{bmatrix} 0 \\ 0 \\ s \end{bmatrix} \tag{1.6}$$

This formulation has some key features in that it can be applied to *any* circuit in a systematic fashion, the equations can be assembled *directly* from the input (circuit specification), as we will see later on, and the coefficients matrix is *very* sparse (has mostly zero elements), although it is larger in dimension than the MNA matrix. Modified nodal analysis, described in Ho et al. (1975), involves the following steps:

1. Write KCL as $Ai = 0$.
2. Use the element equations to eliminate as many current variables as possible from KCL, leading to equations in terms of mostly branch voltages.
3. Use KVL to replace all the branch voltages by nodal voltages to ground.
4. Append element equations of those elements whose current variables could not be eliminated as additional equations of the MNA system.

We will see the details of this process later on, and it leads to the MNA system:

$$\begin{bmatrix} Y & B \\ C & Z \end{bmatrix} \begin{bmatrix} v \\ i \end{bmatrix} = \begin{bmatrix} s_v \\ s_i \end{bmatrix} \tag{1.7}$$

As with STA, this formulation can be applied to *any* circuit in a systematic fashion and the equations can be assembled *directly* from the input (circuit specification). As well, the coefficient matrix is sparse, but often not as sparse as the STA matrix, although it is smaller in dimension. The MNA matrix can become singular during the numerical solution process and, therefore, requires careful pivoting.

With the larger matrix size, STA can take longer to formulate the equations than MNA, but it solves them faster; it is well suited for repeated use as in statistical analysis. Most modern circuit simulators use the MNA approach.

As an example of the MNA formulation, consider the linear resistive circuit in Fig. 1.5. We write KCL at every node and then eliminate the current variables using the element equations, as follows:

$$\text{KCL at node 2:} \quad \frac{V - v_2}{R_1} = \frac{v_2 - v_3}{R_2} \quad \Rightarrow \quad \left(\frac{1}{R_1} + \frac{1}{R_2} \right) v_2 - \frac{1}{R_2} v_3 = \frac{V}{R_1}$$

$$\text{KCL at node 3:} \quad \frac{v_2 - v_3}{R_2} = \frac{v_3}{R_3} \quad \Rightarrow \quad -\frac{1}{R_2} v_2 + \left(\frac{1}{R_2} + \frac{1}{R_3} \right) v_3 = 0$$

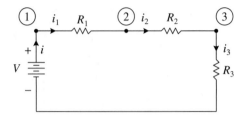

Figure 1.5: A linear circuit, used to demonstrate the MNA formulation.

This leads to the MNA matrix equation:

$$\left[\begin{array}{cc} \left(\dfrac{1}{R_1}+\dfrac{1}{R_2}\right) & \dfrac{-1}{R_2} \\[2ex] \dfrac{-1}{R_2} & \left(\dfrac{1}{R_2}+\dfrac{1}{R_3}\right) \end{array}\right] \left[\begin{array}{c} v_2 \\ v_3 \end{array}\right] = \left[\begin{array}{c} \dfrac{V}{R_1} \\[2ex] 0 \end{array}\right] \tag{1.8}$$

Notice that the system matrix can be written as the sum of three matrices:

$$\left[\begin{array}{cc} \left(\dfrac{1}{R_1}+\dfrac{1}{R_2}\right) & \dfrac{-1}{R_2} \\[2ex] \dfrac{-1}{R_2} & \left(\dfrac{1}{R_2}+\dfrac{1}{R_3}\right) \end{array}\right] = \left[\begin{array}{cc} \dfrac{1}{R_1} & 0 \\[1ex] 0 & 0 \end{array}\right] + \left[\begin{array}{cc} \dfrac{1}{R_2} & \dfrac{-1}{R_2} \\[1ex] \dfrac{-1}{R_2} & \dfrac{1}{R_2} \end{array}\right] + \left[\begin{array}{cc} 0 & 0 \\[1ex] 0 & \dfrac{1}{R_3} \end{array}\right] \tag{1.9}$$

each of which relates to a specific element. These contributions of the various elements are called *element stamps*, as we will see later on. In general, a resistor like R_2, which is not connected to ground, has the following element stamp:

$$
\begin{array}{ccccc}
 & v^+ & & v^- & \\
 & \vdots & & \vdots & \\
n^+ & \cdots \;\; +G_2 & \cdots & -G_2 & \cdots \\
 & \vdots & & \vdots & \\
n^- & \cdots \;\; -G_2 & \cdots & +G_2 & \cdots \\
 & \vdots & & \vdots &
\end{array}
$$

where $G_2 = 1/R_2$. This and similar element stamps are used to directly build the required MNA matrix as the simulator is reading (parsing) the circuit description file.

1.3 SOLUTION TECHNIQUES

As seen in the above examples, such as in (1.8), solving linear resistive circuits reduces to solving the *linear system*:

$$Ax = b \tag{1.10}$$

This is a classical problem that is basic to many engineering disciplines and has a variety of solution techniques. Direct methods of solution include matrix inversion, Gaussian elimination, and LU factorization. Indirect methods (relaxation methods) of solution include Gauss-Jacobi and Gauss-Seidel, successive over-relaxation, etc. The most common method is LU factorization, which proceeds as follows:

1. Factor A as $A = LU$, where L is lower-diagonal and U is upper-diagonal.
2. Solve $Lz = b$ for z, by forward substitution.
3. Solve $Ux = z$ for x, by backward substitution.

We will see the details of this process later on and we will recognize that a most desirable property throughout all this is *matrix sparsity*.

1.3.1 Nonlinear Circuits

Solving nonlinear circuits is typically done using *Newton's method*. We will see that this means that we repeatedly, until convergence, perform the following two steps:

1. *Linearize* the circuit equations around a candidate solution point.
2. *Solve* the resulting linear circuit using LU factorization to discover a better solution point.

In this way, the MNA formulation for linear resistive circuits turns out to be *sufficient*, because the solution of a nonlinear circuit is reduced to repeated solutions of linearized versions of that circuit.

As an example of the process of linearization around a candidate solution point, consider a nonlinear resistor with the element equation $i = f(v)$, as depicted in Fig. 1.6. The equation of the tangent line at the point (v_0, i_0) is:

$$i = f'(v_0)[v - v_0] + i_0 = f'(v_0)v + I_{eq} \qquad (1.11)$$

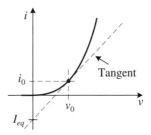

Figure 1.6: An i-v characteristic for a nonlinear resistor, showing linearization around a candidate solution point.

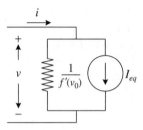

Figure 1.7: A linear circuit that has the same current–voltage characteristic as the tangent line in Fig. 1.6.

where $I_{eq} = i_0 - f'(v_0)v_0$. This equation is also the i-v characteristic of the sub-circuit shown in Fig. 1.7. This sub-circuit is called a *companion model*, of the nonlinear resistor. The element stamp of the companion model is then used to build the matrix of the linearized circuit, and the resulting linear system is solved. This process is repeated until the successive candidate solution points have converged to their final value.

1.3.2 Dynamic Circuits

All the preceding has been for resistive circuits and we have focused on the solution at a single point in time, a so-called DC Analysis. In general, circuits include dynamic (L, C) elements, and we are interested in the response over time, a so-called Transient Analysis. This is done by using a *finite difference approximation* of the derivative, such as:

$$i = C\frac{dv}{dt} \approx C\frac{(v(t + \Delta t) - v(t))}{\Delta t} \tag{1.12}$$

By replacing all derivatives by their finite difference approximations, the circuit equations effectively become *algebraic*, rather than differential, and possibly nonlinear. Given the solution at time t, i.e., $v(t)$ and $i(t)$, these equations are then solved for $v(t + \Delta t)$ and $i(t + \Delta t)$. Thus, by this operation of *time discretization*, the problem is reduced to solving a possibly nonlinear resistive network at every time-point, based on the use of another kind of *companion model* for the dynamic elements.

1.4 CIRCUIT SIMULATION FLOW

The flow-chart shown in Fig. 1.8 is useful to visualize the overall simulation flow inside a circuit simulator. The simulator repeatedly applies time discretization, element linearization, and matrix equation solution. In the following chapters, we will describe the many details, and pitfalls, of these various activities.

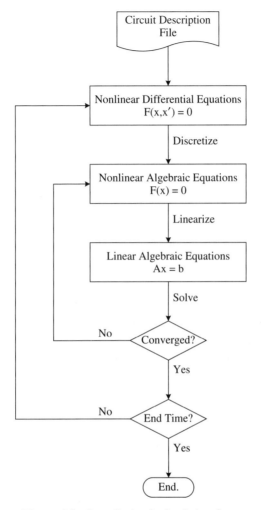

Figure 1.8: Overall circuit simulation flow.

1.4.1   Analysis Modes

Circuit simulators offer different *analysis modes*. Berkeley's *The Spice Page*, at the web site:

`http://bwrc.eecs.berkeley.edu/Classes/IcBook/SPICE`

lists the following analysis modes for `SPICE3`:

- `DC Analysis`: Determines the DC operating point of the circuit with inductors shorted and capacitors opened. It is automatically performed prior to a `Transient Analysis`, to determine the initial conditions, and prior to an

AC Small-Signal Analysis, to determine the linearized, small-signal models for nonlinear devices. It can also be used to get the DC transfer curves by means of a DC sweep.

- Transient Analysis: Computes the output variables, voltages and currents, as functions of time over a user-specified time interval.

- AC Small-Signal Analysis: Finds the AC output variables as functions of frequency (transfer function), over a user-specified range of frequencies.

- Pole-Zero Analysis: Computes the poles and/or zeros in the small-signal AC transfer function. It is time-consuming for large circuits.

- Small-Signal Distortion Analysis: Computes steady-state harmonic and intermodulation products for small input signal magnitudes.

- Sensitivity Analysis: Finds the sensitivity of an output variable with respect to all circuit variables, including model parameters.

- Noise Analysis: Studies device-generated noise for the given circuit, providing the noise contributions of each device to the output voltage.

In addition, analysis at different temperatures is allowed; further details are available in Vladimirescu (1994). Other simulators offer additional capabilities, such as statistical analysis, switched capacitor circuit analysis, etc. However, perhaps the most common usage of circuit simulation involves running, first, a DC Analysis, followed by a Transient Analysis. Thus, our study will focus on only these two analysis modes.

Notes   Additional background on circuit simulation is available in various texts, such as in Chua and Lin (1975), sections 2.1–2.5 and 3.1–3.5, in Vladimirescu (1994), in Vlach and Singhal (1994), sections 1.1–1.5, and in Pillage et al. (1995), chapter 1.

Problems

1.1. (Computer Project) Write a parser, in C or C++, that can read a circuit specification in terms of a simple "language" that we now describe. The language is quite limited and restrictive, and represents the bare minimum that will be needed for subsequent projects in this book. The parser should not be case-sensitive, and should interpret any contiguous sequence of spaces or tabs as equivalent to a single space. Every line of the input file should describe a single circuit element, and the description of every circuit element should be given wholly within a single input line. The order of lines in the input file is immaterial and any characters following a % in an input line should be considered as *comments* and ignored. Circuit node names should be non-negative integers, from the set {0, 1, 2, . . .}, and the node name 0 should be reserved and used for the *ground* or *reference* node.

The accepted circuit elements and their specifications are given below. In this, the symbol <node.*>, where * can be any single alphanumeric character, denotes a node name. Specifically, <node.+> denotes the node that is the positive voltage reference point for the element and <node.-> denotes the negative reference node. The positive direction of current in any element is assumed to be from <node.+> to <node.->. The symbol <int> denotes a non-negative integer, and <value> denotes a non-negative real number. The <value> given for a circuit parameter, like resistance or capacitance, should be in the standard units: Volt, Ampere, Ohm, Farad, or Henry, but it should *not* include the corresponding unit. Finally, anything inside brackets, such as [G2] or [<value>] is an *optional* field.

• Voltage source: Only independent DC voltage sources are allowed, specified as:

    V<int>   <node.+>   <node.->   <value>

• Current source: Only independent DC current sources are allowed, specified as:

    I<int>   <node.+>   <node.->   <value>   [G2]

• Resistor: Only linear resistors are allowed, specified as:

    R<int>   <node.+>   <node.->   <value>   [G2]

• Capacitor: Only linear capacitors are allowed, specified as:

    C<int>   <node.+>   <node.->   <value>   [G2]

• Inductor: Only linear inductors are allowed, and they should be specified as:

    L<int>   <node.+>   <node.->   <value>

• Diode: The diode model, and its parameter values, will not be part of the input description. Instead, the model will be built into any subsequent simulation code that you will write and only the terminals should be specified here. Optionally, a scale factor can also be included so as to allow the specification of diodes that are larger than minimum size. The specification is:

    D<int>   <node.+>   <node.->   [<value>]

- BJT: Similar to the diode model, only the terminals and an optional scale factor are given. Let QN denote an npn device and QP denote pnp; the specification is:

```
QN<int>    <node.C>    <node.B>    <node.E>    [<value>]
QP<int>    <node.C>    <node.B>    <node.E>    [<value>]
```

where the nodes represent the collector, base, and emitter terminals, respectively.

- MOSFET: Similar to the above, we give only the terminals and an optional scale factor, and the body terminal is to be ignored:

```
MN<int>    <node.D>    <node.G>    <node.S>    [<value>]
MP<int>    <node.D>    <node.G>    <node.S>    [<value>]
```

where MN denotes an n-channel device and MP is p-channel, and the nodes represent the drain, gate, and source terminals, respectively.

The parser should create a data structure, as a linked list of records, where each record describes a separate circuit element, including its terminals and parameter values. Test your parser on circuits of your choice. An example is given in Fig. 2.35.

CHAPTER 2

Network Equations

2.1 ELEMENTS AND NETWORKS

For our purposes, an *element* is a two-terminal electrical device. An electrical *network*, or *circuit* is a system consisting of a set of elements and a set of *nodes*, where every element terminal is identified with a unique node, and every node is identified with at least one element terminal. A network is completely connected, i.e., there is always at least one path from one node to another. Thus, an electrical network can be represented by a *graph* whose *vertices* correspond to circuit nodes and *edges* correspond to circuit elements. We distinguish between two types of elements, *passive* and *active*.

2.1.1 Passive Elements

A passive element is an element with the property that the voltage across it, $v(t)$, and the current through it, $i(t)$, may be related through a functional relationship of the form:

$$v = f(i, i') \quad \text{or} \quad i = f(v, v') \tag{2.1}$$

where $f : \mathbb{R} \to \mathbb{R}$, and where $i'(t)$ and $v'(t)$ are the first-order derivatives of $i(t)$ and $v(t)$, respectively. The functional relationship (2.1) is called the *element equation*. When the element equation does not depend on the derivative, so that $i = f(v)$ or $v = f(i)$, then the element is said to be *resistive*, otherwise it is *dynamic*. An element is said to be *linear* if $f(\cdot)$ is a *linear* function, otherwise it is *nonlinear*.

A resistive passive element is called a *resistor*; it can be linear or nonlinear, and has the symbols shown in Fig. 2.1. A linear resistor is characterized by the element equation $v = Ri$, where R is its *resistance*, or $i = Gv$, where G is its *conductance*. A nonlinear resistor is said to be voltage-controlled when its element equation is in the form $i = f(v)$, and current-controlled when in the form $v = f(i)$.

A *capacitor* is a passive element whose electrical charge, $q(t)$, is a function of the applied voltage across it, $v(t)$, so that $q = f(v)$, where $f : \mathbb{R} \to \mathbb{R}$. With current being equal to the rate of change of charge, $i(t) = dq/dt$, and with

Circuit Simulation, by Farid N. Najm
Copyright © 2010 John Wiley & Sons, Inc.

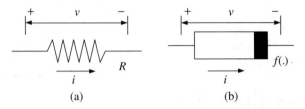

Figure 2.1: The symbols for (a) a linear resistor and (b) a nonlinear resistor.

$C(v) \triangleq dq/dv$ (the symbol $\triangleq$ denotes a definition) as the *capacitance*, we have that:

$$i(t) = \frac{dq}{dv}\frac{dv}{dt} = C(v)\frac{dv}{dt} \qquad (2.2)$$

When the capacitance is fixed, independent of voltage, then the capacitor is said to be *linear*, in which case:

$$q(t) = Cv(t) \quad \text{and} \quad i(t) = C\frac{dv}{dt} \qquad (2.3)$$

In general, the capacitance can be a function of voltage, such as illustrated in Fig. 2.2, in which case the capacitor is said to be *nonlinear*.

An *inductor* is a passive element whose magnetic flux, $\phi(t)$, is a function of the current applied through it, $i(t)$, so that $\phi = f(i)$, where $f : \mathbb{R} \to \mathbb{R}$. By Faraday's law, the induced voltage across the inductor is equal to the rate of change of flux, $v(t) = d\phi/dt$, so that, with $L(i) \triangleq d\phi/di$ defined as the *inductance*, we have that:

$$v(t) = \frac{d\phi}{di}\frac{di}{dt} = L(i)\frac{di}{dt} \qquad (2.4)$$

When the inductance is fixed, independent of current, then the inductor is said to be *linear*, in which case:

$$\phi(t) = Li(t) \quad \text{and} \quad v(t) = L\frac{di}{dt} \qquad (2.5)$$

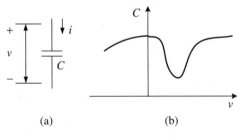

Figure 2.2: The symbol for a capacitor (a) and an illustrative plot of capacitance as a function of voltage (b) for an on linear capacitor.

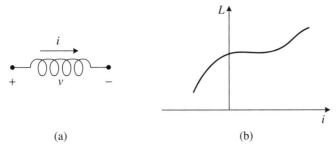

(a)   (b)

Figure 2.3: The symbol for an inductor (a) and an illustrative plot of inductance as a function of current (b) for a nonlinear inductor.

In general, the inductance can be a function of current, such as illustrated in Fig. 2.3, in which case the inductor is said to be *nonlinear*.

2.1.2 Active Elements

An active element is an element with the property that either the voltage across it, $v(t)$, or the current through it, $i(t)$, can be expressed in any one of the following forms:

1. Voltage is either a constant or a function of time, $v = f(t)$, in which case the element is called an *independent voltage source* and has the symbols shown in Fig. 2.4a and Fig. 2.4b.
2. Current is either a constant or a function of time, $i = f(t)$, in which case the element is called an *independent current source* and has the symbol shown in Fig. 2.4c.
3. Voltage is a function of the current i_x in some *other* network element, $v = f(i_x)$, or a function of the voltage v_x across some *other* network element, $v = f(v_x)$, in which case the element is called a *controlled voltage source* (CVS) and has the symbols shown in Fig. 2.5. More generally, v_x may be the voltage difference across any pair of nodes in the network.
4. Current is a function of the current i_x in some *other* network element, $i = f(i_x)$, or a function of the voltage v_x across some *other* network element, $i = f(v_x)$, in which case the element is called a *controlled current source*

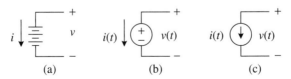

(a)   (b)   (c)

Figure 2.4: Symbols for (a) a constant independent voltage source, (b) a time-varying independent voltage source, and (c) an independent current source.

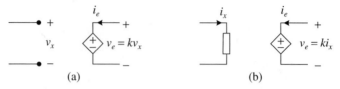

Figure 2.5: Controlled voltage sources, showing a linear voltage-controlled source (a) and a linear current-controlled source (b).

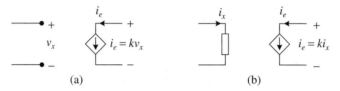

Figure 2.6: Controlled current sources, showing a linear voltage-controlled source (a) and a linear current-controlled source (b).

(CCS) and has the symbols shown in Fig. 2.6. More generally, v_x may be the voltage difference across any pair of nodes in the network.

A controlled source is said to be *linear* if the associated function $f(\cdot)$ is linear, otherwise it is *nonlinear*. The two varieties of controlled voltage sources are the voltage-controlled voltage source (VCVS) and the current-controlled voltage source (CCVS). Likewise, the two varieties of controlled current sources are the voltage-controlled current source (VCCS) and the current-controlled current source (CCCS). Often, a simple nonlinear controlled source can be replaced by a sub-circuit consisting of a nonlinear resistor and only *linear* controlled sources. For example, a VCVS implementing $v_e = f(v_x)$ can be replaced by the equivalent sub-circuit shown in Fig. 2.7. Therefore, in such cases, it is possible to formulate the circuit equations in a way that the *only* nonlinear elements are nonlinear passive elements. However, more complex controlled sources cannot be so simplified. On the other hand, it is always possible to represent a nonlinear

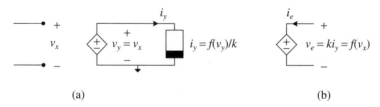

Figure 2.7: Replacement of a nonlinear VCVS by an equivalent circuit containing only *linear* controlled sources.

resistor by a nonlinear controlled source, and this is often done in practical simulators in order to simplify the implementation.

The above element types are referred to as *lumped* elements, so as to distinguish them from *distributed* devices, such as transmission lines. In this text, we will be only concerned with *lumped networks*, i.e., with networks composed of only lumped elements.

2.1.3 Equivalent Circuit Model

In most practical cases, a circuit component that does not fit the above types can be replaced by an *equivalent circuit* in terms of the above types. This is often an approximation, but can be a very good one. The resulting equivalent circuit is a *circuit model* of that original component. The component and its model are (approximately) indistinguishable when examined from their external terminals. Thus, any circuit can be represented in terms of a network composed of only the above element types, which are therefore referred to as the *minimal basic set*.

The usefulness of a circuit simulator is greatly enhanced by the availability of *accurate* (i.e., realistic) circuit models for a wide variety of possible components. We will very briefly take a look at simple diode and transistor models.

Diode Model The pn-diode, as in Fig. 2.8, has the following element equation:

$$i = I_s \left(e^{v/\eta V_T} - 1 \right) \tag{2.6}$$

where $\eta \approx 1$ and $V_T \triangleq (kT/q) \approx 26\,\mathrm{mV}$ at $T = 298\,\mathrm{K}$, and has the equivalent circuit model shown in Fig. 2.9, where R_s and R_c are linear resistors that are often neglected, with $R_s \approx 0$ and $R_c \approx \infty$, and where R_d, C_d, and C_j are nonlinear elements.

Transistor Model The n-channel MOSFET is shown in Fig. 2.10, along with a simple equivalent circuit model. This is the simplest (SPICE level-1) MOSFET model, inspired by the 1968 model for the FET by Shichman and Hodges of AT&T Bell Labs. The capacitors C_{gd} and C_{gs} have a linear part (due to overlap capacitance) and a nonlinear part (due to channel charge). The other capacitors are nonlinear. The nonlinear resistors represent the reverse biased drain and source pn-junctions. The controlled source is a VCCS, controlled by the transistor terminal voltages via a nonlinear expression.

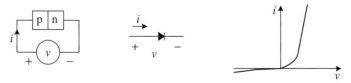

Figure 2.8: The pn-junction diode, showing its structure, circuit symbol, and current-voltage characteristic.

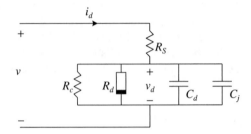

Figure 2.9: An equivalent circuit model for a diode.

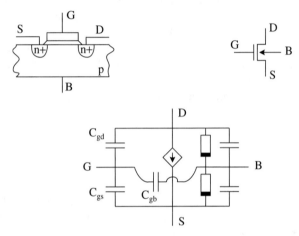

Figure 2.10: The n-channel MOSFET, showing its structure, circuit symbol, and a simple version of its equivalent circuit model.

2.1.4 Network Classification

As we saw above, capacitors and inductors are referred to as *dynamic elements*. A network is referred to as a *dynamic network* if it contains dynamic elements, otherwise it is called a *resistive network*. A network with only linear elements is referred to as a *linear network*, otherwise it is a *nonlinear network*. It is also useful to classify networks according to whether or not they have controlled sources. Thus, networks can be linear or nonlinear, resistive or dynamic, with or without controlled sources.

The basic minimal set of lumped elements, seen above, are *time-invariant* in the sense that their i-v functional characteristics are invariant under a time-shift. A network composed of only time-invariant elements is said to be time-invariant. We will only be interested in *time-invariant* networks. Furthermore, we will only be concerned with time-domain analysis, not frequency-domain analysis.

2.2   TOPOLOGICAL CONSTRAINTS

When concerned with the *topology* of a network, an element is referred to as a *branch*. A network may be characterized by a set of *network constraints*, which capture the network behavior in a system of equations. There are two types of network constraints:

1. *Branch constraints*, also called branch equations or element equations.
2. *Topological constraints*, arising from Kirchoff's Current Law (KCL) and Voltage Law (KVL). These are linear algebraic constraints arising from the structure of the network itself, not due to the properties of any elements.

The study of network topology and topological constraints is best done by appealing to *graph theory*.

2.2.1   Network Graphs

A natural and simple way to study a network N is to define a *directed graph* G_d associated with it:

1. Create a graph vertex corresponding to every network node. Graph vertices will also be occasionally referred to as nodes.
2. Create a graph edge corresponding to every network element. Graph edges will also be occasionally referred to as elements or branches.

The edge *direction* is in the same direction as the positive reference current direction of that element. By convention, we also let the edge direction indicate the positive reference direction for voltage across that element, so that the + voltage terminal is always at the tail of the arrow that defines the edge direction, as in Fig. 2.11. Thus, a network N induces a directed graph G_d which completely captures the topology of the network. When we are not concerned with the reference directions of current and voltage, we remove the edge directions, leading to an *undirected graph* G_n. A network, along with its corresponding two graphs G_d and G_n, is shown in Fig. 2.12. For brevity, we will refer to either G_d or G_n as the *network graph* corresponding to the network N.

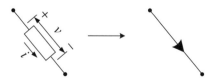

Figure 2.11: A circuit element and the corresponding graph edge.

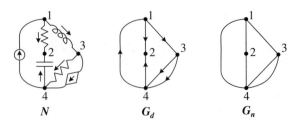

Figure 2.12: A circuit, its directed graph, and its undirected graph.

Notation For a network graph with n vertices (or nodes) and m edges (or elements), one node will be identified as the *reference node* and will be denoted by the integer 0, and all other nodes will be numbered $1, 2, \ldots, n - 1$. These form the set of graph nodes $V = \{0, 1, 2, \ldots, n - 1\}$. The reference node is also called the *datum* or, in most cases, simply *ground*. All edges will be denoted $e_1, e_2, e_3, \ldots, e_m$. These form the set of graph edges $E = \{e_1, e_2, e_3, \ldots, e_m\}$. This labeling scheme is illustrated with the graphs in Fig. 2.13. Finally, all network graphs G_n and G_d under study will be assumed to have some basic properties:

1. They are *connected*.
2. They have *no self-loops* (they are, so-called, loopless graphs).
3. The undirected graph G_n has cycles, so that $m \geq n$ (i.e., it is not a tree).

Incidence Matrix For a node j, let $\mathcal{E}_j^{\text{out}}$ be the set of (directed) edges whose *tail* is connected to j, and let $\mathcal{E}_j^{\text{in}}$ be the set of (directed) edges whose *head* is connected to j. We define the $n \times m$ *incidence matrix* M, according to:

$$M_{ik} = \begin{cases} +1, & \text{if } e_k \in \mathcal{E}_{i-1}^{\text{out}}, \\ -1, & \text{if } e_k \in \mathcal{E}_{i-1}^{\text{in}}, \\ 0, & \text{otherwise.} \end{cases} \qquad (2.7)$$

where $i = 1, 2, \ldots, n$ and $k = 1, 2, \ldots, m$. As an example, the directed graph in Fig. 2.13 has the following incidence matrix:

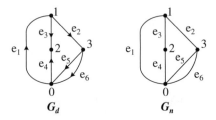

Figure 2.13: Edge labels, shown on a directed and an undirected graph.

$$M = \begin{bmatrix} +1 & 0 & 0 & +1 & -1 & -1 \\ -1 & +1 & +1 & 0 & 0 & 0 \\ 0 & 0 & -1 & -1 & 0 & 0 \\ 0 & -1 & 0 & 0 & +1 & +1 \end{bmatrix} \tag{2.8}$$

Notice some general properties of *any* incidence matrix M:

1. Every column k has a single $+1$ entry and a single -1 entry, corresponding to the terminals of edge e_k, and all other entries are 0.
2. In every row $i = j + 1$, the number of $+1$ entries is equal to the out-degree of the vertex j, i.e., the number of out-going edges.
3. In every row $i = j + 1$, the number of -1 entries is equal to the in-degree of the vertex j, i.e., the number of in-coming edges.

As a result, the element-wise sum of all the rows of M is the 0 row vector. Thus, if a row of M is missing, we can easily find it, by taking the element-wise sum of all the other rows and multiplying by -1. This means that the rows of M, viewed as vectors, are *not linearly independent*!

Voltage Assignments   Considering the directed graph $G_d = (V, E)$, let p_0, $p_1, \ldots, p_{n-1}$ be the electric *potentials* at every node $0, 1, \ldots, n - 1$. Let $u_1, u_2, \ldots, u_m$ be the voltages across every edge $e_1, e_2, \ldots, e_m$, and define the vectors $p = [p_0 \quad p_1 \quad \cdots \quad p_{n-1}]^T$ and $u = [u_0 \quad u_1 \quad \cdots \quad u_m]^T$. For every directed edge e_j, let $t(e_j)$ be the node at the *tail* of e_j and $h(e_j)$ be the node at the *head* of e_j. The voltage assignment u would satisfy Kirchoff's Voltage Law (KVL) if and only if there exists a potential vector p such that:

$$u_j = p_{t(e_j)} - p_{h(e_j)}, \quad \forall e_j \in E \tag{2.9}$$

which can be compactly expressed as:

$$u = M^T p \tag{2.10}$$

Define $v_j = p_j - p_0$, for every node $j = 0, 1, 2, \ldots, (n - 1)$, as the *voltage of every node relative to the reference node*. Therefore, $v_0 = 0$ and $p_j = p_0 + v_j$, so that:

$$u = M^T \begin{bmatrix} p_0 \\ p_0 \\ p_0 \\ \vdots \\ p_0 \end{bmatrix} + M^T \begin{bmatrix} 0 \\ v_1 \\ v_2 \\ \vdots \\ v_{n-1} \end{bmatrix} = M^T \begin{bmatrix} p_0 \\ p_0 \\ p_0 \\ \vdots \\ p_0 \end{bmatrix} + A^T v \tag{2.11}$$

where $v = [v_1 \quad v_2 \quad \cdots \quad v_{n-1}]^T$, and A is the $(n - 1) \times m$ matrix obtained from M by removing its first row, called the *reduced incidence matrix*. By convention, *let* $p_0 = 0$ *at the reference node*, so that:

$$\text{KVL} \quad \Longleftrightarrow \quad u = A^T v \tag{2.12}$$

Current Assignment If $i_1, i_2, \ldots, i_m$ are the currents in $e_1, e_2, \ldots, e_m$, then by Kirchoff's Current Law (KCL), at every node $j \in V$, we have:

$$\sum_{e_k \in \mathcal{E}_j^{\text{out}}} i_k - \sum_{e_l \in \mathcal{E}_j^{\text{in}}} i_l = 0 \tag{2.13}$$

Thus, if $i = [i_1 \quad i_2 \quad \cdots \quad i_m]^T$ is the vector of all branch currents, we can compactly capture KCL for the whole network as:

$$Mi = 0 \tag{2.14}$$

Because one row of M can always be written in terms of all the rest, then one equation of $Mi = 0$ is *redundant*. Thus, it is sufficient to express KCL using the reduced incidence matrix:

$$\text{KCL} \quad \Longleftrightarrow \quad Ai = 0 \tag{2.15}$$

It can be shown that, for a connected, loopless, directed graph G_d, the set of rows of the reduced incidence matrix A are *linearly independent*. Notice that every row of the equation $Ai = 0$ simply expresses KCL at every node, *other than the reference node*, as:

$$\sum_{e_k \in \mathcal{E}_j^{\text{out}}} i_k - \sum_{e_l \in \mathcal{E}_j^{\text{in}}} i_l = 0 \tag{2.16}$$

This observation will be useful later on.

Example For the network in Fig. 2.12, and using the edge labels adopted in Fig. 2.13, the reduced incidence matrix A is as shown:

$$A = \begin{bmatrix} -1 & +1 & +1 & 0 & 0 & 0 \\ 0 & 0 & -1 & -1 & 0 & 0 \\ 0 & -1 & 0 & 0 & +1 & +1 \end{bmatrix} \tag{2.17}$$

which leads to KVL in matrix form, as shown:

$$\begin{bmatrix} u_1 \\ u_2 \\ u_3 \\ u_4 \\ u_5 \\ u_6 \end{bmatrix} = \begin{bmatrix} -1 & 0 & 0 \\ +1 & 0 & -1 \\ +1 & -1 & 0 \\ 0 & -1 & 0 \\ 0 & 0 & +1 \\ 0 & 0 & +1 \end{bmatrix} \begin{bmatrix} v_1 \\ v_2 \\ v_3 \end{bmatrix} \tag{2.18}$$

and a KCL in matrix form, as shown:

$$
\begin{bmatrix}
-1 & +1 & +1 & 0 & 0 & 0 \\
0 & 0 & -1 & -1 & 0 & 0 \\
0 & -1 & 0 & 0 & +1 & +1
\end{bmatrix}
\begin{bmatrix}
i_1 \\ i_2 \\ i_3 \\ i_4 \\ i_5 \\ i_6
\end{bmatrix}
=
\begin{bmatrix}
0 \\ 0 \\ 0
\end{bmatrix}
\tag{2.19}
$$

Combined Form   Combining KCL (2.15) and KVL (2.12), then the *topological constraints*, due to the network topology only, and independent of any element characteristics, become:

$$
\begin{bmatrix}
A & 0 & 0 \\
0 & I & -A^T
\end{bmatrix}
\begin{bmatrix}
i \\ u \\ v
\end{bmatrix}
=
\begin{bmatrix}
0 \\ 0
\end{bmatrix}
\tag{2.20}
$$

This gives $(m + n - 1)$ independent equations in $(2m + n - 1)$ unknowns. The element equations provide another m independent equations in the same unknowns. With a total of $(2m + n - 1)$ independent equations in $(2m + n - 1)$ unknowns, we can solve the network.

The above representation (2.20) is highly *sparse*, which turns out to be a major advantage for solving large circuits. Technically, an $n \times n$ matrix is *sparse* if it has $\mathcal{O}(n)$ non-zero entries. The matrix A has no more than $2m - 1$ non-zero entries. Thus, the above form of the constraints is *provably sparse*. We will see later on that the above form of the topological constraints leads directly to the *sparse tableau analysis (STA)* and the *modified nodal analysis (MNA)* equations.

2.3 CYCLE SPACE AND BOND SPACE

The above form of the constraints (2.20) is not unique; it is possible to capture KCL and KVL more compactly, with fewer equations and variables. We will now digress briefly to see these other ways of capturing network topology by appealing to the notions of *cycle space* and *bond space*. This will lead to a way to write the constraints more *compactly*, as m equations in $2m$ unknowns. However, this will not be as sparse as the above form, therefore not as useful in practice for large circuits. Nevertheless, the study of the cycle space and bond space will give us useful insight into network topology.

2.3.1 Current Assignments

In general, a *current assignment* $i = [i_1 \quad i_2 \quad \cdots \quad i_m]^T$ is a mapping from the set of edges, E, to $\mathbb{R}$. A current assignment i that satisfies KCL, $Ai = 0$, is said to be a *valid current assignment*, also called a *circulation*. Circulations have some key properties:

1. If $i^{(1)}$ and $i^{(2)}$ are circulations, then $i^{(1)} + i^{(2)}$ is also a circulation.
2. If i is a circulation, then αi is also a circulation, $\forall \alpha \in \mathbb{R}$.

Therefore, the set of all circulations is a *sub-space* of $\mathbb{R}^m$, denoted by $\mathfrak{C}$:

$$\mathfrak{C} = \left\{ i : i \in \mathbb{R}^m \text{ and } i \text{ is a circulation} \right\} \tag{2.21}$$

The sub-space $\mathfrak{C}$ is referred to as the *cycle space* of the network.

2.3.2 Voltage Assignments

In general, a *voltage assignment* $u = [u_1 \quad u_2 \quad \cdots \quad u_m]^T$ is a mapping from the set of edges, E, to $\mathbb{R}$. A voltage assignment u that satisfies KVL, $u = A^T v$, is said to be a *valid voltage assignment*. Valid voltage assignments have some key properties:

1. If $u^{(1)}$ and $u^{(2)}$ are valid voltage assignments, then $u^{(1)} + u^{(2)}$ is also a valid voltage assignment.
2. If u is a valid voltage assignment, then αu is also a valid voltage assignment, $\forall \alpha \in \mathbb{R}$.

Therefore, the set of all valid voltage assignments is a *sub-space* of $\mathbb{R}^m$, denoted by $\mathfrak{B}$:

$$\mathfrak{B} = \left\{ u : u \in \mathbb{R}^m \text{ and } u \text{ is a valid voltage assignment} \right\} \tag{2.22}$$

The sub-space $\mathfrak{B}$ is referred to as the *bond space* of the network.

2.3.3 Orthogonal Spaces

The following is an important result related to the cycle and bond spaces.

Theorem 2.1. *Given a directed graph G_d, with incidence matrix M, then:*

1. *$\mathfrak{B}$ is spanned by the rows of M, i.e.:*

$$\forall u \in \mathfrak{B}, \exists \alpha \in \mathbb{R}^n : u = M^T \alpha \quad \underline{and} \quad \forall \alpha \in \mathbb{R}^n, u = M^T \alpha \in \mathfrak{B} \tag{2.23}$$

2. *$\mathfrak{B}$ is orthogonal to $\mathfrak{C}$, denoted $\mathfrak{B} \perp \mathfrak{C}$, i.e., every $u \in \mathfrak{B}$ is orthogonal to every $i \in \mathfrak{C}$, so that $u^T i = 0$. This is, in fact, a special case of* Tellegen's *theorem.*
3. *$\mathfrak{B} \oplus \mathfrak{C} = \mathbb{R}^m$, so that $dim(\mathfrak{B}) + dim(\mathfrak{C}) = m$, i.e., together, $\mathfrak{B}$ and $\mathfrak{C}$ span the whole of $\mathbb{R}^m$.*

A proof of this result, and further details regarding these spaces, may be found in advanced texts on graph theory, as well as, in a circuits context, in Chua et al. (1987). In the following, we let $\dim(\mathfrak{B}) = r$, so that $\dim(\mathfrak{C}) = m - r$. When G is connected, it can be shown that $r = n - 1$, so that, for our work:

$$\dim(\mathfrak{B}) = n - 1 \tag{2.24}$$

and

$$\dim(\mathfrak{C}) = m - n + 1 \tag{2.25}$$

Let $b_1, b_2, \ldots, b_r$ be $m \times 1$ basis vectors of $\mathfrak{B}$. Let $c_1, c_2, \ldots, c_{m-r}$ be $m \times 1$ basis vectors of $\mathfrak{C}$. We use these basis vectors to define two basis matrices, as follows. Let B be the $r \times m$ matrix whose rows are $b_1^T, b_2^T, \ldots, b_r^T$, i.e., it is a *basis matrix* for the bond space $\mathfrak{B}$. Let C be the $(m - r) \times m$ matrix whose rows are $c_1^T, c_2^T, \ldots, c_{m-r}^T$, i.e., it is a *basis matrix* for the cycle space $\mathfrak{C}$.

2.3.4 Topological Constraints

The above leads to two key results (*topological constraints*), as follows:

1. If $i \in \mathfrak{C}$ is a valid current assignment (obeys KCL), then $i \perp \mathfrak{B}$ and:

$$Bi = 0 \tag{2.26}$$

   which is a compact form of KCL, with r independent equations in m unknowns.
2. If $u \in \mathfrak{B}$ is a valid voltage assignment (obeys KVL), then $u \perp \mathfrak{C}$ and:

$$Cu = 0 \tag{2.27}$$

   which is a compact form of KVL, with $m - r$ independent equations in m unknowns.

This gives m independent equations in $2m$ unknowns. The element equations provide another m independent equations in the same $2m$ unknowns. With a total of $2m$ independent equations in $2m$ unknowns, we can solve the network. It remains to explain how to construct the basis matrices B and C, which we do in the next two sections.

2.3.5 Fundamental Circulation

Let $\mathcal{C}$ be any *undirected* cycle in G_d, with a given fixed *orientation* around the cycle. Let $\mathcal{C}^+$ be the set of edges in $\mathcal{C}$ whose directions in G_d agree with the orientation of $\mathcal{C}$. Let $\mathcal{C}^-$ be the set of edges in $\mathcal{C}$ whose directions in G_d do not agree with the orientation of $\mathcal{C}$. Consider the following current assignment, which can be shown to be a *valid current assignment*, i.e., it is a circulation (it obeys KCL):

$$i_j = \begin{cases} +1, & \text{if } e_j \in \mathcal{C}^+, \\ -1, & \text{if } e_j \in \mathcal{C}^-, \\ 0, & \text{if } e_j \notin \mathcal{C}. \end{cases} \qquad (2.28)$$

This particular circulation is called the *fundamental circulation* corresponding to the cycle $\mathcal{C}$. It can be shown that the following procedure gives a basis matrix C of the cycle space. Start by choosing a spanning tree of G_d, which would have $(n-1)$ edges, then:

1. For each non-tree edge, consider the *unique* cycle that results (in G_d) from adding that edge to the tree.
2. Use the fundamental circulation (in G_d) corresponding to that cycle as a row of the basis matrix C.

There are $m - (n-1)$ non-tree edges, leading to the required $m - n + 1$ rows of the basis matrix C.

Example   Using our running example of Fig. 2.12 and Fig. 2.13, we choose a spanning tree composed of e_2, e_3, and e_5 and we identify the three cycles shown in Fig. 2.14. For each cycle, we write the fundamental circulation using a clockwise orientation:

$$i^{(1)} = \begin{bmatrix} 0 & 0 & 0 & 0 & -1 & +1 \end{bmatrix}^T$$
$$i^{(2)} = \begin{bmatrix} 0 & +1 & -1 & +1 & +1 & 0 \end{bmatrix}^T$$
$$i^{(3)} = \begin{bmatrix} +1 & +1 & 0 & 0 & +1 & 0 \end{bmatrix}^T$$

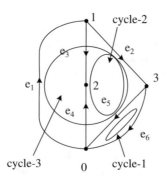

Figure 2.14: Cycles in a graph, corresponding to edges that are not part of the chosen spanning tree.

so that the basis matrix C is as shown:

$$C = \begin{bmatrix} 0 & 0 & 0 & 0 & -1 & +1 \\ 0 & +1 & -1 & +1 & +1 & 0 \\ +1 & +1 & 0 & 0 & +1 & 0 \end{bmatrix} \tag{2.29}$$

2.3.6 Fundamental Potential Difference

The basis matrix B can be constructed based on the same spanning tree, in a similar fashion. The method operates by identifying a so-called *fundamental potential difference* corresponding to each tree edge. There are $n - 1$ tree edges, leading to the $n - 1$ rows of B. However, we will not study this process because there is an easier way of finding B, as follows.

Recall that, using the reduced incidence matrix A, we can capture KCL with the compact form $Ai = 0$. Indeed, when G is connected, it can be shown that A is a basis matrix for $\mathcal{B}$, so that one can simply choose:

$$B = A \tag{2.30}$$

In summary, then, the *topological constraints*, due to the network topology only, and independent of any element characteristics, become:

$$\begin{bmatrix} A & 0 \\ 0 & C \end{bmatrix} \begin{bmatrix} i \\ u \end{bmatrix} = \begin{bmatrix} 0 \\ 0 \end{bmatrix} \tag{2.31}$$

which are m independent equations in $2m$ unknowns. As mentioned earlier, however, although (2.31) is more compact than (2.20), it is *not* preferred in practice for handling large circuits, because it is not as sparse.

2.4 FORMULATION OF LINEAR ALGEBRAIC EQUATIONS

We have seen how the network *topological constraints* can be expressed as (2.20), repeated here for convenience:

$$\begin{bmatrix} A & 0 & 0 \\ 0 & I & -A^T \end{bmatrix} \begin{bmatrix} i \\ u \\ v \end{bmatrix} = \begin{bmatrix} 0 \\ 0 \end{bmatrix} \tag{2.32}$$

We need to augment these equations with the *element equations* (also called *branch equations*) in order to get a solvable system. In this section, we restrict our attention to *linear networks with no dynamic elements*, i.e., linear resistive networks, so that we only allow linear resistors, independent sources, and linear controlled sources. This is actually sufficient; we will see later on that a general nonlinear dynamic network can be reduced, for the purpose of time-domain simulation, to a linear resistive one. For such networks, the element equations are *linear algebraic equations*, the same as the above topological constraints. Combining the two sets of equations will yield a system of simultaneous linear equations whose solution gives the network voltages and currents. We will consider two approaches: the *sparse tableau analysis* (STA) formulation and the *modified nodal analysis* (MNA) formulation.

2.4.1   Sparse Tableau Analysis

The element equations can be easily written in linear algebraic form, as follows. If edge e_k is a resistor, then $u_k - Ri_k = 0$. If edge e_k is an independent source, then $u_k = V$ or $i_k = I$. If edge e_k is a controlled source, then, for a VCVS, $u_k = \alpha u_x$, for a CCVS, $u_k = \alpha i_x$, for a VCCS, $i_k = \alpha u_x$, and for a CCCS, $i_k = \alpha i_x$. Therefore, the element/branch equations can be compactly expressed as:

$$Zi + Yu = s \qquad (2.33)$$

or:

$$\begin{bmatrix} Z & Y \end{bmatrix} \begin{bmatrix} i \\ u \end{bmatrix} = s \qquad (2.34)$$

where Z and Y are $m \times m$ matrices, and they are sparse, with no more than m non-zero elements each, and s is a known $m \times 1$ vector. Combining this with the topological constraints (2.32), leads to the complete system:

$$\begin{bmatrix} A & 0 & 0 \\ 0 & I & -A^T \\ Z & Y & 0 \end{bmatrix} \begin{bmatrix} i \\ u \\ v \end{bmatrix} = \begin{bmatrix} 0 \\ 0 \\ s \end{bmatrix} \qquad (2.35)$$

which has $(2m + n - 1)$ independent equations in $(2m + n - 1)$ unknowns. The system matrix is sparse, with no more than $2(2m - 1) + m + 2m = (7m - 2)$ non-zero entries. This is referred to as the *sparse tableau analysis* (STA) formulation and was described originally in Hachtel et al. (1971).

Reduced Tableau   STA has some key advantages. For one thing, it is very general, allowing one to handle the full set of basic elements under consideration. Furthermore, the tableau matrix is very sparse, which is a key advantage in simulation of large circuits. However, the formulation has some redundancy, because it retains all three sets of variables: i, u, and v. These variables are not independent, and one can easily reduce the number of variables and equations. Indeed, it is trivial to obtain the branch voltages from the node voltages. In most cases, it is also equally trivial to obtain the branch currents from the branch voltages, using the element equations. But, any alternative approach must also be general enough and sparse.

Modified nodal analysis (MNA) is one such approach. It is more compact than STA and is also sparse, although not quite as sparse as STA. Both STA and MNA survive today as viable techniques for large-scale circuit simulation, but the use of MNA is more prevalent. We will focus exclusively on MNA in this text.

As a first step towards *nodal analysis*, which will then lead us to MNA, we will start with a simple reduction of the STA formulation. Substitute the second equation of STA ($u = A^T v$, KVL) into the third equation ($Zi + Yu = s$, the branch equations), and rearrange, to get what one may call the *reduced tableau form*:

$$\begin{bmatrix} YA^T & Z \\ 0 & A \end{bmatrix} \begin{bmatrix} v \\ i \end{bmatrix} = \begin{bmatrix} s \\ 0 \end{bmatrix} \tag{2.36}$$

Thus, we have eliminated all the branch voltage variables, but more can be done, as part of what is called *nodal analysis*, which we study next.

2.4.2 Nodal Analysis

Nodal analysis (NA) is based on the following simplifying assumption. As a result, it is not usable for general circuit simulation, but it is useful to study because it shows the way forward towards MNA.

Assumption 2.1. (Simplifying Assumption) The network contains no voltage sources (neither independent nor controlled).

This is a curious assumption because, obviously, practical circuits typically contain voltage sources! In practice, of course, any physical voltage source would be non-ideal and can be converted to a non-ideal current source using Thévenin's theorem. However, this is not good enough as justification for this assumption, because one would expect a general-purpose circuit simulator to be able to handle ideal voltage sources. Nevertheless, it is useful to see what this assumption leads to.

A key consequence of this assumption is that the branch equations can be written in the following form:

$$i = Yu + s \tag{2.37}$$

It is easy to see how an element equation can be cast into this form, if it is a resistor, an independent current source, or a VCCS. However, the case of a CCCS is a bit harder to see. To see how a CCCS can be included in the analysis, notice that a CCCS can always be modified so that its controlling element e_x is *not* another current source. This is assuming, of course, the network has no circular dependencies, such as $i_e = k_x i_x$, $i_x = k_y i_y$, and $i_y = k_e i_e$. In our case, this would mean that e_x would have to be a resistor, because the network contains no voltage sources. However, when e_x is a resistor, the CCCS can be easily converted to a VCCS, whose branch equation can be cast into the above form. These conversions impede the efficiency of an automated approach, and we will see that MNA overcomes these difficulties as well.

Substitute the branch equation (2.37) into KCL ($Ai = 0$), then use KVL ($u = A^T v$), leading to:

$$AYA^T v = -As \tag{2.38}$$

This is the *nodal analysis* formulation. It is very compact, with an $(n-1) \times (n-1)$ matrix, and features only the node voltages. Sparsity is not yet evident, but we will see later on that it is quite sparse. However, it is an incomplete solution, because it does not allow ideal voltage sources. Notice that the above NA equation effectively expresses KCL at every node, other than the reference node, followed by elimination of the current variables. The matrix:

$$G \triangleq AYA^T \tag{2.39}$$

is called the *nodal admittance matrix*, also called the *conductance matrix*, and its properties are key to an efficient solution.

Regarding the *form* of the element equations, when an element equation is of the form $i = Yu + s$, it is said to be in *admittance form*. If it is in the form $u = Zi + s$, it is said to be in *impedance form*. Nodal analysis requires elements to have equations in admittance form, but we will see that MNA allows element equations in either form.

Notice that the form of the equation $i = Yu + s$ provides an easy way to eliminate *all* the current variables. However, insisting on this form makes it impossible to include voltage sources in the formulation. The equation for an ideal voltage source cannot depend on its current! The desire to eliminate *all* the current variables has led us to this situation. MNA solves this impasse by aiming to eliminate most, but not necessarily all, current variables, as we will see shortly, after a brief digression.

2.4.3 Unique Solvability

Does a network always have a unique solution? In general, allowing for nonlinear elements, dynamic elements, as well as controlled sources, the answer is no! A network may *not* have a solution, or may have *multiple* solutions. In some cases, a unique solution may exist but only for specific values of element parameters. Such cases are only of academic interest and of little value in practice, because practical elements have non-zero tolerances. Loosely speaking, we will refer to a network as being *uniquely solvable* if it has a unique solution under parameter values that are not "too restricted."

Some further definitions are useful at this point. A *cutset* in a connected graph is a set of edges which, if removed, would cause the graph to become disconnected. A *current source cutset* is a cutset whose every edge corresponds to a current source in the network, be it independent or controlled. A *voltage source loop* is a cycle in the graph whose every edge corresponds to a voltage source, whether independent or controlled. It can be shown that a network that is uniquely solvable, must satisfy the following two so-called *consistency requirements*:

1. The network graph *must not contain any current source cutsets*.
2. The network graph *must not contain any voltage source loops*.

These are *necessary, but not sufficient*, conditions for unique solvability.

Although they may appear non-physical and *pathological*, situations that are excluded by these consistency requirements do in fact arise in circuits. This can happen due to overly simplified device models, due to the use of a hypothetical, incomplete, circuit description, or due to the types of companion models used for the nonlinear elements, as we will see later on. In practice, it is possible to remedy such situations, if they arise, by adding some insignificant parasitics, such as large resistors in parallel with some current sources, or small resistors in series with some voltage sources. In any case, in the following, *we will assume throughout that the above consistency requirements are always met*.

If a network is linear, resistive, and has no controlled sources, then it can be shown that the consistency requirements become *sufficient* for unique solvability. Furthermore, if the network is linear and resistive, such as we have assumed for NA, then we can make some useful statements about the G matrix, and solvability. We will now study these issues, by introducing some results from matrix theory and considering how they apply to the conductance matrix G.

Irreducible Matrix   Matrix irreducibility can be expressed in terms of the *matrix graph*. A square $n \times n$ matrix $A = [a_{ij}]$ induces a directed graph G_A, whose vertices are $\{1, 2, \ldots, n\}$, and whose directed edges are (i, j) if $a_{ij} \neq 0$. If there is a directed path in the graph from every vertex to every other vertex, then the graph is said to be *strongly connected*. If G_A is strongly connected, then the matrix A is said to be *irreducible*.

Diagonal Dominance   Another set of useful definitions are the following.

Definition 2.1. (Diagonal Dominance) Let $G = [g_{ij}]$ be an $n \times n$ matrix, then G is said to be *diagonally dominant* if:

$$|g_{ii}| \geq \sum_{\forall j \neq i} |g_{ij}|, \quad \forall i \tag{2.40}$$

Definition 2.2. (Strict Diagonal Dominance) Let $G = [g_{ij}]$ be an $n \times n$ matrix, then G is said to be *strictly diagonally dominant* if:

$$|g_{ii}| > \sum_{\forall j \neq i} |g_{ij}|, \quad \forall i \tag{2.41}$$

Definition 2.3. (Irreducible Diagonal Dominance) Let $G = [g_{ij}]$ be an $n \times n$ matrix, then G is said to be *irreducibly diagonally dominant* if it is *irreducible*, it is *diagonally dominant*, and there is an $i \in \{1, 2, \ldots, n\}$ for which:

$$g_{ii} > \sum_{\forall j \neq i} |g_{ij}| \tag{2.42}$$

If a matrix is either strictly diagonally dominant or irreducibly diagonally dominant, then it is nonsingular. For a linear resistive circuit with no controlled sources, it can be shown that the conductance matrix $G = AYA^T$ is *diagonally dominant*. Such a circuit can be easily transformed, by adding large resistances from every node to ground, so that G becomes *strictly* diagonally dominant. If this is done, then G^{-1} exists and the network is uniquely solvable.

For a connected, linear, resistive circuit, with no controlled sources, that meets the consistency requirements, it can be shown that G is irreducibly diagonally dominant, so that, G^{-1} exists and the network is uniquely solvable.

Positive Definite

Definition 2.4. (Positive Definite) If an $n \times n$ matrix A is symmetric and $x^T Ax > 0$ for all non-zero $x \in \mathbb{R}^n$, then A is said to be *positive definite* or *symmetric positive definite* (SPD).

If A is SPD, then $a_{ii} > 0$, $\forall i$, A is nonsingular, and A^{-1} is also SPD. If A is symmetric, then all its eigenvalues λ_i are real, and:

$$A \text{ is SPD} \quad \Longleftrightarrow \quad \lambda_i > 0, \forall i \tag{2.43}$$

For a connected, linear, resistive circuit, with no controlled sources, that meets the consistency requirements, it can be shown that G is SPD. In this case, G^{-1} exists, and the network is uniquely solvable.

M-matrix

Definition 2.5. (M-matrix) A square matrix G is said to be an *M-matrix* if:

$$g_{ij} \leq 0, \forall i \neq j \quad \text{and} \quad \Re(\lambda_i) > 0, \forall i \tag{2.44}$$

where $\Re(\lambda_i)$ is the real part of the (possibly complex) eigenvalue λ_i.

If G is an M-matrix, then it is nonsingular and $G^{-1} \geq 0$. If a matrix G is either strictly diagonally dominant or irreducibly diagonally dominant, and if:

$$g_{ii} > 0, \forall i \quad \text{and} \quad g_{ij} \leq 0, \forall i \neq j \tag{2.45}$$

then G is an M-matrix. For a connected, linear, resistive circuit, with no controlled sources, that meets the consistency requirements, it can be shown that G is an M-matrix. In this case, $G^{-1} \geq 0$ exists and the network is uniquely solvable.

2.4.4 Modified Nodal Analysis

We are now ready to present modified nodal analysis, originally described in Ho et al. (1975). To handle voltage sources, the key idea is to not insist on eliminating their currents, but to *retain* those currents as *additional variables*. For of these new variables, we add a new equation, namely the branch equation for that (voltage source) element. The size of the matrix equation will grow, by as many equations as we have voltage sources, compared to nodal analysis. In fact, we can retain more currents than just the voltage source currents, and standard MNA retains the following:

1. All voltage source currents, be they independent or controlled.
2. Any current that is a *control variable* for a CCVS or a CCCS.
3. Any current that is a user-specified simulation output.

In addition, when formulating MNA for dynamic circuits, as we will see later on, all inductor currents must be maintained as additional variables. Thus, MNA is a "happy medium" between the reduced form of STA, where no currents were eliminated, and nodal analysis, where all were eliminated. Notice that retaining currents that act as control variables allows us to overcome the need to "convert" a CCCS, as we had to do in nodal analysis.

Definition 2.6. (Element Groups) All elements whose currents are to be eliminated will be referred to as being in *group 1*, while all other elements will be referred to as *group 2*.

We partition the current vector i according to group membership, so that currents of group 1 elements are put into i_1, and the rest are grouped into i_2:

$$i = \begin{bmatrix} i_1 \\ i_2 \end{bmatrix} \tag{2.46}$$

Likewise, we partition the branch voltage vector u, so that all group 1 element voltages are grouped into u_1, and the rest are grouped into u_2:

$$u = \begin{bmatrix} u_1 \\ u_2 \end{bmatrix} \tag{2.47}$$

With this, the general branch equation (2.33), which, as is typical in the MNA literature, we now re-write as $Zi - Yu = s$, can be partitioned as:

$$\begin{bmatrix} I & Z_{12} \\ 0 & Z_{22} \end{bmatrix} \begin{bmatrix} i_1 \\ i_2 \end{bmatrix} - \begin{bmatrix} Y_{11} & Y_{12} \\ Y_{21} & Y_{22} \end{bmatrix} \begin{bmatrix} u_1 \\ u_2 \end{bmatrix} = \begin{bmatrix} s_1 \\ s_2 \end{bmatrix} \tag{2.48}$$

where I, crucially, is the identity matrix.

Recall, the elements in group 1 must be either resistors, independent current sources, VCCS, or CCCS. The above matrix equation means that these elements must have branch equations in the following matrix form:

$$i_1 + Z_{12}i_2 = Y_{11}u_1 + Y_{12}u_2 + s_1 \tag{2.49}$$

It is easy to see how group 1 element equations can be cast into this form, keeping in mind that the controlling element for the CCCS is in group 2. As for elements in group 2, their equations are in the general matrix form:

$$Z_{22}i_2 = Y_{21}u_1 + Y_{22}u_2 + s_2 \tag{2.50}$$

Furthermore, KCL $(Ai = 0)$ is partitioned as well:

$$A_1i_1 + A_2i_2 = 0 \tag{2.51}$$

and KVL $(u = A^T v)$ breaks up into two equations:

$$u_1 = A_1^T v \qquad \text{and} \qquad u_2 = A_2^T v \tag{2.52}$$

Starting with KCL (2.51), plug in i_1 from the first branch equation (2.49), then plug in u_1 and u_2 from KVL (2.52) into the result, to get:

$$\left(A_1 Y_{11} A_1^T + A_1 Y_{12} A_2^T\right) v + (A_2 - A_1 Z_{12}) i_2 = -A_1 s_1 \tag{2.53}$$

Note, this equation is the result of writing KCL $(Ai = 0)$ *at all nodes other than the reference node*, followed by substitutions of i_1, u_1, and u_2. Indeed, row k of this equation is the result of writing KCL at node k, followed by substitutions of i_1, u_1, and u_2.

Then, plug in u_1 and u_2 from KVL (2.52) into the second branch equation (2.50), which leads to:

$$- \left(Y_{21} A_1^T + Y_{22} A_2^T\right) v + Z_{22}i_2 = s_2 \tag{2.54}$$

Here, instead of KCL, the recipe is to start with the element equation of each element in group 2, and then substitute into that u_1 and u_2.

Combining the above two results, (2.53) and (2.54), leads to the full and compact form of the *modified nodal analysis* (MNA) formulation:

$$\begin{bmatrix} \left(A_1 Y_{11} A_1^T + A_1 Y_{12} A_2^T\right) & (A_2 - A_1 Z_{12}) \\ -\left(Y_{21} A_1^T + Y_{22} A_2^T\right) & Z_{22} \end{bmatrix} \begin{bmatrix} v \\ i_2 \end{bmatrix} = \begin{bmatrix} -A_1 s_1 \\ s_2 \end{bmatrix} \tag{2.55}$$

If the network contains no controlled sources, then Y_{11} is diagonal and Z_{12}, Y_{12}, and Y_{21} are all zero, leading to:

$$\begin{bmatrix} A_1 Y_{11} A_1^T & A_2 \\ -Y_{22} A_2^T & Z_{22} \end{bmatrix} \begin{bmatrix} v \\ i_2 \end{bmatrix} = \begin{bmatrix} -A_1 s_1 \\ s_2 \end{bmatrix} \tag{2.56}$$

Furthermore, if group 2 contains no current sources, then $Y_{22} = I$, and the MNA system reduces to:

$$\begin{bmatrix} A_1 Y_{11} A_1^T & A_2 \\ -A_2^T & Z_{22} \end{bmatrix} \begin{bmatrix} v \\ i_2 \end{bmatrix} = \begin{bmatrix} -A_1 s_1 \\ s_2 \end{bmatrix} \tag{2.57}$$

which is the common form of the MNA system that one typically sees in the literature. Finally, if the network is connected and the group 2 elements do not form a cutset, and given the consistency requirements, then it is possible to show that $G \triangleq A_1 Y_{11} A_1^T$ is diagonally dominant, SPD, and an M-matrix.

Assembling the MNA System   In practice, in a simulator, we do not use the above matrix equations to construct the MNA system of equations. Instead, it can be built by inspection, on the fly, in linear time, as the circuit description file is being read in. To see this, recall the two "recipes" for arriving at the MNA equation:

1. Top part (2.53): For every node other than the reference node, write KCL, then:
   (a) Eliminate all currents of group 1 elements using branch equations.
   (b) Replace all branch voltages in terms of node voltages using KVL.
2. Bottom part (2.54): For every group 2 element, write its branch equation, then replace all branch voltages in terms of node voltages using KVL.

A study of the net effect of this process shows that every element has a compact and easy to find *contribution* to the matrix equation. Indeed, it is clear that:

1. The current of every element will appear exactly twice as a KCL current in the top equations, corresponding to KCL at its two terminal nodes, but elements connected to the reference node will appear only once.
2. For a group 2 element, its current and its terminal voltages will also appear as part of its element equation in the bottom equations.

The *contribution* of every element to the matrix equation is described by means of a template, which is called an *element stamp*. The process starts by initializing the matrix and right-hand side (RHS) vector to zero. Then, the element stamps are added to the matrix and RHS vector as the elements are read in. When all the elements have been read in, the matrix equation is complete and ready to be solved.

Element Stamps   Consider a resistor whose terminal nodes are denoted n^+ and n^-, with voltages v^+ and v^-, respectively, and whose current i_R has a reference direction from n^+ to n^-, as shown in Fig. 2.15. Then, $i_R = v^+(1/R) + v^-(-1/R)$, and the element stamps are, for a group 1 resistor in Fig. 2.16 and, for a group 2 resistor in Fig. 2.17. A more compact way to represent these element

Figure 2.15: A resistor.

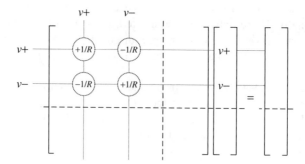

Figure 2.16: Element stamp for a resistor in group 1.

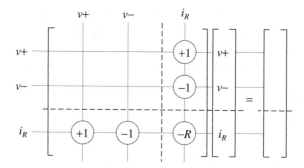

Figure 2.17: Element stamp for a resistor in group 2.

stamps is as shown in Table 2.1 and Table 2.2, where the row and column labels in the tables refer to the corresponding rows and columns of the MNA system.

For an independent current source, with the element equation $i_S = I$, with a reference direction for current from terminal node n^+ to terminal node n^-, whose voltages are v^+ and v^- respectively, as shown in Fig. 2.18, the element stamps

Table 2.1: Element stamp for a resistor in group 1.

	v^+	v^-	RHS
v^+	$1/R$	$-1/R$	
v^-	$-1/R$	$1/R$	

Table 2.2: Element stamp for a resistor in group 2.

	v^+	v^-	i	RHS
v^+			$+1$	
v^-			-1	
i	$+1$	-1	$-R$	

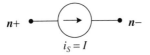

Figure 2.18: An independent current source.

are as follows. When in group 1, the element stamp is as shown in Fig. 2.19. When in group 2, the element stamp is as shown in Fig. 2.20 or, equivalently, in Table 2.3 and Table 2.4.

For an independent voltage source, with the element equation $v^+ - v^- = V$, and a reference direction for current from n^+ to n^-, as shown in Fig. 2.21, the element stamp is as shown in Fig. 2.22 and Table 2.5. For a VCVS, with element equation $v^+ - v^- = k(v_x^+ - v_x^-)$, and a reference direction for current from v^+ to v^-, as shown in Fig. 2.23, the element stamp is as shown in Fig. 2.24 and Table 2.6. For a CCVS, with element equation $v^+ - v^- = ki_x$, and a reference direction for current from v^+ to v^-, as shown in Fig. 2.25, the element stamp is as shown in Fig. 2.26 and Table 2.7. The cases of a VCCS and a CCCS can be found just as easily, and are left as an exercise for the reader. Both the VCCS

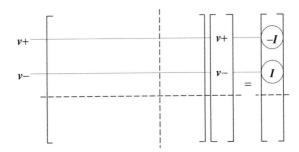

Figure 2.19: Element stamp for an independent current source in group 1.

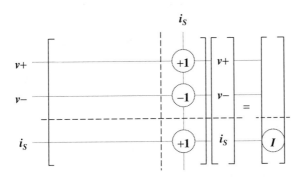

Figure 2.20: Element stamp for an independent current source in group 2.

Table 2.3: Element stamp for an independent current source in group 1.

	v^+	v^-	RHS
v^+			$-I$
v^-			$+I$

Table 2.4: Element stamp for an independent current source in group 2.

	v^+	v^-	i	RHS
v^+			$+1$	
v^-			-1	
i			$+1$	I

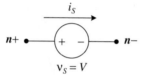

Figure 2.21: An independent voltage source.

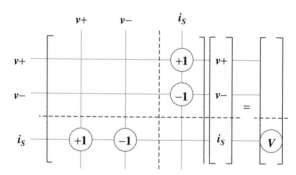

Figure 2.22: Element stamp for an independent voltage source.

and CCCS can be either in group 1 or group 2, and the control element of a VCCS can be in group 1 or group 2. In this way, with a library of element stamps, one can build the MNA equations for any network, "on the fly," while reading its description file.

Example   For the circuit shown in Fig. 2.27, we will use element stamps to construct the full MNA matrix equation. We start by identifying group membership, and we place elements e_1 and e_2 in group 1, and elements e_3 and e_4 in group 2.

Table 2.5: Element stamp for an independent voltage source.

	v^+	v^-	i	RHS
v^+			$+1$	
v^-			-1	
i	$+1$	-1		V

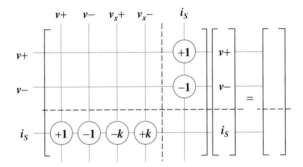

Figure 2.23: A voltage-controlled voltage source.

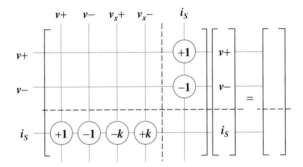

Figure 2.24: Element stamp for a VCVS.

Table 2.6: Element stamp for a VCVS.

	v^+	v^-	v_x+	v_x-	i	RHS
v^+					$+1$	
v^-					-1	
i	$+1$	-1	$-k$	$+k$		

Note that e_3 must be in group 2 because its current is the control variable for the CCCS e_2. The element stamps for the group 1 elements are shown in Table 2.8, and those of the group 2 elements are in Table 2.9. The resulting matrix equation is:

$$
\begin{bmatrix}
0 & 0 & 1 & 1 \\
0 & 1 & -3 & 0 \\
-1 & 1 & 1 & 0 \\
1 & 0 & 0 & 0
\end{bmatrix}
\begin{bmatrix}
v_1 \\
v_2 \\
i_3 \\
i_4
\end{bmatrix}
=
\begin{bmatrix}
0 \\
0 \\
0 \\
3
\end{bmatrix}
\tag{2.58}
$$

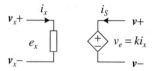

Figure 2.25: A current-controlled voltage source.

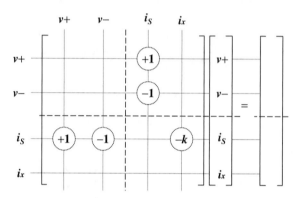

Figure 2.26: Element stamp for a CCVS.

Table 2.7: Element stamp for a CCVS.

	v^+	v^-	i_s	i_x	RHS
v^+			$+1$		
v^-			-1		
i_s	$+1$	-1		$-k$	
i_x					

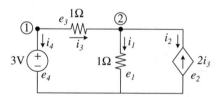

Figure 2.27: A circuit to demonstrate the MNA equation formulation.

Sparsity Considerations in MNA   Recall, an $n \times n$ matrix is said to be technically[1] *sparse* if it has $\mathcal{O}(n)$ non-zero entries. While the sparsity of STA

---

[1]We will see later on that the definition of sparsity can be less strict than this. Effectively, practitioners call a matrix *sparse* if, for the computational task at hand, there is some benefit to taking its pattern of zeros/non-zeros into account. This is not very useful as an *a priori* characterization of sparsity, but it reflects the practical experience that it is hard to "pin down" matrix sparsity with a strict mathematical definition.

Table 2.8: Element stamps for elements e_1 and e_2 in the circuit shown in Fig. 2.27.

(for e_1)			(for e_2)	
	v_2	RHS		i_3	RHS
v_2	1		v_2	-2

Table 2.9: Element stamps for elements e_3 and e_4 in the circuit shown in Fig. 2.27.

(for e_3)					(for e_4)		
	v_1	v_2	i_3	RHS		v_1	i_4	RHS
v_1			$+1$		v_1		$+1$	
v_2			-1		i_4	$+1$		$+3$
i_3	-1	$+1$	$+1$					

was provable, as we saw earlier, sparsity of MNA is not so easily established. Instead, we can offer two arguments in favor of the suggestion that the MNA matrix should in practice be sparse, as follows:

1. Suppose the network has a single voltage source, with one terminal connected to ground, and no controlled sources or dynamic elements. In this case, for a network with n nodes (including ground), the system has n equations in n unknowns, the bottom row has exactly 1 non-zero element, and every other row k has $1 + d(k)$ non-zero elements, where $d(k)$ is the number of neighbors of node k (its degree in the undirected graph G_n). Therefore, the total number of non-zero elements is:

$$1 + \sum_{k=1}^{n-1}(1 + d(k)) = n + 2m \tag{2.59}$$

This is true because, in a graph, the sum of all vertex degrees is equal to twice the number of edges. Therefore, if m is $\mathcal{O}(n)$, which it usually is, then the matrix is sparse.

2. More generally, the MNA matrix has at least $(n - 1)$ rows and, because typical element stamps contribute no more than 6 non-zero entries per stamp, then the total number of non-zero entries is no more than $6m$. Therefore, if m is $\mathcal{O}(n)$, which it usually is, then the MNA matrix is sparse.

Both arguments depend on the condition that m is $\mathcal{O}(n)$, which can be justified as follows. For a connected, loopless, undirected graph with no multiple edges, if the graph is *planar*, then it is known that $m \leq 3n - 6$. Since most circuits are nearly planar, and because the number of parallel elements is certainly bounded, then m should be $\mathcal{O}(n)$ in practice. Indeed, practical experience shows that the MNA matrix is quite sparse, and MNA is the formulation of choice for circuit simulation.

2.5 FORMULATION OF LINEAR DYNAMIC EQUATIONS

We now present the equations in the case when the network is linear but dynamic, i.e., it contains (linear) capacitors and inductors. The equations can be arrived at by a simple extension of the MNA formulation, starting from the branch equation (2.48), and based on the fact that capacitors can be in either group 1 or group 2, but inductors must be in group 2.

In this case, a general branch equation of the form $Zi + Li' - Yu - Cu' = s$ can be written in partitioned matrix form as:

$$
\begin{bmatrix} I & Z_{12} \\ 0 & Z_{22} \end{bmatrix} \begin{bmatrix} i_1 \\ i_2 \end{bmatrix}
$$
$$
+ \begin{bmatrix} 0 & 0 \\ 0 & L_{22} \end{bmatrix} \begin{bmatrix} i'_1 \\ i'_2 \end{bmatrix} - \begin{bmatrix} Y_{11} & Y_{12} \\ Y_{21} & Y_{22} \end{bmatrix} \begin{bmatrix} u_1 \\ u_2 \end{bmatrix} - \begin{bmatrix} C_{11} & 0 \\ 0 & C_{22} \end{bmatrix} \begin{bmatrix} u'_1 \\ u'_2 \end{bmatrix} = \begin{bmatrix} s_1 \\ s_2 \end{bmatrix} \qquad (2.60)
$$

where C_{11}, C_{22}, and L_{22} are all diagonal matrices. Notice that, if there are no capacitors in group 2, then $C_{22} = 0$. Using similar algebraic substitutions, as was done previously, we arrive at the final form for MNA in the dynamic case:

$$
\begin{bmatrix} \left(A_1 Y_{11} A_1^T + A_1 Y_{12} A_2^T\right) & (A_2 - A_1 Z_{12}) \\ -\left(Y_{21} A_1^T + Y_{22} A_2^T\right) & Z_{22} \end{bmatrix} \begin{bmatrix} v \\ i_2 \end{bmatrix}
$$
$$
+ \begin{bmatrix} A_1 C_{11} A_1^T & 0 \\ -C_{22} A_2^T & L_{22} \end{bmatrix} \begin{bmatrix} v' \\ i'_2 \end{bmatrix} = \begin{bmatrix} -A_1 s_1 \\ s_2 \end{bmatrix} \qquad (2.61)
$$

If the network contains no controlled sources, then this MNA system is simplified to:

$$
\begin{bmatrix} A_1 Y_{11} A_1^T & A_2 \\ -Y_{22} A_2^T & Z_{22} \end{bmatrix} \begin{bmatrix} v \\ i_2 \end{bmatrix} + \begin{bmatrix} A_1 C_{11} A_1^T & 0 \\ -C_{22} A_2^T & L_{22} \end{bmatrix} \begin{bmatrix} v' \\ i'_2 \end{bmatrix} = \begin{bmatrix} -A_1 s_1 \\ s_2 \end{bmatrix} \qquad (2.62)
$$

Furthermore, if capacitors are present only in group 1, then some further simplification is possible due to $C_{22} = 0$, leading to:

$$
\begin{bmatrix} A_1 Y_{11} A_1^T & A_2 \\ -Y_{22} A_2^T & Z_{22} \end{bmatrix} \begin{bmatrix} v \\ i_2 \end{bmatrix} + \begin{bmatrix} A_1 C_{11} A_1^T & 0 \\ 0 & L_{22} \end{bmatrix} \begin{bmatrix} v' \\ i'_2 \end{bmatrix} = \begin{bmatrix} -A_1 s_1 \\ s_2 \end{bmatrix} \qquad (2.63)
$$

Finally, if in addition we require that group 2 contains only voltage sources and inductors, then $Z_{22} = 0$ and $Y_{22} = I$, leading to:

$$\begin{bmatrix} A_1 Y_{11} A_1^T & A_2 \\ -A_2^T & 0 \end{bmatrix} \begin{bmatrix} v \\ i_2 \end{bmatrix} + \begin{bmatrix} A_1 C_{11} A_1^T & 0 \\ 0 & L_{22} \end{bmatrix} \begin{bmatrix} v' \\ i_2' \end{bmatrix} = \begin{bmatrix} -A_1 s_1 \\ s_2 \end{bmatrix} \qquad (2.64)$$

which is the common form of the MNA system that one typically sees in the literature.

The dynamic MNA equations can also be built using element stamps, as follows, although, in practice simulators never assemble time-domain dynamic equations in this way. Our examination of this issue is simply an interesting exercise that shows how the dynamic equations can be efficiently set up, if so desired.

2.5.1 Dynamic Element Stamps

For a capacitor, with element equation $i_C = C dv^+/dt - C dv^-/dt$, and a reference direction for current from n^+ to n^-, as shown in Fig. 2.28, it is easy to see that the element stamps are as follows. When in group 1, the capacitor element stamp is as shown in Fig. 2.29. When in group 2, the capacitor element stamp is as shown in Fig. 2.30. For an inductor, with element equation $v^+ - v^- = L di_L/dt$, and a reference direction for current from n^+ to n^-, as shown in Fig. 2.31, the element stamp is as shown in Fig. 2.32. Element stamps for other (resistive) element types are constructed as we saw previously.

Frequency Domain   The above techniques for equation formulation can be applied in the frequency domain, for circuits with linear capacitors and inductors.

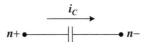

Figure 2.28: A capacitor.

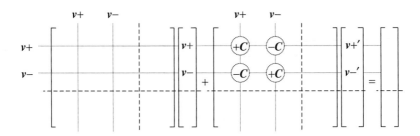

Figure 2.29: Element stamp for a capacitor in group 1.

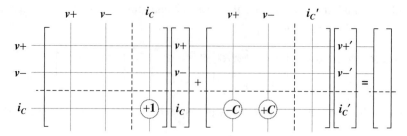

Figure 2.30: Element stamp for a capacitor in group 2.

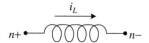

Figure 2.31: An inductor.

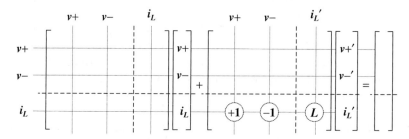

Figure 2.32: Element stamp for an inductor.

A capacitor C has an admittance of $j\omega C$, and can be either in group 1 or group 2, with a corresponding element stamp. An inductor L has an impedance of $j\omega L$, and must be in group 2, with a corresponding element stamp. The corresponding element stamps are very similar to that of a resistor. We will not cover frequency domain analysis in this text.

2.5.2 Unique Solvability

In general, a linear, resistive network with no controlled sources, and given the consistency requirements, is uniquely solvable. We saw the justification for this earlier, in connection with the conductance matrix of nodal analysis, but it also applies when voltage sources are present (i.e., for the MNA system) because one can use superposition to solve a network with voltage sources using multiple applications of nodal analysis (although that would be expensive). Unfortunately, no such result is available for the dynamic case.

For dynamic circuits, the challenge is to make sure that the "order of complexity" or "index" of the differential equations, is no greater than 2. We will briefly touch on this topic later on, but we skip the details for now. Additional constraints are often required in order to achieve this, which relate to CV-loops and LI-cutsets. A CV-loop is a loop consisting of only capacitors and voltage sources. An LI-cutset is a cutset made of only inductors and current sources. Circuits with controlled sources, whether linear or nonlinear, have more problems, and require more constraints, typically of a topological nature.

Notes Additional reading is available in Chua and Lin (1975), sections 2.1–2.5, 3.1–3.5, and 4.1–4.2, in Chua et al. (1987), chapters 5, 8, and 12, as well as in Vlach and Singhal (1994), sections 1.1–1.5, 3.1–3.8, and 4.1–4.4.

The question of unique solvability continues to be an open problem and an active research topic in circuit theory. Classical results are summarized in Chua and Lin (1975). For nonlinear resistive circuits, one can further consult the early work in Nishi and Chua (1984). For linear dynamic networks, some recent results are described in Reißig (1999), while nonlinear dynamic networks are addressed in Estévez Schwarz and Tischendorf (2000) and in Tischendorf (1996). In practice, with an accurate circuit model, sufficient parasitics will typically exist in the circuit description file, which helps avoid pathological behavior, so that most practical circuits are found to be uniquely solvable.

Problems

2.1. Consider the linear network in Fig. 2.33, in which the reference directions for positive current are as indicated.

  (a) Draw the directed and undirected graphs corresponding to this network and give its incidence matrix, M.

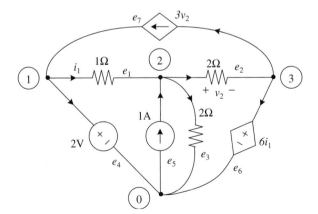

Figure 2.33: Linear network for problem 2.1.

(b) Find the matrix C, which is the basis matrix for the cycle space of this network.

(c) Give the resulting topological constraints for the network, as a system of m equations in $2m$ unknowns, where m is the number of elements.

2.2. If matrix $A = [a_{ij}]$ is symmetric positive definite, prove that $a_{ii} > 0$, $\forall i$.

2.3. For a linear, resistive circuit with no controlled sources and no voltage sources, show that the nodal admittance matrix G is symmetric and diagonally dominant.

2.4. If, in addition to the conditions of problem 2.3, we further require that a resistor is connected from every node to ground, show that G is *strictly* diagonally dominant.

2.5. If, in addition to the conditions of problem 2.3, we further require that the network is connected and meets the consistency requirements, show that G is irreducibly diagonally dominant.

2.6. Suppose a linear resistive network contains a number of voltage sources, each having a grounded terminal. Explain how you would use nodal analysis to formulate the equations for such a network.

2.7. Consider a linear resistive network, that may contain voltage sources. Show how it can be solved by repeated application of nodal analysis, by appealing to the principle of superposition.

2.8. Give the MNA element stamps for a VCCS and a CCCS, considering the case when each element is in group 1 or in group 2.

2.9. Give a full derivation for the MNA equations for a linear circuit with dynamic elements.

2.10. (Computer Project) Using the parser developed previously in problem 1.1 as a front-end, write a C or C++ program that accepts a description of

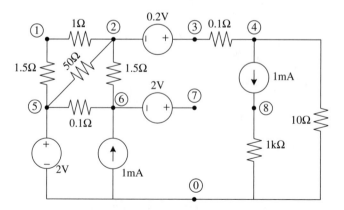

Figure 2.34: A test circuit.

```
V1 5 0 2

V2 3 2 0.2

V3 7 6 2

I1 4 8 1e-3

I2 0 6 1e-3


R1 1 5 1.5

R2 1 2 1

R3 5 2 50    G2    % this is a group 2 element

R4 5 6 0.1

R5 2 6 1.5

R6 3 4 0.1

R7 8 0 1e3

R8 4 0 10    G2    % this is a group 2 element
```

Figure 2.35: Circuit description file for the circuit in Fig. 2.34.

any resistive linear circuit, with no controlled sources, and constructs the corresponding MNA system using element stamps. In other words, your program should accept any network of linear resistors, independent current sources, and independent voltage sources. Your program should make use of the linked-list data structure created by the parser. It should interpret the optional field [G2] introduced earlier, in the specification of the parser, as indicating that an element belongs to group 2. The [G2] flag is not *required* for membership in group 2. Test your program on the circuit shown in Fig. 2.34, where the 10Ω and 50Ω resistors are required to be in group 2. With the circuit description file given in Fig. 2.35, the correct solution is given in (2.65).

$$
\begin{bmatrix}
10 & 0 & -10 & 0 & 0 & 0 & 0 & 0 & 1 & 0 & 0 & 0 & 0 \\
0 & 0.001 & 0 & 0 & 0 & 0 & 0 & 0 & 0 & 0 & 0 & 0 & 0 \\
-10 & 0 & 10 & 0 & 0 & 0 & 0 & 0 & 0 & 0 & 0 & 1 & 0 \\
0 & 0 & 0 & 5/3 & -2/3 & 0 & -1 & 0 & 0 & -1 & 0 & -1 & 0 \\
0 & 0 & 0 & -2/3 & 32/3 & -10 & 0 & 0 & 0 & 0 & -1 & 0 & 0 \\
0 & 0 & 0 & 0 & -10 & 32/3 & -2/3 & 0 & 0 & 1 & 0 & 0 & 1 \\
0 & 0 & 0 & -1 & 0 & -2/3 & 5/3 & 0 & 0 & 0 & 0 & 0 & 0 \\
0 & 0 & 0 & 0 & 0 & 0 & 0 & 0 & 0 & 0 & 1 & 0 & 0 \\
1 & 0 & 0 & 0 & 0 & 0 & 0 & 0 & -10 & 0 & 0 & 0 & 0 \\
0 & 0 & 0 & -1 & 0 & 1 & 0 & 0 & 0 & -50 & 0 & 0 & 0 \\
0 & 0 & 0 & 0 & -1 & 0 & 0 & 1 & 0 & 0 & 0 & 0 & 0 \\
0 & 0 & 1 & -1 & 0 & 0 & 0 & 0 & 0 & 0 & 0 & 0 & 0 \\
0 & 0 & 0 & 0 & 0 & 1 & 0 & 0 & 0 & 0 & 0 & 0 & 0
\end{bmatrix}
\begin{bmatrix}
V(4) \\ V(8) \\ V(3) \\ V(2) \\ V(6) \\ V(5) \\ V(1) \\ V(7) \\ I(R8) \\ I(R3) \\ I(V3) \\ I(V2) \\ I(V1)
\end{bmatrix}
=
\begin{bmatrix}
-0.001 \\ 0.001 \\ 0 \\ 0 \\ 0.001 \\ 0 \\ 0 \\ 0 \\ 0 \\ 0 \\ 2 \\ 0.2 \\ 2
\end{bmatrix}
$$

(2.65)

CHAPTER 3

Solution of Linear Algebraic Circuit Equations

In this chapter, we study the solution methods for linear resistive networks. Once the network equations have been assembled, the solution of a linear resistive network reduces to a solution of the *linear algebraic system*:

$$Ax = b \qquad (3.1)$$

where A is a square $n \times n$ matrix, b is an n-vector, and x is an n-vector of the unknown circuit voltages and currents. This is a classical problem, whose solution has been the subject of decades of research in mathematics and numerical analysis. In the West, early solutions to this problem date back to the work of Gauss in 1809. Our version of this problem has the two key features that the problem size is very big (n can be in the millions) and, luckily, the matrix A is very sparse, typically having no more than about 20 non-zero entries per row, on average. Had circuit matrices been "full," rather than sparse, we would not be able to simulate the circuits that we simulate today. A full matrix is undesirable for reasons of both memory (storage) and execution time. The cardinal rule in circuit simulation is that one must never allow the matrix to become full; sparsity must be maintained.

Of course, the obvious solution to the problem $Ax = b$ is $x = A^{-1}b$, so that one way to find x is to first explicitly compute A^{-1} and then multiply that by b. One way of finding A^{-1} is the classical method of *matrix cofactors*, but that method is very inefficient for large matrices. Another option is to use Cramer's rule, which is based on the notion of the *determinant* of a (square) matrix, denoted $\det(\cdot)$. If we let A_k be the matrix resulting from A once its kth column is replaced by b, then Cramer's rule provides:

$$x_k = \frac{\det(A_k)}{\det(A)} \qquad (3.2)$$

However, Cramer's rule is also extremely inefficient, requiring $2(n+1)!$ multiplications to find the solution, x. When $n = 10$, this evaluates to 79,833,600

Circuit Simulation, by Farid N. Najm
Copyright © 2010 John Wiley & Sons, Inc.

multiplications! In any case, for circuit simulation, we *never* explicitly find the matrix inverse because, even if A is sparse, A^{-1} is typically full. In large numerical analysis problems, in general, one almost never explicitly computes the matrix inverse.

Instead, one of the most efficient and robust solution methods, and the most commonly used, is *Gaussian elimination (GE)*. Gaussian elimination is also easy to implement and can be easily extended to exploit matrix sparsity, which is a *key requirement* for circuit simulation. Taking a broader view, two *classes* of efficient techniques are available:

1. Direct methods: solve the system in a fixed and pre-determined number of steps. This class includes Gaussian elimination, and its many variants, such as *LU* factorization and Cholesky decomposition.

2. Indirect methods: are iterative techniques that approach the solution with (hopefully) gradually improving accuracy. These methods converge only if the matrix A has certain properties. Examples are Gauss-Seidel, Gauss-Jacobi, Conjugate Gradient, etc.

Most modern circuit simulators are based on the use of *LU* factorization. However, we will look at a whole range of solution techniques, including both direct and indirect methods.

3.1 DIRECT METHODS

We will describe basic Gaussian elimination, provide the theoretical justification for it, then describe *LU* factorization as a variant of it. But, first, we will briefly review some matrix basics. The material in this section is based on a number of sources, including Horn and Johnson (1985), Chua and Lin (1975), Ruehli (1986), Vlach and Singhal (1994), and Pillage et al. (1995).

3.1.1 Matrix Preliminaries

As customary, let $\mathbb{R}^{m \times n}$ denote the set of all $m \times n$ real matrices and let $\mathbb{R}^n$ denote the set of all n-dimensional real vectors. If $A \in \mathbb{R}^{m \times n}$, then its entry in the ith row and jth column will be denoted a_{ij}.

Determinant   Let $A \in \mathbb{R}^{n \times n}$ be a square matrix and let $i \in \{1, 2, \ldots, n\}$, then the familiar *determinant* of A is the real number given by:

$$\det(A) \triangleq \begin{cases} \sum_{j=1}^{n}(-1)^{i+j}a_{ij}c_{ij}, & \text{if } n > 1, \\ a_{11}, & \text{if } n = 1. \end{cases} \tag{3.3}$$

where c_{ij} is the determinant of the $(n-1) \times (n-1)$ matrix that results from A by removing its ith row and jth column.

Determinants have a number of interesting properties. If A is a square matrix, then A is nonsingular if and only if $\det(A)$ is non-zero, in which case:

$$\det(A^{-1}) = \frac{1}{\det(A)} \tag{3.4}$$

A matrix and its transpose have the same determinant:

$$\det(A^T) = \det(A) \tag{3.5}$$

If $A, B \in \mathbb{R}^{n \times n}$, then:

$$\det(AB) = \det(A)\det(B) \tag{3.6}$$

A useful corollary, when k is a (positive) integer, is that:

$$\det(A^k) = \det(A)^k \tag{3.7}$$

If $\alpha \in \mathbb{R}$ is any real scalar and $A \in \mathbb{R}^{n \times n}$, then:

$$\det(\alpha A) = \alpha^n \det(A) \tag{3.8}$$

If $\lambda_1, \lambda_2, \ldots, \lambda_n$ are all the (possibly complex) eigenvalues of $A \in \mathbb{R}^{n \times n}$, then:

$$\det(A) = \prod_{i=1}^{n} \lambda_i \tag{3.9}$$

so that a matrix is singular if and only if it has a zero eigenvalue. If $A, B \in \mathbb{R}^{n \times n}$ are *similar*, i.e., if there exists a nonsingular $X \in \mathbb{R}^{n \times n}$ such that $A = X^{-1}BX$, then:

$$\det(A) = \det(B) \tag{3.10}$$

so that the determinant is *similarity-invariant*. As a result, the determinant of a linear transformation $T : V \to V$, where V is a finite-dimensional vector space, is independent of the basis of V.

We will later on be interested in certain *elementary row operations* on matrices, whose effect on the matrix determinant will be based on the following. If B is obtained from A by exchanging two rows or two columns, then:

$$\det(B) = -\det(A) \tag{3.11}$$

Therefore, reordering (i.e., permuting) the matrix rows or columns changes only the sign of the determinant. Thus, $|\det(A)|$ is *invariant to row or column reordering (permutation)*. If B is obtained from A by multiplying one row or column by $\alpha \in \mathbb{R}$, then:

$$\det(B) = \alpha \det(A) \tag{3.12}$$

Finally, if B is obtained from A by adding to one row (or column) the product of any $\alpha \in \mathbb{R}$ and any other row (or column), then, somewhat surprisingly:

$$\det(B) = \det(A) \tag{3.13}$$

Diagonal Matrix   A square matrix D is called a *diagonal matrix* if $d_{ij} = 0$ whenever $i \neq j$, i.e., if all its off-diagonal entries are zero, and is denoted $D = \text{diag}(d_{11}, d_{22}, \ldots, d_{nn})$. A diagonal matrix is called a *scalar matrix* if $d_{ii} = d_{jj}$ for all i and j. The determinant of a diagonal matrix is simply the product of its diagonal entries:

$$\det(D) = \prod_{i=1}^{n} d_{ii} \tag{3.14}$$

Thus, a diagonal matrix is nonsingular if and only if all its diagonal entries are non-zero. Let $D = \text{diag}(d_{11}, d_{22}, \ldots, d_{nn})$, then λ is an eigenvalue of D if and only if $\lambda \in \{d_{11}, d_{22}, \ldots, d_{nn}\}$. Thus, a diagonal matrix carries all its eigenvalues on its diagonal.

If $A, D \in \mathbb{R}^{n \times n}$ and D is diagonal, then DA has the effect of multiplying the *rows* of A by the diagonal entries of D, so that the ith row of A is multiplied by d_{ii}. If $A, D \in \mathbb{R}^{n \times n}$ and D is diagonal, then AD has the effect of multiplying the *columns* of A by the diagonal entries of D, so that the ith column of A is multiplied by d_{ii}.

Finally, if $D, E \in \mathbb{R}^{n \times n}$ are both diagonal, then:

$$DE = ED = \text{diag}(d_{11}e_{11}, d_{22}e_{22}, \ldots, d_{nn}e_{nn}) \tag{3.15}$$

Triangular Matrix   A square matrix T is said to be *upper triangular* if all its entries below the diagonal are 0. Furthermore, if $t_{ii} = 0$ for all i, then T is called *strictly upper triangular*. For example, the matrix on the left in (3.16) is upper triangular and the one on the right is strictly upper triangular.

$$
\begin{bmatrix}
t_{11} & t_{12} & t_{13} & \cdots & t_{1n} \\
0 & t_{22} & t_{23} & \cdots & t_{2n} \\
0 & 0 & t_{33} & \cdots & t_{3n} \\
\vdots & \vdots & \vdots & \ddots & \vdots \\
0 & 0 & \cdots & 0 & t_{nn}
\end{bmatrix}
\quad
\begin{bmatrix}
0 & t_{12} & t_{13} & \cdots & t_{1n} \\
0 & 0 & t_{23} & \cdots & t_{2n} \\
0 & 0 & 0 & \cdots & t_{3n} \\
\vdots & \vdots & \vdots & \ddots & \vdots \\
0 & 0 & \cdots & 0 & 0
\end{bmatrix}
\tag{3.16}
$$

If T is upper triangular and $t_{ii} = 1$ for all i, then T is called *unit upper triangular*; we will also refer to it as being *of unit-type*, for brevity.

A square matrix T is said to be *lower triangular* if all its entries above the diagonal are 0. Furthermore, if $t_{ii} = 0$ for all i, then T is called *strictly lower triangular*. For example, the matrix on the left in (3.17) is lower triangular and the one on the right is strictly lower triangular.

$$\begin{bmatrix} t_{11} & 0 & 0 & \cdots & 0 \\ t_{21} & t_{22} & 0 & \cdots & 0 \\ t_{31} & t_{32} & t_{33} & \cdots & 0 \\ \vdots & \vdots & \vdots & \ddots & \vdots \\ t_{n1} & t_{n2} & \cdots & t_{n,n-1} & t_{nn} \end{bmatrix} \qquad \begin{bmatrix} 0 & 0 & 0 & \cdots & 0 \\ t_{21} & 0 & 0 & \cdots & 0 \\ t_{31} & t_{32} & 0 & \cdots & 0 \\ \vdots & \vdots & \vdots & \ddots & \vdots \\ t_{n1} & t_{n2} & \cdots & t_{n,n-1} & 0 \end{bmatrix} \qquad (3.17)$$

If T is lower triangular and $t_{ii} = 1$ for all i, then T is called *unit lower triangular*; we will also refer to it as being *of unit-type*, for brevity.

A matrix is said to be *triangular* if it is either upper triangular or lower triangular. The determinant of a triangular matrix is simply the product of its diagonal entries:

$$\det(T) = \prod_{i=1}^{n} t_{ii} \qquad (3.18)$$

Thus, *a triangular matrix is nonsingular if and only if all its diagonal entries are non-zero.*

If $T \in \mathbb{R}^{n \times n}$ is a triangular matrix, then λ is an eigenvalue of T if and only if $\lambda \in \{t_{11}, t_{22}, \ldots, t_{nn}\}$. Thus, a triangular matrix carries all its eigenvalues on its diagonal.

Finally, the following algebraic properties of triangular matrices can be useful. Triangular matrices do not necessarily commute under multiplication. The inverse of an upper (lower) triangular matrix is upper (lower) triangular. The inverse of a unit upper (lower) triangular matrix is unit upper (lower) triangular. The product of two upper (lower) triangular matrices is upper (lower) triangular. The product of two unit upper (lower) triangular matrices is unit upper (lower) triangular.

Permutation Matrices   The following class of matrices are useful in the study of Gaussian elimination.

Definition 3.1. (Permutation Matrix) A *permutation matrix* is a square matrix that has a single 1 entry in every row and every column, while all other entries are 0.

If $P \in \mathbb{R}^{n \times n}$ is a permutation matrix, then P can be obtained from I by a permutation of its rows or its columns. Recall, a *permutation* is a mapping from $\{1, 2, \ldots, n\}$ onto itself, which is *one-to-one* and *onto*; informally, a permutation is a "reshuffle." It can be shown that $\det(P) = \pm 1$, so that P is nonsingular, and $P^{-1} = P^T$. Furthermore, if $A \in \mathbb{R}^{n \times n}$, then PA can be obtained from A by a permutation of its *rows*, and AP can be obtained from A by a permutation of its *columns*.

Combinatorics   For assessing computational complexity of algorithms, some basic results from combinatorics are useful, as follows. When the order does not matter, the number of combinations without repetition, of r out of k items, ("k choose r") is the *binomial coefficient*:

$$\binom{k}{r} = \frac{k!}{r!(k-r)!} \tag{3.19}$$

One can show that, for a fixed $r \le n$:

$$\sum_{k=r}^{n} \binom{k}{r} = \binom{n+1}{r+1} \tag{3.20}$$

and this leads to:

$$\sum_{k=1}^{n} k = \frac{n(n+1)}{2} \tag{3.21}$$

and

$$\sum_{k=1}^{n} k^2 = \frac{n(n+1)(2n+1)}{6} \tag{3.22}$$

3.1.2 Gaussian Elimination (GE)

Consider the simple case of a system of two linear equations in two unknowns:

$$x_1 + x_2 = 5$$
$$x_1 - x_2 = 1 \tag{3.23}$$

An elementary method of simplifying this system is to use the first equation to write $x_1 = 5 - x_2$ and substitute that in place of x_1 in the second equation, thereby eliminating x_1 from the second equation. This simple method of variable elimination was extended by Gauss into the systematic procedure that we call *Gaussian elimination* (GE). The basic method consists of two phases, called *forward elimination* (FE) and *backward substitution* (BS).

Forward elimination proceeds in $n - 1$ steps, where each step eliminates one equation and one unknown, until all that remains is one equation in one unknown. It operates on both A and b, producing successively simplified versions of the original system. Effectively, this performs a *triangularization* of the matrix, leading to an *upper triangular* matrix equation:

$$\begin{bmatrix} a_{11} & a_{12} & a_{13} & \cdots & a_{1n} \\ 0 & a_{22} & a_{23} & \cdots & a_{2n} \\ 0 & 0 & a_{33} & \cdots & a_{3n} \\ \vdots & \vdots & \vdots & \ddots & \vdots \\ 0 & 0 & \cdots & 0 & a_{nn} \end{bmatrix} x = b \tag{3.24}$$

Once in this form, the solution of the system becomes simple because, with:

$$a_{11}x_1 + a_{12}x_2 + a_{13}x_3 + \cdots + a_{1n}x_n = b_1$$

$$a_{22}x_2 + a_{23}x_3 + \cdots + a_{2n}x_n = b_2$$

$$a_{33}x_3 + \cdots + a_{3n}x_n = b_3$$

$$\vdots \quad \vdots \qquad \vdots$$

$$a_{n-1,n-1}x_{n-1} + a_{n-1,n}x_n = b_{n-1}$$

$$a_{nn}x_n = b_n$$

we can simply write $x_n = b_n/a_{nn}$, $x_{n-1} = \left(b_{n-1} - a_{n-1,n}x_n\right)/a_{n-1,n-1}$, etc. In general, and using the convention that $\sum_{n+1}^{n}(\cdot) = 0$, we would perform, for $k = n, (n-1), \ldots, 1$:

$$x_k = \frac{b_k - \sum_{j=k+1}^{n} a_{kj}x_j}{a_{kk}} \tag{3.25}$$

and this is precisely what the process of backward substitution is. It can be shown that forward elimination requires $\approx n^3/3$ operations,[1] while backward substitution requires $\approx n^2/2$. Thus, Gaussian elimination is, overall, an $\mathcal{O}(n^3)$ algorithm. If the matrix is sparse, then GE can be done much more efficiently than this, as we will see later on.

We have described backward substitution and it remains to explain forward elimination. To this end, consider the initial state of the system:

$$a_{11}x_1 + a_{12}x_2 + \cdots + a_{1n}x_n = b_1$$

$$a_{21}x_1 + a_{22}x_2 + \cdots + a_{2n}x_n = b_2$$

$$a_{31}x_1 + a_{32}x_2 + \cdots + a_{3n}x_n = b_3$$

$$\vdots \qquad \qquad \vdots$$

$$a_{n1}x_1 + a_{n2}x_2 + \cdots + a_{nn}x_n = b_n$$

If we multiply the first equation by (a_{21}/a_{11}) and subtract it from the second equation, we get:

$$
\begin{array}{llllll}
a_{11}x_1 + & a_{12}x_2 & + \cdots + & a_{1n}x_n & = & b_1 \\
0 \cdot x_1 + & \left[a_{22} - a_{12}\left(\dfrac{a_{21}}{a_{11}}\right)\right]x_2 & + \cdots + & \left[a_{2n} - a_{1n}\left(\dfrac{a_{21}}{a_{11}}\right)\right]x_n & = & b_2 - b_1\left(\dfrac{a_{21}}{a_{11}}\right) \\
a_{31}x_1 + & a_{32}x_2 & + \cdots + & a_{3n}x_n & = & b_3 \\
\vdots & \vdots \quad \vdots & & \vdots & & \vdots \\
a_{n1}x_1 + & a_{n2}x_2 & + \cdots + & a_{nn}x_n & = & b_n
\end{array}
$$

[1]As is common practice, *operations* will refer to only floating point multiplications and divisions, which are significantly more expensive than additions or subtractions.

Thus, we have *eliminated* x_1 from the 2nd equation. Repeating this for equation $i = 2, 3, \ldots, n$, using (a_{i1}/a_{11}) in every case, we would eliminate x_1 from all the other equations, leading to:

$$
\begin{aligned}
a_{11}x_1 + \quad & a_{12}x_2 & + \cdots + & \quad a_{1n}x_n & = & \quad b_1 \\
0 \cdot x_1 + \left[a_{22} - a_{12}\left(\frac{a_{21}}{a_{11}}\right)\right]x_2 & + \cdots + \left[a_{2n} - a_{1n}\left(\frac{a_{21}}{a_{11}}\right)\right]x_n & = & \quad b_2 - b_1\left(\frac{a_{21}}{a_{11}}\right) \\
0 \cdot x_1 + \left[a_{32} - a_{12}\left(\frac{a_{31}}{a_{11}}\right)\right]x_2 & + \cdots + \left[a_{3n} - a_{1n}\left(\frac{a_{31}}{a_{11}}\right)\right]x_n & = & \quad b_3 - b_1\left(\frac{a_{31}}{a_{11}}\right) \\
\vdots & \\
0 \cdot x_1 + \left[a_{n2} - a_{12}\left(\frac{a_{n1}}{a_{11}}\right)\right]x_2 & + \cdots + \left[a_{nn} - a_{1n}\left(\frac{a_{n1}}{a_{11}}\right)\right]x_n & = & \quad b_n - b_1\left(\frac{a_{n1}}{a_{11}}\right)
\end{aligned}
$$

Forward elimination is a repetition of this process, $(n - 1)$ times.

As a result, the core of the basic Gaussian elimination algorithm (without any pivoting for now) is as shown in Fig. 3.1. In forward elimination, we start with $k = 1$, and for every row $i \geq 2$, the first row of A is multiplied by a_{i1}/a_{11} and subtracted from the ith row. The result is saved in place of the original ith row. This eliminates the variable x_1 from all but the first equation, so that the system is transformed as shown:

$$
\begin{bmatrix}
a_{11}^{(1)} & a_{12}^{(1)} & a_{13}^{(1)} & \cdots & a_{1n}^{(1)} \\
a_{21}^{(1)} & a_{22}^{(1)} & a_{23}^{(1)} & \cdots & a_{2n}^{(1)} \\
a_{31}^{(1)} & a_{32}^{(1)} & a_{33}^{(1)} & \cdots & a_{3n}^{(1)} \\
\vdots & \vdots & \vdots & \ddots & \vdots \\
a_{n1}^{(1)} & a_{n2}^{(1)} & a_{n3}^{(1)} & \cdots & a_{nn}^{(1)}
\end{bmatrix}
\begin{bmatrix}
b_1^{(1)} \\
b_2^{(1)} \\
b_3^{(1)} \\
\vdots \\
b_n^{(1)}
\end{bmatrix}
\longrightarrow
\begin{bmatrix}
a_{11}^{(1)} & a_{12}^{(1)} & a_{13}^{(1)} & \cdots & a_{1n}^{(1)} \\
0 & a_{22}^{(2)} & a_{23}^{(2)} & \cdots & a_{2n}^{(2)} \\
0 & a_{32}^{(2)} & a_{33}^{(2)} & \cdots & a_{3n}^{(2)} \\
\vdots & \vdots & \vdots & \ddots & \vdots \\
0 & a_{n2}^{(2)} & a_{n3}^{(2)} & \cdots & a_{nn}^{(2)}
\end{bmatrix}
\begin{bmatrix}
b_1^{(1)} \\
b_2^{(2)} \\
b_3^{(2)} \\
\vdots \\
b_n^{(2)}
\end{bmatrix}
$$

Input: $A^{(1)} = A$, $b^{(1)} = b$
Forward elimination:
   for $(k = 1, \ldots, (n-1))$ **do**
      for $(i = k + 1, \ldots, n)$
        $m_{ik} = a_{ik}^{(k)}/a_{kk}^{(k)}$
        $a_{ik}^{(k+1)} = 0$
        for $(j = k + 1, \ldots, n)$ **do**
           $a_{ij}^{(k+1)} = a_{ij}^{(k)} - m_{ik}a_{kj}^{(k)}$
        $b_i^{(k+1)} = b_i^{(k)} - m_{ik}b_k^{(k)}$
Backward substitution:
   for $(k = n, \ldots, 1)$ **do**
$$
x_k = \frac{1}{a_{kk}^{(k)}}\left(b_k^{(k)} - \sum_{j=k+1}^{n} a_{kj}^{(k)}x_j\right)
$$

Figure 3.1: The core of a Gaussian elimination algorithm.

where we have introduced the superscripts (1), (2), etc. to denote the initial and subsequently updated entries of the system. Notice that the first row remains the same, while all other entries of A and b have been updated. The next step involves applying the same procedure to the sub-matrix with indices from 2 to n and, after $(n-1)$ steps, the matrix becomes upper triangular:

$$\begin{bmatrix} a_{11}^{(1)} & a_{12}^{(1)} & a_{13}^{(1)} & \cdots & a_{1n}^{(1)} \\ 0 & a_{22}^{(2)} & a_{23}^{(2)} & \cdots & a_{2n}^{(2)} \\ 0 & 0 & a_{33}^{(3)} & \cdots & a_{3n}^{(3)} \\ \vdots & \vdots & \vdots & \ddots & \vdots \\ 0 & 0 & \cdots & 0 & a_{nn}^{(n)} \end{bmatrix} x = \begin{bmatrix} b_1^{(1)} \\ b_2^{(2)} \\ b_3^{(3)} \\ \vdots \\ b_n^{(n)} \end{bmatrix} \qquad (3.26)$$

In this final state of the system, notice that some rows have been updated more times than others. The 1st row has remained the same, the 2nd row was updated 1 time, the 3rd row was updated 2 times, etc., and the nth row was updated $(n-1)$ times.

Repeated Solutions With Gaussian elimination, if the right-hand side vector b changes, the process has to be restarted from the beginning. In many practical cases, including some simulation scenarios, the matrix may remain fixed while the right-hand side vector b changes. In that case, one can modify Gaussian elimination so as to reuse previous work. Rather than operating on b with the multipliers m_{ik} during forward elimination, the multipliers are instead *saved* in a matrix. For the given b, the saved multipliers are used to operate on that vector. Then, backward substitution is used as usual to find the solution x. Given another right-hand side vector, the saved matrix of multipliers, and backward substitution, are efficiently re-applied on that vector. When so modified, Gaussian elimination becomes equivalent to LU factorization, as we will see below.

Elementary Row Operations The operations performed at every step of Gaussian elimination are called *elementary row operations* (EROs), and they are of three *types*:

Type 1: Exchange two rows (for row pivoting, as we will see later on).
Type 2: Multiply one row by a constant and add the result to another row.
Type 3: Multiply one row by a constant.

These operations are each equivalent to pre-multiplying the system equation $Ax = b$ with one of three corresponding *types* of so-called *elementary matrices*:

Type 1: Obtained by applying a type 1 ERO to the identity matrix I.
Type 2: Obtained by applying a type 2 ERO to the identity matrix I.
Type 3: Obtained by applying a type 3 ERO to the identity matrix I.

The following matrices are examples of the three types:

$$
E_1 = \begin{bmatrix} 1 & 0 & 0 & 0 \\ 0 & 0 & 1 & 0 \\ 0 & 1 & 0 & 0 \\ 0 & 0 & 0 & 1 \end{bmatrix} \quad
E_2 = \begin{bmatrix} 1 & 0 & 0 & 0 \\ 0 & 1 & 0 & 0 \\ \dfrac{-a_{31}}{a_{11}} & 0 & 1 & 0 \\ 0 & 0 & 0 & 1 \end{bmatrix} \quad
E_3 = \begin{bmatrix} 1 & 0 & 0 & 0 \\ 0 & \dfrac{1}{a_{22}^{(2)}} & 0 & 0 \\ 0 & 0 & 1 & 0 \\ 0 & 0 & 0 & 1 \end{bmatrix}
$$

It can be shown that matrices of these three types are always nonsingular, so that pre-multiplying the system by such matrices does not change its solution; the solution of the final system is the solution of the original system. Thus, the net effect of Gaussian elimination is to pre-multiply the system equation by the product of a number of such elementary matrices:

$$
\left(\prod_{i=1}^{m} E_i \right) Ax = \left(\prod_{i=1}^{m} E_i \right) b \tag{3.27}
$$

leading to the triangulated final form (3.26). Further details on this topic are given in Chua and Lin (1975).

Pivoting Crucially, the element $a_{kk}^{(k)}$, which one divides by at every step of GE, must not be zero; this element is referred to as the *pivot*. The simple algorithm shown above assumes that pivots are all non-zero. In practice, pivots can become zero, even for simple systems. For example, consider the following system:

$$
\begin{aligned}
x_1 &- x_2 &+ x_3 &= 0 \\
x_1 &- x_2 &+ 2x_3 &= 2 \\
x_1 &+ 2x_2 &+ 2x_3 &= 1
\end{aligned} \tag{3.28}
$$

After the first step of Gaussian elimination, we have:

$$
\begin{aligned}
x_1 &- x_2 &+ x_3 &= 0 \\
&0x_2 &+ x_3 &= 2 \\
&3x_2 &+ x_3 &= 1
\end{aligned} \tag{3.29}
$$

No further progress can be made, because the element that was to be used as a pivot in the second step is zero. This problem can be overcome by exchanging rows, columns, or both. This *rearrangement* or *reordering* of the matrix is called *pivoting*. For this example, to use *row exchange*, we would exchange the two equations (rows of the matrix), which eliminates the problem:

$$
\begin{aligned}
x_1 &- x_2 &+ x_3 &= 0 \\
&3x_2 &+ x_3 &= 1 \\
&0x_2 &+ x_3 &= 2
\end{aligned} \tag{3.30}
$$

Obviously, exchanging rows, i.e., permuting the matrix rows, requires a similar exchange in the b vector, and is equivalent to pre-multiplying the original system equation by an appropriate permutation matrix:

$$Ax = b \quad \Longleftrightarrow \quad PAx = Pb \qquad (3.31)$$

Recall that *left-multiplication* of a matrix by a permutation matrix achieves a permutation of its *rows*.

Performing a *column exchange* is slightly more complicated and is in fact equivalent to a variable transformation, as follows. If Q is a permutation matrix and if we let $y = Q^T x$ so that $x = Qy$, then:

$$Ax = b \quad \Longleftrightarrow \quad AQy = b \qquad (3.32)$$

This achieves a permutation (exchange) of the columns of A, which eliminates the problem. For our example, if we choose $x_1 = y_1$, $x_2 = y_3$, $x_3 = y_2$, then:

$$
\begin{array}{rcrcrcl}
y_1 & + & y_2 & - & y_3 & = & 0 \\
& & y_2 & + & 0y_3 & = & 2 \\
& & y_2 & + & 3x_3 & = & 1
\end{array}
\quad \text{with} \quad
Q = \begin{bmatrix} 1 & 0 & 0 \\ 0 & 0 & 1 \\ 0 & 1 & 0 \end{bmatrix}
\qquad (3.33)
$$

Recall that *right-multiplication* of a matrix by a permutation matrix achieves a permutation of its *columns*. Once the temporary solution y is found, we can easily find the required solution x, using $x = Qy$.

It is also possible to exchange both rows *and* columns, i.e., to perform a *row and column exchange*, so that this most general form of pivoting is as shown:

$$Ax = b \quad \Longleftrightarrow \quad (PAQ)y = Pb, \quad \text{where} \quad y = Q^T x \qquad (3.34)$$

Occasionally, pivoting is *absolutely required* so as to avoid dividing by zero. If the matrix is nonsingular, then pivoting will always succeed, i.e., there will always *exist* a set of permutations that leads to a non-zero diagonal. In addition, and this goes beyond simply avoiding a division by zero, pivoting is *highly advisable* for two other reasons:

1. Due to *finite precision* of digital computers, pivoting helps reduce the inevitable *roundoff error* in the result. This would be called *pivoting for accuracy* and, as we will see later on, it calls for one to use the largest-magnitude pivot.

2. When the matrix is *sparse* and if sparse matrix techniques are used, then pivoting helps to drastically reduce the computational effort. This would be called *pivoting for sparsity*. We will see later on that it is indeed possible to select a pivot that helps maintains sparsity.

We will return later on to a detailed discussion of pivoting.

3.1.3 *LU* Factorization

There are many kinds of matrix factorizations or decompositions that have been found useful in numerical methods. For circuit simulation, the *LU factorization* is quite useful and is the basis for most simulator implementations.

Theorem 3.1. *If $A \in \mathbb{R}^{n \times n}$ is nonsingular, then there exists a permutation matrix P such that:*

$$PA = LU \tag{3.35}$$

where L is a nonsingular lower triangular matrix and U is a nonsingular upper triangular matrix.

Thus, the *LU* factorization of a matrix A is the equation:

$$PA = \begin{bmatrix} l_{11} & 0 & 0 & \cdots & 0 \\ l_{21} & l_{22} & 0 & \cdots & 0 \\ \vdots & \vdots & \ddots & & \vdots \\ l_{n1} & l_{n2} & l_{n3} & \cdots & l_{nn} \end{bmatrix} \begin{bmatrix} u_{11} & u_{12} & u_{13} & \cdots & u_{1n} \\ 0 & u_{22} & u_{23} & \cdots & u_{2n} \\ \vdots & \vdots & \ddots & & \vdots \\ 0 & 0 & 0 & \cdots & u_{nn} \end{bmatrix} = LU \tag{3.36}$$

Notice that:

$$\det(A) = (-1)^r \det(PA) = (-1)^r \det(L)\det(U) = (-1)^r \prod_{i=1}^{n} l_{ii} u_{ii} \tag{3.37}$$

where the integer $r \geq 0$ is the number of row exchanges performed by the permutation matrix P.

Note that, in general, the L and U matrices are not unique. However, if the diagonal of either L or U is prespecified, then the factorization becomes unique, as we will see below. For example, if we require U to be unit upper triangular, as in Crout's algorithm, then the factorization is unique. Likewise, if we require L to be unit lower triangular, as in Doolittle's algorithm, then the factorization is unique. The justification of these statements is via the following theorem.

Theorem 3.2. *If $A \in \mathbb{R}^{n \times n}$ is nonsingular, then there exists a permutation matrix P such that:*

$$PA = L'DU' \tag{3.38}$$

where L' is a nonsingular unit lower triangular matrix, D is a nonsingular diagonal matrix, U' is a nonsingular unit upper triangular matrix, and this factorization is unique, for this PA.

The following analysis explores the link between these two theorems. Let A be nonsingular and, using (3.35), let $PA = LU$, where L is lower triangular and U is upper triangular, both nonsingular and not necessarily unique. Let D_L be a diagonal matrix whose entries are the diagonal entries of L:

$$D_L = \text{diag}(l_{11}, l_{22}, \ldots, l_{nn}) \tag{3.39}$$

Since $l_{ii} \neq 0$, then there exists a <u>unit</u> lower triangular L', for which:

$$L = L' D_L \tag{3.40}$$

Likewise, if $D_U = \text{diag}(u_{11}, u_{22}, \ldots, u_{nn})$, then there exists a <u>unit</u> upper triangular U', for which:

$$U = D_U U' \tag{3.41}$$

If we define a diagonal matrix $D \triangleq D_L D_U$, so that $d_{ii} = l_{ii} u_{ii}$, then:

$$P A = L U = (L' D_L)(D_U U') = L'(D_L D_U) U' = L' D U' \tag{3.42}$$

and, by the above theorem, this factorization *is unique*. Working backwards, we can factor D in any desirable way, $d_{ii} = l_{ii} u_{ii}$, to achieve any desired diagonal on U or on L (with only non-zero entries). Thus, once a diagonal for U, or for L, is chosen, the rest of the factorization $P A = L U$ becomes determined and unique.

Solving the System   How would we solve a linear system given an LU factorization of the system matrix? If we start with the equation $Ax = b$ and pre-multiply both sides by the permutation matrix P for which $P A = L U$, then:

$$P A x = L U x = P b \tag{3.43}$$

If we let $z = P b$ be the new right-hand side (RHS) vector, easily found by a number of row exchanges of the original vector b, and let $y = U x$, so that we can write the two *triangular* systems:

$$L y = z \qquad \text{and} \qquad U x = y \tag{3.44}$$

We can now solve the system $L y = z$ for y, by *forward substitution* (FS) (similar to backward substitution, but starting from the top), then solve the system $U x = y$ for x, by backward substitution. The overall solution flow is:

1. LU factorization ($\approx n^3/3$ operations)
2. Forward substitution (FS) ($\approx n^2/2$ operations)
3. Backward substitution (BS) ($\approx n^2/2$ operations)

Thus, overall, solving a system using LU factorization is an $\mathcal{O}(n^3)$ algorithm, similar to Gaussian elimination. However, once the matrix has been LU factored, one can find the solution for any new RHS vector b by using one FS followed by one BS, in $\approx n^2$ operations. This is a key advantage of LU factorization over Gaussian elimination. Another minor advantage is that the system $A^T x = b$ can be easily solved once the system $Ax = b$ has been solved, because $P A = L U \Rightarrow (P A)^T = U^T L^T$. Note, U^T is lower triangular and L^T is upper triangular. This feature can be useful in sensitivity analysis.

As mentioned, modern circuit simulators use LU factorization. It remains to describe how to construct the LU factors, but first, we will briefly consider the implementation of forward and backward substitution.

Forward Substitution   Forward substitution is concerned with solving the system $Ly = z$:

$$
\begin{bmatrix}
l_{11} & 0 & \cdots & 0 \\
l_{21} & l_{22} & \cdots & 0 \\
\vdots & \vdots & \ddots & \vdots \\
l_{n1} & l_{n2} & \cdots & l_{nn}
\end{bmatrix}
\begin{bmatrix}
y_1 \\ y_2 \\ \vdots \\ y_n
\end{bmatrix}
=
\begin{bmatrix}
z_1 \\ z_2 \\ \vdots \\ z_n
\end{bmatrix}
\tag{3.45}
$$

In principle, solving this system is trivial, because L is lower triangular, so that we would simply perform, for $k = 1, \ldots, n$, $y_k = \left(z_k - \sum_{j=1}^{k-1} l_{kj} y_j \right) / l_{kk}$. An algorithm based directly on this equation would be as shown in Fig. 3.2. This implementation is efficient if a sparse L is stored *by-rows*. However, when L is sparse and *storage-by-columns* is used, which is probably more common, then the algorithm shown Fig. 3.3 is faster. This alternate implementation works by pre-subtracting the product terms that contain y_k from all future z entries. When it is time to compute a certain y_k value, all the subtractions for its corresponding z_k value have already been made. In either case, when L is of unit-type, then the divisions by l_{kk} are obviously not required.

Backward Substitution   Backward substitution is concerned with solving the system $Ux = y$:

$$
\begin{bmatrix}
u_{11} & u_{12} & \cdots & u_{1n} \\
0 & u_{22} & \cdots & u_{2n} \\
\vdots & \vdots & \ddots & \vdots \\
0 & 0 & \cdots & u_{nn}
\end{bmatrix}
\begin{bmatrix}
x_1 \\ x_2 \\ \vdots \\ x_n
\end{bmatrix}
=
\begin{bmatrix}
y_1 \\ y_2 \\ \vdots \\ y_n
\end{bmatrix}
\tag{3.46}
$$

In principle, solving this system is trivial, because U is upper triangular, so that we would simply perform, for $k = n, \ldots, 1$, $x_k = \left(y_k - \sum_{j=k+1}^{n} u_{kj} x_j \right) / u_{kk}$.

```
for (k = 1, ..., n) do
    for (j = 1, ..., k − 1) do
        z_k = z_k − l_kj y_j
    y_k = z_k / l_kk
```

Figure 3.2: An implementation of forward substitution that is appropriate when a matrix is stored *by rows*.

```
for (k = 1, ..., n) do
    y_k = z_k / l_kk
    if (y_k ≠ 0) then
        for (i = k + 1, ..., n) do
            z_i = z_i − l_ik y_k
```

Figure 3.3: An implementation of forward substitution that is appropriate when a matrix is stored *by columns*.

$$
\begin{aligned}
&\textbf{for } (k = n, \ldots, 1) \textbf{ do} \\
&\quad \textbf{for } (j = k + 1, \ldots, n) \textbf{ do} \\
&\quad\quad y_k = y_k - u_{kj} x_j \\
&\quad x_k = y_k / u_{kk}
\end{aligned}
$$

Figure 3.4: An implementation of backward substitution that is appropriate when a matrix is stored *by rows*.

$$
\begin{aligned}
&\textbf{for } (k = n, \ldots, 1) \textbf{ do} \\
&\quad x_k = y_k / u_{kk} \\
&\quad \textbf{if } (x_k \neq 0) \textbf{ then} \\
&\quad\quad \textbf{for } (i = k - 1, \ldots, 1) \textbf{ do} \\
&\quad\quad\quad y_i = y_i - u_{ik} x_k
\end{aligned}
$$

Figure 3.5: An implementation of backward substitution that is appropriate when a matrix is stored *by columns*.

An algorithm based directly on this equation would be as shown in Fig. 3.4. This implementation is efficient if a sparse U is stored *by-rows*. However, when U is sparse and *storage-by-columns* is used, which is probably more common, then the algorithm in Fig. 3.5 is faster. This alternate implementation works by pre-subtracting the product terms that contain x_k from all future y entries. When it is time to compute a certain x_k value, all the subtractions for its corresponding y_k value have already been made. In either case, when U is of unit-type, then the divisions by u_{kk} are obviously not required.

Factoring the Matrix A few comparable algorithms are available for LU factorization. Crout's algorithm gives a unit upper triangular U, Doolittle's algorithm gives a unit lower triangular L, while Gauss's algorithm, a modification of GE, also gives a unit lower triangular L.

All three methods are asymptotically (i.e., for large n) equivalent in terms of complexity, with $\approx n^3/3$ operations, but Gauss's algorithm is generally preferred, for the following reasons. Compared to Gauss's, the other two algorithms have fewer memory references and smaller round-off error, but they allow only partial (not full) pivoting. *Partial pivoting* is a pivoting process by which we restrict our search for a pivot to those entries in either the kth column (followed by a row exchange) or in the kth row (followed by a column exchange). In contrast, *full pivoting* is a pivoting process that allows for any pivot from the remaining sub-matrix to be used, thereby possibly requiring both row and column exchanges. Gauss's algorithm for LU factorization allows using a pivot that allows both row and column exchange, which is useful for maintaining sparsity, as we will see later on. Also, Gauss's algorithm has a useful application in connection with block matrices for parallel simulation, as we will see later on. However, it should be also mentioned that, having $u_{ii} = 1$ (Crout), rather than $l_{ii} = 1$ (Doolittle and Gauss), is preferable when the right-hand side vector is highly sparse. This is because, with $u_{ii} = 1$, we would divide by l_{kk} as part of solving $Ly = z = Pb$, and we would have very few of these to do. Although we will show a basic Gauss's algorithm that gives $l_{ii} = 1$, it is easy to modify this to allow for $u_{ii} = 1$. This is left as an exercise for the reader.

In the following, we assume that A is nonsingular and, for clarity of presentation, we assume that the factorization exists with $P = I$ as the permutation matrix. In other words, we assume that no pivoting is required. It is not hard to lift this requirement, by modifying the below algorithms to include pivoting, as we will see later on.

Algorithms for LU Factorization   It is actually easy to derive an algorithm for LU factorization. Consider the matrix equation $A = LU$, or:

$$\begin{bmatrix} l_{11} & 0 & \cdots & 0 \\ l_{21} & l_{22} & \cdots & 0 \\ \vdots & \vdots & \ddots & \vdots \\ l_{n1} & l_{n2} & \cdots & l_{nn} \end{bmatrix} \begin{bmatrix} u_{11} & u_{12} & \cdots & u_{1n} \\ 0 & u_{22} & \cdots & u_{2n} \\ \vdots & \vdots & \ddots & \vdots \\ 0 & 0 & \cdots & u_{nn} \end{bmatrix} = \begin{bmatrix} a_{11} & a_{12} & \cdots & a_{1n} \\ a_{21} & a_{22} & \cdots & a_{2n} \\ \vdots & \vdots & \ddots & \vdots \\ a_{n1} & a_{n2} & \cdots & a_{nn} \end{bmatrix}$$

If we think of the l_{ij} and u_{ij} as unknowns, and the a_{ij} as given constants, then this leads to n^2 nonlinear equations in $(n^2 + n)$ unknowns:

$$\sum_{k=1}^{\min(i,j)} l_{ik} u_{kj} = a_{ij}, \quad \forall i, j \tag{3.47}$$

This system is *under-determined*; it has n more unknowns than equations. Thus, we are free to choose certain values up-front for n of the variables, and the system can then be solved for all the rest. Because the equations (3.47) are nonlinear, this choice cannot be arbitrary. The various LU algorithms typically set the value of either the L or the U diagonal. If we choose $l_{ii} = 1$, we get a factorization with a unit-type L. If we choose $u_{ii} = 1$, we get a factorization with a unit-type U. One could choose a value other than 1, anything other than 0, but a unit-type L or U gives some reduction of the computational effort. Once a diagonal has been chosen, finding the rest of L and U is easy. Notice, for later reference, that (3.47) leads to the following. When $i \leq j$, then $a_{ij} = \sum_{k=1}^{i} l_{ik} u_{kj} = \sum_{k=1}^{i-1} l_{ik} u_{kj} + l_{ii} u_{ij}$, from which:

$$u_{ij} = \left(a_{ij} - \sum_{k=1}^{i-1} l_{ik} u_{kj} \right) / l_{ii} \tag{3.48}$$

When $i \geq j$, then $a_{ij} = \sum_{k=1}^{j} l_{ik} u_{kj} = \sum_{k=1}^{j-1} l_{ik} u_{kj} + l_{ij} u_{ij}$, from which:

$$l_{ij} = \left(a_{ij} - \sum_{k=1}^{j-1} l_{ik} u_{kj} \right) / u_{jj} \tag{3.49}$$

These equations can be combined in different ways to find L and U.

Crout's Algorithm   Let us start by assuming that $u_{ii} = 1$ so that U is unit upper triangular. We use (3.47) and expand the summation, as in Vlach and Singhal (1994), for every entry of the matrix A, to get:

$$A = \begin{bmatrix} l_{11} & l_{11}u_{12} & l_{11}u_{13} & \cdots \\ l_{21} & l_{21}u_{12} + l_{22} & l_{21}u_{13} + l_{22}u_{23} & \cdots \\ l_{31} & l_{31}u_{12} + l_{32} & l_{31}u_{13} + l_{32}u_{23} + l_{33} & \cdots \\ l_{41} & l_{41}u_{12} + l_{42} & l_{41}u_{13} + l_{42}u_{23} + l_{43} & \cdots \\ \vdots & \vdots & \vdots & \ddots \end{bmatrix}$$

Given this, we can proceed in the following way. We start at level 1 and consider the matrix $A(\{1, \ldots, n\}, \{1, \ldots, n\})$, i.e., the full original matrix. For the *first column*, we immediately know that, for all $i \geq 1$:

$$l_{i1} = a_{i1} \tag{3.50}$$

and, for the *first row*, since l_{11} is known, then for all $j \geq 2$:

$$u_{1j} = a_{1j}/l_{11} \tag{3.51}$$

We then consider level 2 and the sub-matrix $A(\{2, \ldots, n\}, \{2, \ldots, n\})$. For the *first column*, since l_{i1} and u_{12} are known, then for all $i \geq 2$:

$$l_{i2} = a_{i2} - l_{i1}u_{12} \tag{3.52}$$

and, for the *first row*, since l_{21} and u_{1j} are known, then for all $j \geq 3$:

$$u_{2j} = (a_{2j} - l_{21}u_{1j})/l_{22} \tag{3.53}$$

Similarly, for every subsequent level $k \geq 3$, we consider the remaining sub-matrix, i.e., $A(\{k, \ldots, n\}, \{k, \ldots, n\})$, and we write:

$$\text{first column:} \quad l_{ik} = a_{ik} - \sum_{m=1}^{k-1} l_{im}u_{mk}, \quad \forall i \geq k \tag{3.54}$$

$$\text{first row:} \quad u_{kj} = \left(a_{kj} - \sum_{m=1}^{k-1} l_{km}u_{mj} \right) \Big/ l_{kk}, \quad \forall j \geq k+1 \tag{3.55}$$

Effectively, this is one way of interleaving the application of (3.48) and (3.49). This leads to Crout's algorithm, which is as shown in Fig. 3.6. Note that the algorithm does not actually compute u_{ii}; it is already known to be 1. One feature of Crout's algorithm is the alternating sequence of column and row evaluations.

Another feature is the option of doing *in-place computation*, as follows. Note that each a_{ij} is needed only in order to determine the corresponding entry of

> **Input:** A, a nonsingular matrix
> **for** $(k = 1, \ldots, n)$ **do**
>   **for** $(i = k, \ldots, n)$ **do**
> $l_{ik} = a_{ik} - \sum_{m=1}^{k-1} l_{im} u_{mk}$
>   **for** $(j = k + 1, \ldots, n)$ **do**
> $u_{kj} = \left(a_{kj} - \sum_{m=1}^{k-1} l_{km} u_{mj} \right) \Big/ l_{kk}$

Figure 3.6: Crout's algorithm for LU factorization.

either L or U, depending on whether $i \geq j$ or $i < j$. Then, since the 0 entries of L and U, as well as all the u_{ii}, are *known*, and do not need to be stored, we can let l_{ij} or u_{ij} rewrite a_{ij}. Thus, we can have L and U over-write A, to end up with a so-called *auxiliary matrix* S consisting of (only) the initially unknown elements of L and U:

$$S = \begin{bmatrix} l_{11} & u_{12} & u_{13} & \cdots & u_{1n} \\ l_{21} & l_{22} & u_{23} & \cdots & u_{2n} \\ l_{31} & l_{32} & l_{33} & \cdots & u_{3n} \\ \vdots & \vdots & \vdots & \ddots & \vdots \\ l_{n1} & l_{n2} & l_{n3} & \cdots & l_{nn} \end{bmatrix} = L + U - I \tag{3.56}$$

The computational cost (number of multiplications and divisions) of the implementation of Crout's algorithm, as shown in Fig. 3.6, is:

$$\text{cost} = \sum_{k=1}^{n} \left(\sum_{i=k}^{n} (k - 1) + \sum_{j=k+1}^{n} (k - 1 + 1) \right) = \frac{n^3}{3} - \frac{n}{3} \tag{3.57}$$

Two Variants on Crout   There are two variants on the basic Crout's algorithm that can be easily derived, as follows. Starting again from $u_{ii} = 1$, consider once more the expanded matrix view:

$$A = \begin{bmatrix} l_{11} & l_{11}u_{12} & l_{11}u_{13} & \cdots \\ l_{21} & l_{21}u_{12} + l_{22} & l_{21}u_{13} + l_{22}u_{23} & \cdots \\ l_{31} & l_{31}u_{12} + l_{32} & l_{31}u_{13} + l_{32}u_{23} + l_{33} & \cdots \\ l_{41} & l_{41}u_{12} + l_{42} & l_{41}u_{13} + l_{42}u_{23} + l_{43} & \cdots \\ \vdots & \vdots & \vdots & \ddots \end{bmatrix}$$

Given this, we can solve for the variables one row at a time, as follows. Start at row 1 and perform $l_{11} = a_{11}$ and, for all $j \geq 2$, perform:

$$u_{1j} = a_{1j}/l_{11} \tag{3.58}$$

Then, consider row 2 and perform $l_{21} = a_{21}, l_{22} = a_{22} - l_{21}u_{12}$, and, for all $j \geq 3$, perform:

$$u_{2j} = (a_{2j} - l_{21}u_{1j})/l_{22} \qquad (3.59)$$

This process is repeated, giving the solution one row at a time. This gives a variant on Crout that produces one row at a time, as opposed to interleaving the column and row solutions as in the original Crout algorithm. Similarly, one can produce another variant that gives one column at a time. When dealing with large sparse matrices, there may be an advantage to solving the system one way or another, either by rows or by columns. When applied to a full matrix, these variants perform the same number of operations as Crout's original algorithm, but in a different order.

Doolittle's Algorithm   In this case, we start with the choice of L as unit lower triangular, and the expanded matrix view $A = LU$ becomes as follows:

$$A = \begin{bmatrix} u_{11} & u_{12} & u_{13} & \cdots \\ l_{21}u_{11} & l_{21}u_{12} + u_{22} & l_{21}u_{13} + u_{23} & \cdots \\ l_{31}u_{11} & l_{31}u_{12} + l_{32}u_{22} & l_{31}u_{13} + l_{32}u_{23} + u_{33} & \cdots \\ l_{41}u_{11} & l_{41}u_{12} + l_{42}u_{22} & l_{41}u_{13} + l_{42}u_{23} + l_{43}u_{33} & \cdots \\ \vdots & \vdots & \vdots & \ddots \end{bmatrix}$$

We proceed similarly to Crout's algorithm, but in this case we process a row before we process a column, leading to Doolittle's algorithm as shown in Fig. 3.7.

The computational complexity is exactly the same as for Crout's algorithm, because this is simply a reshuffling of the same operations. Here, too, one can consider similar variants like the ones we considered for Crout's algorithm, so as to obtain one row at a time or one column at a time.

Gauss's Algorithm   Gauss's algorithm for LU factorization is much more interesting because of its links to Gaussian elimination (GE). Recall that Gaussian

---

Input: A, a nonsingular matrix
  for $(k = 1, \ldots, n)$ **do**
    for $(j = k, \ldots, n)$ **do**
      $u_{kj} = a_{kj} - \sum_{m=1}^{k-1} l_{km}u_{mj}$
    for $(i = k + 1, \ldots, n)$ **do**
      $l_{ik} = \left(a_{ik} - \sum_{m=1}^{k-1} l_{im}u_{mk} \right) \Big/ u_{kk}$

Figure 3.7: Doolittle's algorithm for LU factorization.

elimination leads to the upper triangular final form:

$$\begin{bmatrix} a_{11}^{(1)} & a_{12}^{(1)} & a_{13}^{(1)} & \cdots & a_{1n}^{(1)} \\ 0 & a_{22}^{(2)} & a_{23}^{(2)} & \cdots & a_{2n}^{(2)} \\ 0 & 0 & a_{33}^{(3)} & \cdots & a_{3n}^{(3)} \\ \vdots & \vdots & \vdots & \ddots & \vdots \\ 0 & 0 & \cdots & 0 & a_{nn}^{(n)} \end{bmatrix} x = \begin{bmatrix} b_1^{(1)} \\ b_2^{(2)} \\ b_3^{(3)} \\ \vdots \\ b_n^{(n)} \end{bmatrix} \tag{3.60}$$

How is the final RHS vector related to the original b? We can find out if we unravel the b computation in the inner-loop of GE, by reference to Fig. 3.1. Notice that the *last* time that b_i is updated is when $k + 1 = i$, and the sequence of updates to b_i is as follows:

$$b_i^{(1)} = b_i$$

$$k = 1 : \qquad b_i^{(2)} = b_i^{(1)} - m_{i1}b_1^{(1)}$$

$$k = 2 : \qquad b_i^{(3)} = b_i^{(2)} - m_{i2}b_2^{(2)}$$

$$\vdots$$

$$b_i^{(k+1)} = b_i^{(k)} - m_{ik}b_k^{(k)}$$

$$\vdots$$

$$k = i - 2 : \qquad b_i^{(i-1)} = b_i^{(i-2)} - m_{i,i-2}b_{i-2}^{(i-2)}$$

$$k = i - 1 : \qquad b_i^{(i)} = b_i^{(i-1)} - m_{i,i-1}b_{i-1}^{(i-1)}$$

If we add all the above equations, we get the expression for $b_i^{(i)}$:

$$b_i^{(i)} = b_i - \sum_{j=1}^{i-1} m_{ij}b_j^{(j)} \tag{3.61}$$

This can be arranged as $\sum_{j=1}^{i-1} m_{ij}b_j^{(j)} + b_i^{(i)} = b_i$, or, in matrix form:

$$\begin{bmatrix} 1 & 0 & 0 & \cdots & 0 \\ m_{21} & 1 & 0 & \cdots & 0 \\ m_{31} & m_{32} & 1 & \cdots & 0 \\ \vdots & \vdots & \vdots & \ddots & \vdots \\ m_{n1} & m_{n2} & m_{n3} & \cdots & 1 \end{bmatrix} \begin{bmatrix} b_1^{(1)} \\ b_2^{(2)} \\ b_3^{(3)} \\ \vdots \\ b_n^{(n)} \end{bmatrix} = b \tag{3.62}$$

The unit lower triangular matrix in (3.62) is nonsingular, so we can multiply both sides of (3.60) by it without affecting the solution, leading to:

$$
\begin{bmatrix}
1 & 0 & 0 & \cdots & 0 \\
m_{21} & 1 & 0 & \cdots & 0 \\
m_{31} & m_{32} & 1 & \cdots & 0 \\
\vdots & \vdots & \vdots & \ddots & \vdots \\
m_{n1} & m_{n2} & m_{n3} & \cdots & 1
\end{bmatrix}
\begin{bmatrix}
a_{11}^{(1)} & a_{12}^{(1)} & a_{13}^{(1)} & \cdots & a_{1n}^{(1)} \\
0 & a_{22}^{(2)} & a_{23}^{(2)} & \cdots & a_{2n}^{(2)} \\
0 & 0 & a_{33}^{(3)} & \cdots & a_{3n}^{(3)} \\
\vdots & \vdots & \vdots & \ddots & \vdots \\
0 & 0 & \cdots & 0 & a_{nn}^{(n)}
\end{bmatrix}
x = b \qquad (3.63)
$$

Thus, it is clear that GE implicitly performs an LU factorization, in which the L matrix is of unit-type and is simply a matrix of the multiplier entries:

$$
L =
\begin{bmatrix}
1 & 0 & 0 & \cdots & 0 \\
m_{21} & 1 & 0 & \cdots & 0 \\
m_{31} & m_{32} & 1 & \cdots & 0 \\
\vdots & \vdots & \vdots & \ddots & \vdots \\
m_{n1} & m_{n2} & m_{n3} & \cdots & 1
\end{bmatrix}
\qquad (3.64)
$$

and U is the final triangulated system matrix. Indeed, one can show, similarly to the derivation of (3.62), that $LU = A$, the original system matrix. Notice, finally that the final state of the system is as follows:

$$
Ux =
\begin{bmatrix}
a_{11}^{(1)} & a_{12}^{(1)} & a_{13}^{(1)} & \cdots & a_{1n}^{(1)} \\
0 & a_{22}^{(2)} & a_{23}^{(2)} & \cdots & a_{2n}^{(2)} \\
0 & 0 & a_{33}^{(3)} & \cdots & a_{3n}^{(3)} \\
\vdots & \vdots & \vdots & \ddots & \vdots \\
0 & 0 & \cdots & 0 & a_{nn}^{(n)}
\end{bmatrix}
x =
\begin{bmatrix}
b_1^{(1)} \\
b_2^{(2)} \\
b_3^{(3)} \\
\vdots \\
b_n^{(n)}
\end{bmatrix}
= L^{-1} b \qquad (3.65)
$$

So that L^{-1} is nothing but the product of all the elementary matrices by which the system was (implicitly) pre-multiplied as part of GE.

For better comparison with other LU methods, we can rewrite the GE algorithm in a form that is similar to Crout's and Doolittle's algorithms. We do this by first writing the expanded form of the auxiliary matrix S, in the case when L is unit-type, as follows. The expanded form of $A = LU$, based on (3.47) when L is unit-type, is:

$$
A =
\begin{bmatrix}
u_{11} & u_{12} & u_{13} & \cdots \\
l_{21}u_{11} & l_{21}u_{12} + u_{22} & l_{21}u_{13} + u_{23} & \cdots \\
l_{31}u_{11} & l_{31}u_{12} + l_{32}u_{22} & l_{31}u_{13} + l_{32}u_{23} + u_{33} & \cdots \\
l_{41}u_{11} & l_{41}u_{12} + l_{42}u_{22} & l_{41}u_{13} + l_{42}u_{23} + l_{43}u_{33} & \cdots \\
\vdots & \vdots & \vdots & \ddots
\end{bmatrix}
$$

> **Input:** A, a nonsingular matrix
>   **for** $(k = 1, \ldots, n)$ **do**
>   **for** $(j = k, \ldots, n)$ **do**
> $u_{kj} = a_{kj}$
>   **for** $(i = k + 1, \ldots, n)$ **do**
> $l_{ik} = a_{ik}/u_{kk}$
>   **for** $(i = k + 1, \ldots, n)$ **do**
>   **for** $(j = k + 1, \ldots, n)$ **do**
> $a_{ij} = a_{ij} - l_{ik}u_{kj}$

Figure 3.8: Gauss's algorithm for LU factorization.

And we can write an expanded form of $S = L + U - I$, when L is unit-type, based on (3.48) and (3.49), as:

$$
S = \begin{bmatrix}
a_{11} & a_{12} & a_{13} & \cdots \\
a_{21}/u_{11} & a_{22} - l_{21}u_{12} & a_{23} - l_{21}u_{13} & \cdots \\
a_{31}/u_{11} & (a_{32} - l_{31}u_{12})/u_{22} & a_{33} - l_{31}u_{13} - l_{32}u_{23} & \cdots \\
a_{41}/u_{11} & (a_{42} - l_{41}u_{12})/u_{22} & (a_{43} - l_{41}u_{13} - l_{42}u_{23})/u_{33} & \cdots \\
\vdots & \vdots & \vdots & \ddots
\end{bmatrix}
$$

Thus, we can compute the LU factorization as follows. From the first row of S, find $u_{1j} = a_{1j}$, $\forall j \geq 1$. Then, from the first column of S, now that we know u_{11}, we can find $l_{i1} = a_{i1}/u_{11}$, $\forall i \geq 2$. With l_{21} known and u_{1j} known, we then use the 2nd row of S to compute the 2nd row of U, etc.

The process sounds exactly like Doolittle's algorithm. However, we can also organize the computation in a different way, as follows. Note that all the entries of S involve starting with entries of A, and performing a set of subtractions (empty for the first row and column) and a set of divisions (for entries below the diagonal). There are more subtractions to be done as we get deeper into the matrix. One option is to sweep across the whole remaining sub-matrix and do a subtraction before moving down to the next level. Thus, when finding $u_{22} = s_{22} = a_{22} - l_{21}u_{12}$, we can choose to perform one subtraction for the whole sub-matrix $A(\{2, \ldots, n\}, \{2, \ldots, n\})$. The process can be organized as shown in Fig. 3.8 and, being identical to Gaussian elimination, it is referred to as Gauss's algorithm.

Notice that the double for-loop at the bottom performs one subtraction across the whole remaining sub-matrix. On the next iteration, the first column has had all the required subtractions performed and is ready for the final division. The complexity is the same as before, $n^3/3 - n/3$, and one can also develop a variant of this algorithm that produces a unit-type U matrix. Because it accesses the whole sub-matrix, Gauss's algorithm performs more memory referencing than Crout, but it does allow full pivoting. It is often implemented so that it uses in-place computation, storing the L matrix in the bottom left of the triangulated A matrix.

In terms of its operations on the system matrix, Gauss's method for LU factorization is identical to the forward elimination phase of Gaussian elimination, as one can easily see by comparing the two algorithms. In fact, in the literature, it is often the case that the term "Gaussian elimination" actually refers to Gauss's method for LU factorization. If a RHS vector b is available, Gauss's method can also apply similar operations on it, in which case it becomes identical to the full GE algorithm.

3.1.4 Block Gaussian Elimination

Gauss's method for LU factorization is also useful in another context, as we now explain. Recall that the effect of GE on the system $Ax = b$ is to pre-multiply both sides by L^{-1}, so that:

$$Ax = b \quad \xrightarrow{GE} \quad Ux = L^{-1}b \tag{3.66}$$

This is usually expressed by forming an *augmented matrix* $[A|b]$ and pre-multiplying that with L^{-1}:

$$[A|b] \quad \xrightarrow{GE} \quad [U|L^{-1}b] \tag{3.67}$$

Now consider a system expressed in *block matrix* form, as shown:

$$\begin{bmatrix} A & B \\ C & D \end{bmatrix} \begin{bmatrix} x_1 \\ x_2 \end{bmatrix} = \begin{bmatrix} b_1 \\ b_2 \end{bmatrix} \tag{3.68}$$

where x_1, x_2, b_1, and b_2 are vectors. Let A be nonsingular with $A = LU$, where L is of unit-type. Suppose we apply Gauss's method only *partially*, proceeding down the diagonal of A, but stopping once A has been triangulated. Then, the effect on the top equation in (3.68) is to pre-multiply it by L^{-1}, yielding:

$$Ux_1 + L^{-1}Bx_2 = L^{-1}b_1 \tag{3.69}$$

What happens to the bottom equation in the process? Obviously, C is transformed to 0, but what happens to D and b_2?

To answer this question, recall that GE is simply a *variable elimination* process, a substitution process, so that the real question is: what becomes of the bottom equation in (3.68) if we eliminate x_1 from it by making use of the top equation? We can easily answer this by using the top equation to get:

$$x_1 = A^{-1}(b_1 - Bx_2) \tag{3.70}$$

and substituting this into the bottom equation, leading to:

$$(D - CA^{-1}B)x_2 = b_2 - CA^{-1}b_1 \tag{3.71}$$

This "elimination by substitution" is at the heart of GE, so that this is exactly the new form of the bottom equation. Thus, the resulting half-way transformed system is as follows:

$$\begin{bmatrix} U & L^{-1}B \\ 0 & (D - CA^{-1}B) \end{bmatrix} \begin{bmatrix} x_1 \\ x_2 \end{bmatrix} = \begin{bmatrix} L^{-1}b_1 \\ b_2 - CA^{-1}b_1 \end{bmatrix} \tag{3.72}$$

This is called *block Gaussian elimination*, and it can be understood as pre-multiplying the system by a certain matrix, as shown:

$$\begin{bmatrix} L^{-1} & 0 \\ -CA^{-1} & I \end{bmatrix} \begin{bmatrix} A & B & | & b_1 \\ C & D & | & b_2 \end{bmatrix} = \begin{bmatrix} U & L^{-1}B & | & L^{-1}b_1 \\ 0 & (D - CA^{-1}B) & | & b_2 - CA^{-1}b_1 \end{bmatrix}$$

Or, simply focusing on the action on the system matrix:

$$\begin{bmatrix} L^{-1} & 0 \\ -CA^{-1} & I \end{bmatrix} \begin{bmatrix} A & B \\ C & D \end{bmatrix} = \begin{bmatrix} U & L^{-1}B \\ 0 & (D - CA^{-1}B) \end{bmatrix} \tag{3.73}$$

Effectively, this matrix transformation leads to a *"partial U"* matrix, which is a partial version of the final triangulated system matrix:

$$U' \triangleq \begin{bmatrix} U & L^{-1}B \\ 0 & (D - CA^{-1}B) \end{bmatrix} \tag{3.74}$$

It is convenient to perform this "partial GE" using Gauss's method for LU factorization, because it also yields the *"partial L"* matrix on the fly. What does the partial L matrix, i.e., L', look like? We know that the top part of L' has an L and a 0, but we need to determine X and Y:

$$L' \triangleq \begin{bmatrix} L & 0 \\ X & Y \end{bmatrix} \tag{3.75}$$

We can easily *solve* for X because, if the full $[A, B, C, D]$ matrix were to be LU-factorized, we would have:

$$\begin{bmatrix} L & 0 \\ X & \cdot \end{bmatrix} \begin{bmatrix} U & L^{-1}B \\ 0 & \cdot \end{bmatrix} = \begin{bmatrix} A & B \\ C & D \end{bmatrix} \tag{3.76}$$

where the "$\cdot$" denotes unknown final content, so that $XU = C$, $X = CU^{-1}$, and therefore:

$$L' = \begin{bmatrix} L & 0 \\ CU^{-1} & Y \end{bmatrix} \tag{3.77}$$

As for Y, it is convenient to *assume* that the matrix L' is initialized, at the start of GE, as the identity matrix, so that $Y = I$ conveniently leads to:

$$\begin{bmatrix} L & 0 \\ CU^{-1} & I \end{bmatrix} \begin{bmatrix} U & L^{-1}B \\ 0 & (D - CA^{-1}B) \end{bmatrix} = \begin{bmatrix} A & B \\ C & D \end{bmatrix} \tag{3.78}$$

Thus, adopting this initialization assumption on L', we have:

$$L' = \begin{bmatrix} L & 0 \\ CU^{-1} & I \end{bmatrix} \quad \text{and} \quad U' = \begin{bmatrix} U & L^{-1}B \\ 0 & (D - CA^{-1}B) \end{bmatrix} \qquad (3.79)$$

A common shorthand notation is to capture both L' and U' in a single so-called *auxiliary matrix* $S = L' + U' - I$, so that:

$$S = \begin{bmatrix} (L|U) & L^{-1}B \\ CU^{-1} & (D - CA^{-1}B) \end{bmatrix} \qquad (3.80)$$

where $(L|U)$ is a packaging of L (less its unit diagonal) and U into the same square matrix, whose diagonal consists of the U diagonal. Finally, one can also combine this with the right-hand side vector, as:

$$\begin{bmatrix} (L|U) & L^{-1}B \\ CU^{-1} & (D - CA^{-1}B) \end{bmatrix} \begin{bmatrix} L^{-1}b_1 \\ b_2 - CA^{-1}b_1 \end{bmatrix} \qquad (3.81)$$

3.1.5 Cholesky Decomposition

In a certain special case, which often arises in simulation, the factorization can be simplified significantly, as follows. First, a brief review of matrix positive definiteness is useful.

Positive Definite   Recall, if $A \in \mathbb{R}^{n \times n}$ is symmetric, and if $x^T A x > 0$ for all non-zero $x \in \mathbb{R}^n$, then A is said to be *positive definite*, or *symmetric positive definite* (SPD). If A is SPD then it is nonsingular, and its inverse is also SPD. Recall that a matrix is singular if and only if it has a zero eigenvalue. A symmetric matrix is SPD if and only if all its eigenvalues are (strictly) positive. If A is symmetric and strictly diagonally dominant and if all its diagonal entries are strictly positive, then it is SPD. If $A = CC^T$, where $C \in \mathbb{R}^{n \times n}$ is nonsingular, then A is SPD.

If $A \in \mathbb{R}^{n \times n}$ is symmetric, and if $x^T A x \geq 0$ for all non-zero $x \in \mathbb{R}^n$, then A is said to be *positive semi-definite*. In this case, there is no guarantee of nonsingularity. A matrix is positive semi-definite if and only if all its eigenvalues are non-negative. If $A = CC^T$, where $C \in \mathbb{R}^{n \times n}$, then A is positive semi-definite.

Theorem 3.3. *A matrix $A \in \mathbb{R}^{n \times n}$ is positive definite if and only if there exists a nonsingular, lower triangular matrix $L \in \mathbb{R}^{n \times n}$, with (strictly) positive diagonal entries, such that:*

$$A = LL^T \qquad (3.82)$$

and this decomposition is unique, *i.e., there is only one such L with (strictly) positive diagonal entries.*

> **Input:** A, a symmetric, positive definite matrix
> **for** $(k = 1, \ldots, n)$ **do**
>
> $$l_{kk} = \sqrt{a_{kk} - \sum_{j=1}^{k-1} l_{kj}^2}$$
>
> **for** $(i = k + 1, \ldots, n)$ **do**
>
> $$l_{ik} = \frac{1}{l_{kk}} \left(a_{ik} - \sum_{j=1}^{k-1} l_{ij} l_{kj} \right)$$

Figure 3.9: Cholesky decomposition for LU factorization, where $U = L^T$.

Theorem 3.4. *A matrix $A \in \mathbb{R}^{n \times n}$ is positive semi-definite if and only if there exists a lower triangular matrix $L \in \mathbb{R}^{n \times n}$, such that:*

$$A = LL^T \tag{3.83}$$

but this decomposition is not *unique, in general.*

The decomposition of a matrix A into the product $A = LL^T$ is called the *Cholesky decomposition*.

Cholesky Decomposition   We will be interested in the Cholesky decomposition only for the case of a *symmetric positive definite* (SPD) matrix A, which is the case for circuit matrices in certain special cases. In this case, one can easily derive an algorithm for performing the Cholesky decomposition, $A = LL^T$, along the lines of what we have done above. There is no need for permutation or pivoting in this case. The *Crout-Cholesky* algorithm starts from the upper left corner of the matrix and proceeds one column at a time, as shown in Fig. 3.9. Note that the quantity under the square root sign is guaranteed to be positive because of the positive definiteness of the matrix A. The algorithm requires $\approx n^3/6 - n/6$ operations, half the LU case. There are other algorithms, including a variant of Gauss's method, i.e., a *Gauss-Cholesky* algorithm, and a *Cholesky-Banachiewicz* algorithm that works similarly but one row at a time, instead of one column at a time. All these methods have the same computational complexity on a full matrix, but some may be more preferred for sparse matrices, depending on the implementation.

3.2   ACCURACY AND STABILITY OF GE

Is Gaussian elimination (equivalently, LU factorization) a *good* algorithm to use on digital computers, or does it suffer too much from numerical errors? This seemingly simple question hides a lot of complexity and is not easy to answer. In order to appreciate the available answers, resulting from work in numerical analysis of algorithms over the last half century, we must develop some deeper understanding of error, including backward error and forward error, some knowledge of floating point number systems, and some understanding of the general notions of *stability* and *accuracy* of numerical algorithms. It will turn out that the answer can be stated in simple terms: *GE and LU factorization are not immune to numerical error problems, but in practice they work very well if pivoting is used to avoid small pivots*. The rest of this section aims to provide the technical justification for this statement.

The material below is based on a number of sources, including Higham (2002), Muller (2006), Duff et al. (1986), and Golub and Van Loan (1989). The classical reference in this area is the pioneering work in Wilkinson (1965). Our study will also require some review of vector and matrix norms, based in-part on the excellent Horn and Johnson (1985), and we will study the notion of conditioning, leading up finally to techniques of iterative refinement.

It is useful to start the discussion at the very beginning, with the obvious statement that digital computers can use only a *finite* number of *bits* to represent numbers. This is expressed by saying that digital computers have *finite precision*. Thus, one cannot represent arbitrarily large numbers (very large numbers lead to *overflow*) and cannot represent arbitrarily small non-zero real numbers (very small numbers lead to *underflow*). As well, one can represent only a finite subset of the real numbers. Real numbers that cannot be represented exactly must be *rounded off* to the nearest number that *can*, leading to so-called *roundoff error*. In general, and setting aside issues of overflow and underflow, computer arithmetic on real numbers suffers from two types of roundoff errors: those resulting from *storing* a number in memory, and those resulting from applying arithmetic *operations* on numbers. Thus, we must distinguish *exact arithmetic* from *finite precision arithmetic* or *computer arithmetic*.

As we will see below, one can define a *unit roundoff*, denoted u, as the accuracy with which the basic arithmetic operations $(+, -, \times, /)$ can be performed. In general, the value of u depends on the computer hardware, the operating system, the compiler, and data types used in the program. We will define u more formally below, after some preliminary discussion.

In an algorithm, roundoff errors will partially accumulate/grow, partially cancel out, and lead, in complex ways, to the final error in the result. Numerical analysis of algorithms aims to understand the impact of roundoff error, in storage and in operations, on the final algorithm output error. One can, at best, hope that the error in the output is $\approx u$, but in practice one often suffers from growth in the error. Loosely speaking, an algorithm is said to be *numerically accurate and stable* if it does not suffer from very large growth in the roundoff error. Numerical analysis aims to develop practical algorithms that are *accurate and stable*.

3.2.1 Error

Let $f : \mathbb{R} \to \mathbb{R}$ be a real scalar function of a real scalar variable x, and let us consider the errors involved in computing $y \triangleq f(x)$ on a computer. Let Δx be the roundoff error in storing x, so that we denote the *stored* value of x as $\tilde{x} \triangleq x + \Delta x$. Let the *computed* value of y be $\tilde{y} \triangleq \tilde{f}(x)$, where $\tilde{f}$ represents an algorithm for computing f on a computer with the (unavoidable) roundoff errors, due to storing of x as $\tilde{x}$, and due to operating on $\tilde{x}$ with finite precision arithmetic.

Thus, we distinguish between two computations, the *problem* $f(x)$ which is exact but unattainable on a digital computer, and the *algorithm* implementation $\tilde{f}(x)$ which is attainable but inexact. One way that we may be satisfied that $\tilde{y}$ is close enough to y is to ensure that the following two conditions hold, for every x:

1. **Forward error analysis:** By analysis of the *problem* $f(x)$, we establish that its output is not excessively sensitive to perturbations in its input. For example, we may establish that any input perturbation of within ϵ would lead to an output perturbation of within an acceptable δ. Formally, we would write that there exist $\epsilon > 0$ and $\delta > 0$, such that:

$$\forall \hat{x} \in \mathbb{R}, \quad \left| \frac{\hat{x} - x}{x} \right| \le \epsilon \quad \Longrightarrow \quad \left| \frac{f(\hat{x}) - f(x)}{f(x)} \right| \le \delta \qquad (3.84)$$

   The ratio δ/ϵ can be used to reflect sensitivity. In general, ϵ and δ may depend on x, but we ignore such technicalities for now.

2. **Backward error analysis:** By analysis of the *algorithm* $\tilde{f}(x)$, we establish that there exists an $\hat{x} \in \mathbb{R}$, for which:

$$\left| \frac{\hat{x} - x}{x} \right| \le \epsilon \quad \underline{\text{and}} \quad f(\hat{x}) = \tilde{f}(x) \qquad (3.85)$$

Combining (3.84) and (3.85), we get that $f(\hat{x}) = \tilde{f}(x) = \tilde{y}$ is within δ of $f(x) = y$, so that, for every x, we have:

$$\left| \frac{\tilde{y} - y}{y} \right| \le \delta \qquad (3.86)$$

and the roundoff errors in the algorithm outputs are thereby bounded. The situation is illustrated in Fig. 3.10, where the direction of the one-way arrows suggests the reason for the names: the study of the second condition (3.85) is called *backward error analysis*, while the study of the first (3.84) is called *forward error analysis*. We now discuss these notions in more detail.

Backward Error   Consider an input perturbation Δx, for which $\hat{x} \triangleq x + \Delta x$ is such that:

$$f(\hat{x}) = \tilde{y} = \tilde{f}(x) \qquad (3.87)$$

In other words, $\tilde{f}(x) = f(\hat{x})$, so that the computed solution turns out to be the *exact* solution of the problem with *perturbed input data*. The *smallest* such $|\Delta x|$ is called the *backward error* at that value of x. Backward error may also be expressed as a relative error, $\min(|\Delta x|/|x|)$. An algorithm $\tilde{f}$ for computing a function f is said to be *stable* if, for any x, there is a *small* value of $|\Delta x|$,

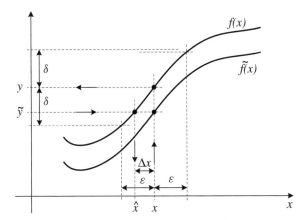

Figure 3.10: Illustration of relations among the various error terms, including forward and backward error.

or of $|\Delta x|/|x|$, for which $\tilde{f}(x) = f(x + \Delta x)$. In other words, if, for any x, the backward error is small.

Backward error may also be understood in terms of data uncertainties. As Higham (2002) puts it, backward error "*interprets rounding errors as being equivalent to perturbations in the data. The data frequently contains uncertainties due to measurements, previous computations, or errors committed in storing numbers on the computer ...*" He continues, "*if the backward error is no larger than these uncertainties then the computed solution can hardly be criticized—it may be the solution we are seeking, for all we know.*" It is certainly the exact solution of a "nearby problem," as is often expressed in the field. The computed solution is the answer we would get if we made small perturbations to the problem and then solved the problem *exactly*, and therefore one has no reason to reject it. This is the essence of the notion of stability.

Forward Error   We denote the *forward error* by $\Delta y \triangleq |\tilde{y} - y|$ or, alternatively, as $|\tilde{y} - y|/|y|$. Suppose algorithm $\tilde{f}$ is stable, so that, for any x, there is a small value of $|\Delta x|$ for which $\tilde{f}(x) = f(x + \Delta x)$. This is not enough. We still need to know whether the difference between $f(x + \Delta x)$ and $f(x)$ is small enough, for small $|\Delta x|$. Notice that this question relates to properties of the *problem* $f(\cdot)$ and not of any *algorithm* $\tilde{f}(\cdot)$. It has nothing to do with the algorithm or its implementation, but with the intrinsic *sensitivity* of the problem to its input data.

Thus, forward error analysis asks the question: if small changes are made to the input data, are the resulting changes in the *exact* solution also small? This requires a so-called *perturbation analysis* of the problem. Again, this boils down to a question about the problem, not about a computer implementation of an algorithm. If the answer to the question is *yes*, then the problem is said to be *well-conditioned*, otherwise it is said to be *ill-conditioned*.

For illustration, suppose that f is differentiable and sufficiently close to linear over Δx, so that we can write a truncated Taylor series expansion as:

$$\tilde{y} - y = f(x + \Delta x) - f(x) \approx f'(x)\Delta x \tag{3.88}$$

where, recall, the algorithm $\tilde{f}$ is assumed *stable*, leading to:

$$\frac{\Delta y}{y} \approx \left(\frac{xf'(x)}{f(x)}\right)\frac{\Delta x}{x} \tag{3.89}$$

and we refer to the following quantity as a *condition number*:

$$c_f(x) = \left|\frac{xf'(x)}{f(x)}\right| \tag{3.90}$$

Thus, even though the algorithm is stable, the error in the output can be large if the problem is ill-conditioned, i.e., if it has a large condition number. As a general rule of thumb, for a stable method, we have:

$$\text{(forward error)} \approx \text{(condition number)} \times \text{(backward error)} \tag{3.91}$$

Exactly what an appropriate condition number is in any given case will depend on the specific problem being considered, as we will see later on in connection with the problem of solving a linear system $Ax = b$.

3.2.2 Floating Point Numbers

Early computers used *fixed point* number representations, in which one operates on numbers as integers, keeping in mind the position of the decimal point, possibly scaling numbers so as to fit within the required range. Later on, it was found that a *floating point* number system is more efficient. It requires more hardware resources but makes programming easier. The *IEEE standard 754-1985* is the predominant standard today for floating point computer arithmetic. Apart from specifying the number of bits in different modes, this standard requires specific handling for exceptions and other salient details. A brief study of floating point number systems is useful in order to appreciate error analysis of algorithms.

A *floating point number system* $F \subset \mathbb{R}$ is a subset of the real numbers whose elements have the form:

$$y = \pm m_t \times \beta^{e-t} \tag{3.92}$$

where:

- t is the *precision*. In IEEE single precision, $t = 24$, while in double precision it is 53.
- β is the *base*, also called the *radix*; $\beta = 2$ in IEEE arithmetic.
- m_t is a t-digit non-negative integer, in the base β, referred to as the *mantissa* or the *significand*, satisfying $0 \le m_t \le \beta^t - 1$.

- e is the *exponent*, in the range $e_{min} \le e \le e_{max}$. In IEEE single precision, $e_{min} = -125$ and $e_{max} = 128$, while in double precision $e_{min} = -1021$ and $e_{max} = 1024$.

The *range* of the non-zero floating point numbers in F is:

$$\beta^{e_{min}-1} \le |y| \le \beta^{e_{max}(1-\beta^{t-1})} \tag{3.93}$$

In IEEE single precision $1.1755 \times 10^{-38} \le |y| \le 3.4028 \times 10^{38}$ and in double precision $2.2251 \times 10^{-308} \le |y| \le 1.7977 \times 10^{308}$, where the end-points of these ranges have been rounded off.

Machine Epsilon   Floating point numbers are not equally spaced on the real line; in fact, the spacing increases by a factor of 2 at every power of 2. The spacing can be characterized by the *machine epsilon*, which is the distance ϵ_m from 1.0 to the next larger floating point number, given by:

$$\epsilon_m = \beta^{1-t} \tag{3.94}$$

In IEEE single precision, $\epsilon_m = 2^{-23} \approx 1.1921 \times 10^{-7}$ and in double precision $\epsilon_m = 2^{-52} \approx 2.2204 \times 10^{-16}$.

Rounding and the Unit Roundoff   If $x \in \mathbb{R}$ lies in the range of F, then we define *rounding* as follows. Rounding a real number x produces a real number $fl(x)$, which is the element of F that is nearest to x. The question of resolving a tie immediately comes up, and is actually quite important. There are various ways to break a tie, such as to round to the value with even mantissa (the most common IEEE mode), to round towards 0, or away from 0, or towards $+\infty$, or towards $-\infty$, etc. It can be shown that, for any x in the range of F, there exists a $\delta \in \mathbb{R}$, such that:

$$fl(x) = x(1 + \delta), \quad \text{with } |\delta| < u \tag{3.95}$$

where $u \triangleq \beta^{1-t}/2 = \epsilon_m/2$ is called the *unit roundoff*. In IEEE single precision, $u \approx 5.9605 \times 10^{-8}$, while in double precision $u \approx 1.1102 \times 10^{-16}$. This u is very useful to express numerical roundoff errors. For one thing, we have:

$$\left| \frac{fl(x) - x}{x} \right| = |\delta| < u \tag{3.96}$$

This relative error actually varies with x, from close to u down to u/β.

The machine epsilon is sometimes defined as the smallest positive real τ which, when added to 1.0 yields a result other than 1.0, i.e., $1 + \tau > 1$. For example, one might find such a τ using this C code fragment:

```
double tau = 1.0;
do{tau /= 2.0;} while((double)(1.0 + tau) != 1.0);
```

However, this definition is imprecise because the result depends on the rounding mode. For the common mode of rounding to even, this gives a τ that is just slightly larger than $\epsilon_m/2$, while, for rounding towards $-\infty$, or towards 0, this gives $\tau = \epsilon_m$. Thus, without further information on the machine, the operating system, the compiler, etc., one can use this method to only get an idea of the order of magnitude of ϵ_m. For our work, we will use the more reliable definition of machine epsilon given by (3.94).

IEEE Arithmetic One advantage of IEEE arithmetic is that it guarantees the following:

Theorem 3.5. *If* $\circ \in \{+, -, \times, /\}$, *then* $\forall x$, y, *and* $z = x \circ y$, *all in the range of* F, *there exists a* $\delta \in \mathbb{R}$, *such that:*

$$\tilde{z} = (x \circ y)(1 + \delta), \quad \text{where} \quad |\delta| \leq u \tag{3.97}$$

i.e.:

$$\left| \frac{\tilde{z} - x \circ y}{x \circ y} \right| = |\delta| \leq u \tag{3.98}$$

so that the computed value $\tilde{z}$ *of a basic arithmetic operation is as good as the rounded exact answer* $fl(z) = z(1 + \delta)$.

This provides a "model" of computer arithmetic that forms the basis for error analysis of numerical algorithms. If $\tilde{z}$ is the computed value of $z = x \pm y$, then notice that, by (3.97):

$$\tilde{z} = (x \pm y)(1 + \delta) = x(1 + \delta) \pm y(1 + \delta) \tag{3.99}$$

Therefore, the atomic algorithmic steps of *addition* and *subtraction* are stable, because a backward error of $|\delta| \approx u$ is certainly small enough. More complicated computations (algorithms) require more detailed error analysis, which is beyond the scope of our study.

Cautionary note: In the literature, it is common to denote the computed value $\tilde{g}(\cdot)$ of a computation or function $g(\cdot)$ as simply $fl(g(\cdot))$. This can be misleading because one can read $fl(g(\cdot))$ as meaning the rounded value of the result of the *exact* computation $g(\cdot)$. This is not the intended meaning! However, it happens to be true for the atomic $\{+, -, \times, /\}$ arithmetic operations, which is often expressed as $fl(x \circ y) = (x \circ y)(1 + \delta)$. In general, one should always understand $fl(g(\cdot))$ to simply mean $\tilde{g}(\cdot)$.

3.2.3 Norms

A brief review of vector and matrix norms will be required before we can discuss the stability and accuracy of Gaussian elimination.

Vector Norms   A *vector norm* over $\mathbb{R}^n$ is a function $\|\cdot\| : \mathbb{R}^n \to \mathbb{R}$ such that, $\forall x, y \in \mathbb{R}^n$,

1. $\|x\| \geq 0$.
2. $\|x\| = 0$ if and only if $x = 0$.
3. $\|cx\| = |c| \cdot \|x\|$ for all scalars $c \in \mathbb{R}$.
4. $\|x + y\| \leq \|x\| + \|y\|$.

Vector norms provide a means of measuring the "size" of a vector and, for any norm and any $x, y \in \mathbb{R}^n$, it can be shown that:

$$\left| \|x\| - \|y\| \right| \leq \|x - y\| \tag{3.100}$$

There is an infinite number of possible vector norms on $\mathbb{R}^n$, but a commonly used class are the p-norms (p need not be an integer):

$$\|x\|_p \triangleq \left(\sum_{i=1}^{n} |x_i|^p \right)^{1/p}, \quad p \geq 1 \tag{3.101}$$

also called the l_p norms, which include some important special cases, such as the l_1 norm, $\|x\|_1 \triangleq |x_1| + |x_2| + \cdots + |x_n|$, also known as the sum norm, the l_2 norm, also known as the Euclidean norm, $\|x\|_2 \triangleq \sqrt{|x_1|^2 + |x_2|^2 + \cdots + |x_n|^2}$, and the l_∞ norm, $\|x\|_\infty \triangleq \max\{|x_1|, |x_2|, \ldots, |x_n|\}$, also known as the max norm. It can be shown that $\|x\|_\infty = \lim_{p \to \infty} \|x\|_p$ and that, for any $x \in \mathbb{R}^n$,

$$\frac{1}{\sqrt{n}} \|x\|_2 \leq \|x\|_\infty \leq \|x\|_2 \leq \|x\|_1 \leq \sqrt{n} \|x\|_2 \tag{3.102}$$

Another useful result is the *Cauchy-Schwarz inequality*:

$$|x^T y| \leq \|x\|_2 \|y\|_2 \tag{3.103}$$

Norms are used to measure accuracy or errors in vector quantities, such as the *absolute error* $\|\tilde{x} - x\|$ and the *relative error* $\|\tilde{x} - x\| / \|x\|$ (if $x \neq 0$).

Finally, the l_∞ norm helps determine the number of correct significant digits. For example, and as a general rule of thumb, if:

$$\frac{\|\tilde{x} - x\|_\infty}{\|x\|_\infty} \approx 10^{-k} \tag{3.104}$$

then the largest component of $\tilde{x}$ has about k correct significant digits.

Matrix Norms   One motivation for defining a matrix norm is as follows. It is known that the computed solution of a linear system has poor quality if the system matrix is "nearly singular." In the theoretical study of "near-singularity," we need a measure of distance on the space of matrices, and matrix norms provide such a measure. Another motivation is that matrix norms help quantify the "size" of a matrix and, therefore, its impact as a linear transformation $y = Ax$.

A *matrix norm* is a function $\| \cdot \| : \mathbb{R}^{m \times n} \rightarrow \mathbb{R}$ such that, $\forall A, B \in \mathbb{R}^{m \times n}$,

1. $\|A\| \geq 0$
2. $\|A\| = 0$ if and only if $A = 0$
3. $\|cA\| = |c| \cdot \|A\|$ for all scalars $c \in \mathbb{R}$
4. $\|A + B\| \leq \|A\| + \|B\|$

One useful norm is the *Frobenius* norm, defined for $A \in \mathbb{R}^{m \times n}$, as:

$$\|A\|_F = \sqrt{\sum_{i=1}^{m} \sum_{j=1}^{n} |a_{ij}|^2} = \sqrt{\operatorname{tr}(A^T A)} \qquad (3.105)$$

where $\operatorname{tr}(B) \triangleq \sum_{i=1}^{n} b_{ii}$ is called the *trace* of a matrix $B \in \mathbb{R}^{n \times n}$. An important class of matrix norms are those *induced* by a vector norm, defined as follows. If $\| \cdot \|$ is a vector norm, then the corresponding *induced* matrix norm is:

$$\|A\| = \max_{x \neq 0} \frac{\|Ax\|}{\|x\|} = \max_{\|x\|=1} \|Ax\| \qquad (3.106)$$

Note that $\|I\| = 1$, where $I \triangleq \operatorname{diag}(1, 1, \ldots, 1)$ is the identity matrix. Note also that, $\forall x \in \mathbb{R}$, $\|Ax\| \leq \|A\| \|x\|$, and one can show that, in general:

$$\|AB\| \leq \|A\| \|B\| \qquad (3.107)$$

where A is $m \times n$ and B is $n \times q$. Thus, for induced norms, the norm is a measure of the largest "gain" that the transformation $y = Ax$ applies to the "size" of the vector x. In the following, whenever a matrix norm is used without further qualification, it will be assumed to be an *induced* norm.

Corresponding to the vector p-norms, one can define the *induced* matrix p-norms, as:

$$\|A\|_p = \max_{x \neq 0} \frac{\|Ax\|_p}{\|x\|_p} \qquad (3.108)$$

For $A \in \mathbb{R}^{m \times n}$, it can be shown that:

$$\|A\|_1 = \max_{1 \leq j \leq n} \sum_{i=1}^{m} |a_{ij}| \quad \text{(max column sum)} \qquad (3.109)$$

$$\|A\|_\infty = \max_{1 \leq i \leq m} \sum_{j=1}^{n} |a_{ij}| \quad \text{(max row sum)} \qquad (3.110)$$

If $\rho(B)$ is the largest $|\lambda|$, where λ is an eigenvalue of the square matrix B ($\rho(B)$ is called the *spectral radius*), then, with $A \in \mathbb{R}^{m \times n}$, one can show:

$$\|A\|_2 = \sqrt{\rho(A^T A)} \quad \text{(spectral norm)} \qquad (3.111)$$

which is not easy to compute, so that the following is often more useful:

$$\|A\|_2 \le \sqrt{\|A\|_1 \|A\|_\infty} \qquad (3.112)$$

(note: the eigenvalues of $A^T A$ are always real and non-negative). For any induced norm, it can be shown that $\rho(A) \le \|A\|$. With $A \in \mathbb{R}^{m \times n}$, the following relations are also useful:

$$\|A\|_2 \le \|A\|_F \le \sqrt{n}\|A\|_2 \qquad (3.113)$$

$$\frac{1}{\sqrt{n}}\|A\|_\infty \le \|A\|_2 \le \sqrt{m}\|A\|_\infty \qquad (3.114)$$

$$\frac{1}{\sqrt{m}}\|A\|_1 \le \|A\|_2 \le \sqrt{n}\|A\|_1 \qquad (3.115)$$

$$\max_{\forall i,j} |a_{ij}| \le \|A\|_2 \le \sqrt{mn} \max_{\forall i,j} |a_{ij}| \qquad (3.116)$$

$$\frac{1}{n}\|A\|_2 \le \frac{1}{\sqrt{n}}\|A\|_\infty \le \|A\|_2 \le \sqrt{n}\|A\|_1 \le n\|A\|_2 \qquad (3.117)$$

Finally, because matrix p-norms are induced norms, then:

$$\|AB\|_p \le \|A\|_p \|B\|_p \qquad (3.118)$$

3.2.4 Stability of GE and *LU* Factorization

How would we characterize the stability of a numerical algorithm for solving the linear system $Ax = b$? Note, the *input* data to the algorithm is the matrix A and the vector b, and the algorithm *output* is x. From basic backward error principles, if, for every A and b, there exist "small" $\|\Delta A\|$ and $\|\Delta b\|$ such that the computed solution $\tilde{x}$ satisfies:

$$(A + \Delta A)\tilde{x} = b + \Delta b \qquad (3.119)$$

then the algorithm is stable—it solves exactly a *nearby* problem. Note, $\|\Delta b\|$ is a *vector norm* and $\|\Delta A\|$ is the *matrix norm* induced by it. Thus, a useful stability metric is the *normwise relative backward error*, defined as:

$$\eta(\tilde{x}) \triangleq \min \left\{ \epsilon : (A + \Delta A)\tilde{x} = b + \Delta b, \quad \frac{\|\Delta A\|}{\|A\|} \le \epsilon, \quad \frac{\|\Delta b\|}{\|b\|} \le \epsilon \right\}$$

where the set notation refers to the set of all ϵ for which *there exist* ΔA and Δb with the stated properties. It can be shown that:

$$\eta(\tilde{x}) = \frac{\|r\|}{\|A\|\|\tilde{x}\| + \|b\|} \tag{3.120}$$

where $r \triangleq b - A\tilde{x}$ is called the *residual*. As Higham (2002) puts it, this result "*makes precise the intuitive feeling that if the residual is small then we have a "good" approximate solution.*" Thus, one way to check the quality of a computed solution $\tilde{x}$ is to find the residual $r = b - A\tilde{x}$ and check if it is small relative to $\|A\|\|\tilde{x}\|$ or $\|b\|$. Of course, we can only compute $\tilde{r}$ and compare it to $\|\tilde{A}\|\|\tilde{x}\|$ and $\|b\|$, but the roundoff error in computing $\tilde{r}$ should be small and acceptable.

Is Gaussian elimination (GE) (equivalently, LU factorization) stable? The short answer is *not necessarily*, but in most practical cases it can be made stable by proper pivoting. Loosely speaking, it can be shown that GE is stable if the matrix terms in the sequence $a_{ij}^{(1)}, a_{ij}^{(2)}, \cdots a_{ij}^{(k)}, \cdots$ do not "grow" very large, relative to a_{ij}. We can be more specific about this, as follows.

GE/LU   Given the equivalence between GE and Gauss's method for LU factorization, then, in the following, we will use the shorthand notation GE/LU to refer to both LU factorization and the GE algorithm when implemented along the lines of Gauss's method for LU factorization, so that the multipliers are saved as an L matrix, and the final triangulated matrix becomes the U matrix. When using GE/LU to solve $Ax = b$, we effectively solve:

$$\tilde{L}\tilde{U}x = \tilde{b} \tag{3.121}$$

using forward and backward substitution. Therefore, error can be incurred 1) during LU factorization, *and* 2) during forward/backward substitution. However, it is possible to show that forward/backward substitution are in fact stable operations, as follows.

Forward/Backward Substitution Is Stable   It is useful at this point to define the coefficient:

$$\gamma_k \triangleq \frac{ku}{1 - ku} \tag{3.122}$$

where k is an integer that will typically be equal to n, the size of the problem, or a small multiple of n, and where u is the unit roundoff. In virtually all cases when using IEEE arithmetic, and for practically solvable problems, it is found that $nu < 1$, so that typically $ku < 1$ and $\gamma_k > 0$. This will be immediately useful. If the triangular system $Tx = b$, where $T \in \mathbb{R}^{n \times n}$ is nonsingular, is solved by substitution, then it can be shown that the computed solution $\tilde{x}$ satisfies:

$$(T + \Delta T)\tilde{x} = b, \quad \text{where} \quad |\Delta T| \leq \gamma_n |T| \tag{3.123}$$

where $|T|$ denotes the matrix whose entries are $|t_{ij}|$. This effectively means that the backward errors in every term of ΔT are quite small; they are $\mathcal{O}(u)$.

Thus, forward/backward substitution is quite stable. As a result, we consider that the solution of $\tilde{L}\tilde{U}x = b$ is stable, and we can focus on stability of the LU factorization part of the process.

Stability of the Factorization   Thus, stability monitoring of GE/LU can be done by checking the size of H, defined as:

$$H \triangleq \tilde{L}\tilde{U} - A \tag{3.124}$$

and, one can show that, as a result of GE/LU, we have:

$$|H| \leq \gamma_n |\tilde{L}||\tilde{U}| \tag{3.125}$$

As for the solution of $Ax = b$, one can prove that GE/LU leads to:

$$(A + \Delta A)\tilde{x} = b, \quad \text{where} \quad |\Delta A| \leq \gamma_{3n} |\tilde{L}||\tilde{U}| \tag{3.126}$$

These results don't quite say that GE/LU is stable or not, but we do learn that the stability of GE/LU is related to the size of the matrix $|\tilde{L}||\tilde{U}|$.

The Need for Pivoting   Looking at (3.126), we are motivated to monitor the value of the ratio:

$$\frac{\left\| |\tilde{L}||\tilde{U}| \right\|}{\|A\|} \tag{3.127}$$

If this ratio is small, then $\|\Delta A\| / \|A\|$ would be small, and GE/LU would be stable. If pivoting is *not* used, and assuming that a divide-by-0 is not encountered, this ratio can be arbitrarily large. For example, as pointed out in Higham (2002), for the matrix $\left[\begin{smallmatrix} \epsilon & 1 \\ 1 & 1 \end{smallmatrix}\right]$, where $\epsilon > 0$, this ratio is of the order of $1/\epsilon$, which can be very large. This gives us a hint that pivoting may be required *for accuracy*. One would hope that, with pivoting, the entries in $\tilde{L}$ and $\tilde{U}$ can be made small enough that the ratio in (3.127) becomes small. If pivoting is used in a way that the pivot used is at least as large as the largest entry in the column at the present level of the GE algorithm, then we get $|l_{ij}| \leq 1$ because the l_{ij} are the multipliers that arise during GE. Thus, the "size" of L can be made small, but the "size" of U is not as easy to bound. One can show that the size of U is, in fact, bounded relative to the size of A, but the coefficient is exponential in n, so that the bound is too loose and not useful.

The Growth Factor   Given the above, it should be clear that, with a pivoting strategy that limits the size of L, the stability of the GE/LU algorithm now rests on the issue of whether the size of U is small or not. This leads us to the classical results of stability of GE, which are in terms of the *growth factor*:

$$\rho \triangleq \frac{\max_{i,j,k} \left| a_{ij}^{(k)} \right|}{\max_{i,j} \left| a_{ij} \right|} \tag{3.128}$$

and one can see why this is useful because it leads to a bound on U, as:

$$|u_{ij}| = \left| a_{ij}^{(i)} \right| \leq \rho \max_{i,j} |a_{ij}| \tag{3.129}$$

This should be expected because U is in fact the final triangular A matrix at the end of the GE algorithm, so that growth of elements of A should be a concern. Indeed, the growth factor directly determines the backward error, according to this classical theorem given in Higham (2002) by reference to early work by Wilkinson:

Theorem 3.6. *(Wilkinson (1963)) Let $A \in \mathbb{R}^{n \times n}$ and suppose that GE with partial pivoting gives a computed solution $\tilde{x}$ to $Ax = b$, then:*

$$(A + \Delta A)\tilde{x} = b, \quad with \quad \frac{\|\Delta A\|_\infty}{\|A\|_\infty} \leq \rho \gamma_{3n} n^2 \tag{3.130}$$

It is also possible to show that, according to Golub and Van Loan (1989):

$$\|\Delta A\|_\infty \leq 8\rho \|A\|_\infty u n^3 + \mathcal{O}(u^2) \tag{3.131}$$

which, ignoring the $\mathcal{O}(u^2)$ term, leads to the approximate upper bound on the relative backward error of $8\rho u n^3$. In any case, these results show that the growth factor ρ must be kept small, and one way to try and do this, though it is not guaranteed to succeed, is by pivoting, as we now explain.

3.2.5   Pivoting for Accuracy

Given the above results, the general strategy in practice is to use pivoting to make sure that the matrix entries do not grow too large. This requires that the pivot element should not be too small, relative to the other elements in the column at the present (kth) level of the GE algorithm. Such a strategy would ensure that the multipliers m_{ik} in the GE algorithm in Fig. 3.1 are small in magnitude, which helps avoid large growth in the updated matrix entries. Ideally, we would like the pivot element to have the highest available absolute value in that column, or in the remaining sub-matrix. There are two options for choosing the pivot element $a_{kk}^{(k)}$:

1. **Partial pivoting:** search only the kth column and pick the largest element (in absolute value) as the pivot by exchanging rows[2]. This is sometimes referred to as *row pivoting*, but this term is not very common.

---

[2]Note that it is possible to do "partial pivoting" of sorts by finding the largest entry in the kth *row*, followed by a column exchange, but that approach would not help limit the size of the

2. **Full pivoting:** search the whole remaining sub-matrix (from k to n) for the largest magnitude pivot and exchange possibly rows and columns.

For efficiency, we do not actually exchange whole rows or columns; instead, we only change the memory pointers to get the desired permutation.

Practical experience shows that GE/LU with partial pivoting is almost always stable, and so it is a very commonly used approach. We will use the acronym GEPP to refer to the use of GE/LU with partial pivoting. One should keep in mind, however, that there is *no guarantee* that GEPP would always be stable. In fact, with a carefully contrived matrix, the following upper-bound on the growth factor under GEPP is achievable, i.e., it is *tight*:

$$\rho \leq 2^{n-1} \max_{i,j} |a_{ij}| \qquad (3.132)$$

so that exponential element size growth *can* occur with GEPP. There is a smaller upper-bound than this when using full pivoting but, in practice, partial pivoting works quite well and there is no need for full pivoting. In terms of computational cost, partial pivoting is $\mathcal{O}(n^2)$, which is not trivial but bearable, while full pivoting is $\mathcal{O}(n^3)$, which is too expensive, and is rarely required.

For matrices that are not necessarily sparse, common practice is to use only GE with partial pivoting (GEPP), with any/all flavors of GE (classical GE, or *LU* factorization). For sparse matrices, we will see later on that one must allow for both row and column exchange, but we will study ways of choosing a pivot that are cheaper than full pivoting.

Pivoting Example The following hypothetical scenario, from Duff et al. (1986), provides an excellent illustration of the effects of pivoting. Suppose we are working with a floating point system that allows only *three* decimal digits—a hypothetical 3-digit floating-point computer—and consider the problem:

$$\begin{bmatrix} 0.001 & 2.42 \\ 1.00 & 1.58 \end{bmatrix} \begin{bmatrix} x_1 \\ x_2 \end{bmatrix} = \begin{bmatrix} 5.20 \\ 5.47 \end{bmatrix} \qquad (3.133)$$

Notice that both the matrix and the right-hand side (RHS) vector use only 3-digit values, so that no roundoff error occurs in storing the data. Any roundoff error problems we may discover are due to the algorithm. Since $a_{11} \neq 0$, the basic GE proceeds with a_{11} as the pivot, and we compute:

$$a_{22}^{(2)} = fl\left(a_{22} - a_{12}\left(\frac{a_{21}}{a_{11}}\right)\right) = fl(1.58 - 2420) = -2420 \qquad (3.134)$$

multipliers, although it can help avoid a zero pivot as we saw earlier. Therefore, in this text, partial pivoting will always mean one where we select a pivot from the kth *column*, followed by a *row* exchange.

where the use of the small pivot 0.001 led to a growth in the value of a_{22}. The resulting triangulated system is:

$$\begin{bmatrix} 0.001 & 2.42 \\ 0 & -2420 \end{bmatrix} \begin{bmatrix} x_1 \\ x_2 \end{bmatrix} = \begin{bmatrix} 5.20 \\ -5200 \end{bmatrix} \tag{3.135}$$

The resulting *computed* solution is:

$$\begin{bmatrix} \tilde{x}_1 \\ \tilde{x}_2 \end{bmatrix} = \begin{bmatrix} 0.00 \\ 2.15 \end{bmatrix} \tag{3.136}$$

while the 3-digit approximation to the *true* solution is:

$$\begin{bmatrix} x_1 \\ x_2 \end{bmatrix} \approx \begin{bmatrix} 1.18 \\ 2.15 \end{bmatrix} \tag{3.137}$$

so that the error is clearly large, and and the residual is:

$$\begin{bmatrix} r_1 \\ r_2 \end{bmatrix} \approx \begin{bmatrix} -0.003 \\ 1.17 \end{bmatrix} \tag{3.138}$$

whose norm is clearly not very small relative to $\|b\|$. If we use partial pivoting, we would interchange the two rows of A so as to use 1.00 as the pivot:

$$\begin{bmatrix} 1.00 & 1.58 \\ 0.001 & 2.42 \end{bmatrix} \begin{bmatrix} x_1 \\ x_2 \end{bmatrix} = \begin{bmatrix} 5.47 \\ 5.20 \end{bmatrix} \tag{3.139}$$

leading to the reduced system:

$$\begin{bmatrix} 1.00 & 1.58 \\ 0 & 2.42 \end{bmatrix} \begin{bmatrix} x_1 \\ x_2 \end{bmatrix} = \begin{bmatrix} 5.47 \\ 5.19 \end{bmatrix} \tag{3.140}$$

showing no growth in the term a_{22}, and the computed solution:

$$\begin{bmatrix} \tilde{x}_1 \\ \tilde{x}_2 \end{bmatrix} = \begin{bmatrix} 2.09 \\ 2.14 \end{bmatrix} \tag{3.141}$$

with a residual of:

$$\begin{bmatrix} r_1 \\ r_2 \end{bmatrix} \approx \begin{bmatrix} 0.01911 \\ -0.0012 \end{bmatrix} \tag{3.142}$$

which is much smaller than before, and has small norm compared to $\|b\|$.

Remarks   With sparse matrices, as we will see later on, pivoting can destroy sparsity. A general strategy becomes to pivot for sparsity when the pivot is larger than some threshold, and otherwise to pivot for accuracy. This is called *threshold pivoting*. When this is done, stability can suffer and backward error can increase, but *iterative refinement* can be help combat this, as we will see later on. When the matrix is *diagonally dominant*, then no pivoting is required.

3.2.6 Conditioning of $Ax = b$

Even with partial pivoting, so that GE is stable, with a small backward error, an ill-conditioned problem can still suffer from large forward error. Recall the rule of thumb for a stable algorithm:

$$(\text{forward error}) \approx (\text{condition number}) \times (\text{backward error}) \qquad (3.143)$$

We are now interested in the question: what is an appropriate condition number for a linear system problem $Ax = b$, and how can we detect ill-conditioning?

As an example, consider the system:

$$\begin{bmatrix} 1.0 & 2.0 \\ 2.0 & 3.999 \end{bmatrix} \begin{bmatrix} x_1 \\ x_2 \end{bmatrix} = \begin{bmatrix} 4.0 \\ 7.999 \end{bmatrix} \qquad (3.144)$$

The exact solution of the system is:

$$\begin{bmatrix} x_1 \\ x_2 \end{bmatrix} = \begin{bmatrix} 2.0 \\ 1.0 \end{bmatrix} \qquad (3.145)$$

If we only slightly perturb the system by changing the RHS vector:

$$\begin{bmatrix} 1.0 & 2.0 \\ 2.0 & 3.999 \end{bmatrix} \begin{bmatrix} x_1 \\ x_2 \end{bmatrix} = \begin{bmatrix} 4.001 \\ 7.998 \end{bmatrix} \qquad (3.146)$$

then, the exact solution becomes:

$$\begin{bmatrix} x_1 \\ x_2 \end{bmatrix} = \begin{bmatrix} -3.999 \\ 4.0 \end{bmatrix} \qquad (3.147)$$

Thus, a minor change to the problem data has produced a large change in the solution. This system is ill-conditioned.

In general, for a perturbation Δb in the RHS vector, but no perturbation in A for now, it can be shown that:

$$\frac{\|\Delta x\|}{\|x\|} \le \|A\| \|A^{-1}\| \frac{\|\Delta b\|}{\|b\|} \qquad (3.148)$$

Likewise, for a perturbation ΔA in the system matrix, but no perturbation in b for now, it can be shown that:

$$\frac{\|\Delta x\|}{\|x + \Delta x\|} \le \|A\| \|A^{-1}\| \frac{\|\Delta A\|}{\|A\|} \qquad (3.149)$$

Condition Number This motivates the definition of a *condition number* for the matrix A, as:

$$\kappa(A) \triangleq \|A\| \|A^{-1}\| \qquad (3.150)$$

with $\kappa(A) \triangleq \infty$ when A is singular. If $\kappa(A)$ is large, then the problem is said to be *ill-conditioned* and the forward error can be large, as is clear from (3.148) and (3.149). If $\kappa(A) \approx 1$, then $\mathcal{O}(u)$ perturbations in the data (A or b) lead to $\mathcal{O}(u)$ perturbations in the solution, and the problem is said to be *well-conditioned*. In general, it can be shown that $\kappa(A) \geq 1$, and that $\forall c \in \mathbb{R}$, with $c \neq 0$, we have:

$$\kappa(cA) = \kappa(A) \tag{3.151}$$

We are interested in the case where there are perturbations in both A and b, and the following result is useful.

Theorem 3.7. *Let $Ax = b$ and $(A + \Delta A)(x + \Delta x) = (b + \Delta b)$, and $0 < \epsilon < 1$, with:*

$$\frac{\|\Delta A\|}{\|A\|} \leq \epsilon, \qquad \frac{\|\Delta b\|}{\|b\|} \leq \epsilon, \quad and \quad \epsilon < \frac{1}{\kappa(A)} \tag{3.152}$$

then:

$$\frac{\|\Delta x\|}{\|x\|} \leq \frac{\epsilon\kappa(A)}{1 - \epsilon\kappa(A)}\left(1 + \frac{\|b\|}{\|A\|\|x\|}\right) \leq \frac{2\epsilon\kappa(A)}{1 - \epsilon\kappa(A)} \tag{3.153}$$

Thus, again, a large $\kappa(A)$ denotes ill-conditioning. One should keep in mind the requirements (3.152) before using this result. In practice, one would hope that $\epsilon \approx u$ and that $u\kappa(A) < 1$. Indeed, as we will see below, if $u\kappa(A) \not< 1$, then the solution of the system is entirely unreliable.

Singular Value Decomposition   A brief digression to review the singular value decomposition will be useful at this point. A matrix $Q \in \mathbb{R}^{n \times n}$ is said to be *orthogonal* if $Q^T Q = I$. If $A \in \mathbb{R}^{m \times n}$, then there exist *orthogonal* matrices $U \in \mathbb{R}^{m \times m}$ and $V \in \mathbb{R}^{n \times n}$ leading to the following so-called *singular value decomposition* (SVD):

$$A = U\Sigma V^T \tag{3.154}$$

where $\Sigma = \text{diag}(\sigma_1, \sigma_2, \ldots, \sigma_p) \in \mathbb{R}^{m \times n}$, with $p = \min(m, n)$, and where $\sigma_1 \geq \sigma_2 \geq \cdots \geq \sigma_p \geq 0$ are called the *singular values* of A. It is easy to see that $\Sigma = U^T AV$, so that A is *diagonalizable* by means of the transformation $U^T AV$. One can show that the singular values σ_i are the non-negative square roots of the eigenvalues of AA^T, and hence are uniquely determined. The largest and the smallest singular values of A are denoted $\sigma_{max}(A)$ and $\sigma_{min}(A)$, respectively:

$$\sigma_{max}(A) = \sqrt{\lambda_{max}(AA^T)} \quad and \quad \sigma_{min}(A) = \sqrt{\lambda_{min}(AA^T)} \tag{3.155}$$

where $\lambda_{max}(AA^T)$ and $\lambda_{min}(AA^T)$ are the largest and the smallest eigenvalues of AA^T, respectively, both guaranteed real and non-negative. Finally, it can be shown that the columns of U are eigenvectors of AA^T and the columns of V are eigenvectors of $A^T A$. Singular values also have a geometric meaning: they are the lengths of the semi-axes of the hyperellipsoid defined by $\{Ax : \|x\|_2 = 1\}$.

Conditioning   The condition number is sometimes denoted $\kappa_p(A)$ to signify the fact that the matrix p-norm of A and A^{-1} is being used; it can be shown that:

$$\frac{1}{n}\kappa_2(A) \leq \kappa_1(A) \leq n\kappa_2(A) \tag{3.156}$$

$$\frac{1}{n}\kappa_\infty(A) \leq \kappa_2(A) \leq n\kappa_\infty(A) \tag{3.157}$$

$$\frac{1}{n^2}\kappa_1(A) \leq \kappa_\infty(A) \leq n^2\kappa_1(A) \tag{3.158}$$

If A is ill-conditioned in the α-norm, then it is also ill-conditioned in the β-norm, because constants c_1 and c_2 can always be found such that:

$$c_1\kappa_\alpha(A) \leq \kappa_\beta(A) \leq c_2\kappa_\alpha(A) \tag{3.159}$$

For the 2-norm, it can be shown based on an SVD of A that:

$$\kappa_2(A) = \frac{\sigma_{max}(A)}{\sigma_{min}(A)} \tag{3.160}$$

Intuitively, as given in Ruehli (1986), a problem $Ax = b$ is ill-conditioned if some of the n-dimensional hyperplanes representing the n equations are nearly-parallel, as shown in Fig. 3.11.

Conditioning and Pivoting   Going back to our previous ill-conditioned example:

$$\begin{bmatrix} 1.0 & 2.0 \\ 2.0 & 3.999 \end{bmatrix} \begin{bmatrix} x_1 \\ x_2 \end{bmatrix} = \begin{bmatrix} 4.0 \\ 7.999 \end{bmatrix} \tag{3.161}$$

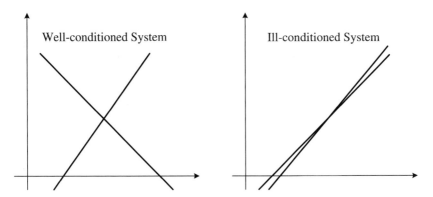

Figure 3.11: The geometric interpretation of ill-conditioning, in the case of a two-dimensional system [after Ruehli (1986)].

we have:

$$A = \begin{bmatrix} 1.0 & 2.0 \\ 2.0 & 3.999 \end{bmatrix} \quad A^{-1} = \begin{bmatrix} 3,999 & -2,000 \\ -2,000 & 1,000 \end{bmatrix} \tag{3.162}$$

$$\|A\|_\infty = 5.999 \quad \|A^{-1}\|_\infty = 5,999 \tag{3.163}$$

and

$$\kappa_\infty(A) = 35,988 \tag{3.164}$$

which is a very large condition number. Notice that $\det(A) = -0.001$, so that A is nearly singular, which is usually a sign of "trouble," and it explains the large values of the inverse matrix. However, A^{-1} is not directly computed by GE, and so this does not necessarily reflect element size growth during GE. In fact, the matrix A has no particularly small elements, so that GE on it would not run into any small pivot issues or any growth problems. This reinforces the point that conditioning is an "orthogonal" issue to that of stability, and cannot be "fixed" by pivoting. It is a property of the problem itself, not of the algorithm.

Indeed, it can be shown that *the condition number is invariant to pivoting, be it partial or full*. Specifically, for any $A \in \mathbb{R}^{n \times n}$, and if $P, Q \in \mathbb{R}^{n \times n}$ are permutation matrices, one can show that:

$$\kappa_p(PAQ) = \kappa_p(A) \tag{3.165}$$

for any p-norm. It can also be shown that $\kappa_2(A)$ is *similarity-invariant*. But, do we conclude from this example that a small determinant (near singularity) is a sign of ill-conditioning? The answer is *no*, as we will see.

Conditioning and the Determinant   Note that having $\det(A) \approx 0$ is not required for ill-conditioning, nor is it indicative of ill-conditioning, as we illustrate. Consider the following matrix:

$$A = \begin{bmatrix} 100.01 & 100 \\ 100 & 100.01 \end{bmatrix} \quad \text{with} \quad A^{-1} = \frac{1}{2.0001} \begin{bmatrix} 100.01 & -100 \\ -100 & 100.01 \end{bmatrix} \tag{3.166}$$

We have $\det(A) = 2.0001$, which does not seem problematic; however, $\kappa_\infty = \|A\| \|A^{-1}\| = 20,001$, so that the matrix is ill-conditioned. On the other hand, consider a real symmetric 10×10 matrix with all eigenvalues equal to 0.1, then $\det(A) = 0.1^{10} = 10^{-10}$, while:

$$\kappa_2(A) = \sigma_{max}(A)/\sigma_{min}(A) = 1.0 \tag{3.167}$$

because the singular values of a real symmetric matrix are equal to the absolute values of its eigenvalues.

Conditioning and Uneven Scaling   If a matrix A is such that, for some $\|v\| = \|w\|$ we have $\|Av\| \gg \|Aw\|$, then, with $y = Aw$ we have:

$$1 \ll \frac{\|Av\|}{\|Aw\|} \frac{\|w\|}{\|v\|} = \frac{\|Av\|}{\|v\|} \frac{\|A^{-1}y\|}{\|y\|} \le \|A\| \|A^{-1}\| = \kappa(A) \tag{3.168}$$

so that $\kappa(A) \gg 1$. Thus, an informal characterization of conditioning is that well-conditioned matrices would scale the norms of all vectors about equally. For example, consider the same matrix we saw above:

$$A = \begin{bmatrix} 100.01 & 100 \\ 100 & 100.01 \end{bmatrix} \tag{3.169}$$

The following two vectors are the same "size" but scale quite differently as a result of the transformation $y = Ax$:

$$v = \begin{bmatrix} 1 \\ -1 \end{bmatrix} \quad w = \begin{bmatrix} 1 \\ 1 \end{bmatrix} \tag{3.170}$$

with:

$$\|Av\|_\infty = 0.01 \quad \|Aw\|_\infty = 200.01 \tag{3.171}$$

and this uneven scaling of equally sized vectors is another indicator of possible ill-conditioning. The above v and w are eigenvectors, corresponding to the two eigenvalues (in this case, equal to the singular values) 0.01 and 200.01. The large range is indicative of a problem. In general, it can be shown that the condition number is the ratio of the largest to the smallest "gain" applied by A to the vector norms:

$$\kappa(A) = \frac{\max\limits_{x \ne 0} \frac{\|Ax\|}{\|x\|}}{\min\limits_{x \ne 0} \frac{\|Ax\|}{\|x\|}} \tag{3.172}$$

Conditioning and Pivoting   Another useful example, based on a comment in Higham (2002), is this:

$$\begin{bmatrix} \epsilon & -1 \\ 1 & 1 \end{bmatrix} \begin{bmatrix} x_1 \\ x_2 \end{bmatrix} = \begin{bmatrix} 1 \\ 0 \end{bmatrix} \tag{3.173}$$

in which $\epsilon > 0$ is "small," and where the exact solution is:

$$\begin{bmatrix} x_1 \\ x_2 \end{bmatrix} = \begin{bmatrix} 1/(1+\epsilon) \\ -1/(1+\epsilon) \end{bmatrix} \approx \begin{bmatrix} 1 \\ -1 \end{bmatrix} \tag{3.174}$$

The condition number for the system matrix is:

$$\kappa_\infty = \frac{4}{1+\epsilon} \approx 4.0 \tag{3.175}$$

which is small, so that the matrix is well-conditioned. However, if no pivoting is performed, the new value of a_{22} is:

$$a'_{22} = fl\left(1 + \frac{1}{\epsilon}\right) = \frac{1}{\epsilon} \tag{3.176}$$

for small enough ϵ, due to roundoff, leading to the triangular system:

$$\begin{bmatrix} \epsilon & -1 \\ 0 & 1/\epsilon \end{bmatrix}\begin{bmatrix} x_1 \\ x_2 \end{bmatrix} = \begin{bmatrix} 1 \\ -1/\epsilon \end{bmatrix} \tag{3.177}$$

whose computed solution is:

$$\begin{bmatrix} \tilde{x}_1 \\ \tilde{x}_2 \end{bmatrix} = \begin{bmatrix} 0 \\ -1 \end{bmatrix} \tag{3.178}$$

which is obviously highly inaccurate, with a residual of:

$$r = b - A\tilde{x} = \begin{bmatrix} 0 \\ 1 \end{bmatrix} \tag{3.179}$$

which is of the same size as b, normwise.

Thus, we see that GE, applied to a very well-conditioned matrix, can turn out to be highly unstable and lead to big errors, so that pivoting is definitely required. But, no amount of pivoting can remedy an ill-conditioned problem!

Significant Digits   As mentioned, in practice, GEPP is almost always found to be a stable algorithm with small backward error (and a small residual $r = b - A\tilde{x}$). In some problems, it is enough that the residual is small! In other words, in some problems, the residual is an acceptable measure of solution error. But, often, we are interested in the total (forward) error and in the high accuracy of the solution itself, and that depends on the condition number. What is the impact of ill-conditioning on accuracy?

With a stable GEPP (so that the backward error is $\approx u$), it can be shown that the forward error is:

$$\frac{\|\tilde{x} - x\|_\infty}{\|x\|_\infty} \approx u\kappa_\infty(A) \tag{3.180}$$

so that, again, (forward error) $\approx$ (backward error)$\times$(condition number). Thus, a condition number may be judged as large, or not, *in relation to* the unit roundoff u of the machine. If $\kappa(A) \geq (1/u)$, then the solution is certainly unreliable! In practice, it is generally felt that if $\kappa(A) > (1/\sqrt{u})$ then the computed solution may not be trustworthy.

The above result leads to the following useful *heuristic* which is commonly employed: if $u\kappa_\infty(A) \approx 10^{-k}$, then GEPP gives $\approx k$ correct significant digits. Two important questions remain: how do we efficiently detect ill-conditioning, and can we do anything about it, so as to improve accuracy?

Detecting Ill-Conditioning   Duff et al. (1986) report that there are several possible ways to detect ill-conditioning, many of which turn out to be either unreliable or too expensive. These include finding A^{-1}, finding the singular values or the eigenvalues of A, finding the determinant of A, and checking the pivot values. All of these are failed options, for one reason or another. Instead, here are two methods that have been found to be reliable and cheap in practical experience:

1. Solve a problem with a known solution and check the result. Choose a y and compute $c = Ay$, then solve $Ax = c$ and check the result $\tilde{x}$ against the known y. This is workable but the next method is preferred in practice.
2. The so-called LINPACK estimate: This method *estimates* $\|A^{-1}\|_1$ by first solving $A^T v = c$ for a specially constructed vector c, and then solving $Aw = v$ and computing $\|A^{-1}\|_1 \approx \|w\|_1/\|v\|_1$. This is justifiable using the singular value decomposition of A.

For further details, the interested reader should consult Duff et al. (1986).

Fixing an Ill-Conditioned Problem   A matrix is said to be *poorly scaled* if the range of magnitudes of its entries is very large—it has some very large and very small entries. Equivalently, a wide range in b leads to a poorly scaled problem. One can scale each equation to equilibrate the RHS vector entries. We will assume that this is always done, so that a problem is poorly scaled only if the A matrix is poorly scaled. With such a matrix the quality of the solution suffers, due to various effects:

1. If the normwise backward error $\|\Delta A\|_\infty/\|A\|_\infty$ is small, the backward error in the small magnitude entries can be large. The larger matrix elements "mask" these errors due to the ∞-norm. One could use *componentwise* metrics rather than normwise metrics, but the theoretical results are not as easily applicable in practice.
2. Roundoff error gets worse because, during GEPP, one is often adding or subtracting some very large to very small numbers.
3. The solution comes out poorly scaled as well, so that an ∞-norm error metric for the solution vector is of little value for its smaller components. For example, if $x^T = \begin{bmatrix} 2.1 & 1.3 \times 10^7 \end{bmatrix}$, then an error bound such as $\|\Delta x\|_\infty \leq 3$ means that x_2 is accurate but does not convey much accuracy in x_1.

In such cases, one would do well to *scale* the problem first before solving it, and this can be done by multiplication by diagonal matrices D_1 and D_2:

$$D_1 A D_2 y = D_1 b \quad \text{with} \quad D_2 y = x \tag{3.181}$$

It turns out that *scaling can reduce the condition number*:

$$\exists D_1, D_2 : \quad \kappa(D_1 A D_2) < \kappa(A) \tag{3.182}$$

However, the theory for this is incomplete, and there are no useful general purpose strategies for scaling—it tends to be problem-specific. During modeling, one should try to ensure that the system matrix is not poorly scaled, by using similar units, normalization, and similar devices.

3.2.7 Iterative Refinement

Suppose $\tilde{x}$ is the *computed* solution to $Ax = b$, with the residual $r = b - A\tilde{x}$, so that:

$$A\tilde{x} = b - r \tag{3.183}$$

Let z be the *exact* solution to $Az = r$, then it is clear that:

$$A(\tilde{x} + z) = b \tag{3.184}$$

so that z is the required *correction* for $\tilde{x}$. Thus, with $PA = LU$ available, one possible approach for improvement is:

1. Compute $\tilde{r} = fl(b - A\tilde{x})$.
2. Solve $LUz = P\tilde{r}$, for $\tilde{z}$.
3. Set $\tilde{x}' = fl(\tilde{x} + \tilde{z})$.

However, since with a stable GEPP the residual is already "small," it turns out that $\tilde{r} = fl(b - A\tilde{x})$ has few, if any, correct significant digits. And the same becomes true for $\tilde{z}$ and $\tilde{x}'$ ("garbage in, garbage out"). But, if true partial pivoting is not being used, such as in order to preserve sparsity, then the backward error (and residual) may not be too small. In this case, the above steps *can* be useful to improve accuracy, and they can be applied iteratively a few times, such as, for example, as shown in Fig. 3.12, where $\epsilon_{rel}, \epsilon_{abs} > 0$ are user-specified *relative* and *absolute error tolerances*. Hence the name, *iterative refinement*, also called *iterative improvement*.

Mixed Precision Iterative Refinement Even if true partial pivoting is being used, there is a way to get an improved solution, provided $r = b - A\tilde{x}$ is computed with *increased precision*. For example, all the computations would be in single precision, except that computing $r = b - A\tilde{x}$ would be done in

---

Input: L, U, P, and $\tilde{x}$ as the found solution of $LUx = Pb$
repeat
    $r = b - A\tilde{x}$
    Solve $LUz = Pr$
    $\tilde{x} := \tilde{x} + z$
until $(\|z\| \le \epsilon_{rel}\|\tilde{x}\| + \epsilon_{abs})$

Figure 3.12: Iterative refinement algorithm.

double precision. Notice that the original A matrix would need to be retained for this, so that it is not enough to have only L and U available. This is referred to as *mixed precision iterative refinement*.

The following heuristic is commonly employed. If the (single precision) unit roundoff is $u \approx 10^{-d}$ and if $\kappa_\infty(A) \approx 10^q$, then, after k iterations, x has $\approx \min(d, k(d - q))$ correct significant digits. For example, if $u \approx 10^{-8}$ and $\kappa_\infty(A) \approx 10^4$, then $u\kappa_\infty(A) \approx 10^{-4}$ and GEPP should give us ≈ 4 correct significant digits, but we can do better, as follows. After $k = 2$ iterations of mixed precision iterative refinement, with $(d - q) = 4$, then $\approx \min(8, 8) = 8$ significant digits should be correct. Roughly speaking, if $u\kappa_\infty(A) < 1$, then it can be shown that mixed precision iterative refinement can ultimately produce a solution with *full* (single) precision, so that:

$$\frac{\|\tilde{x} - x\|_\infty}{\|x\|_\infty} \approx u \tag{3.185}$$

which really is the most we can hope for.

One drawback of mixed precision iterative refinement is that its implementation is machine-dependent and may not be very portable. Another limitation, of course, is that we cannot use it in cases when all computations are already being done in the highest available precision. Most circuit simulators perform all computations in double precision. In circuit simulation, the basic (fixed precision) iterative refinement *can* be used to improve accuracy when true partial pivoting is not used.

3.3   INDIRECT/ITERATIVE METHODS

Much research has been done to improve the efficiency of circuit simulation. One such research direction has been the use of *relaxation methods*. Relaxation methods are *iterative* techniques which *relax* the accuracy requirements during early iterations, but *tighten* it in later iterations. A wide range of relaxation methods have been proposed, and a good reference for this area is Ruehli (1987), but we will focus on a class of methods which are applicable for solving the system of linear algebraic equations, $Ax = b$. Specifically, we will study the two methods: Gauss-Jacobi and Gauss-Seidel. Solution methods like GE and LU factorization take a predetermined number of steps before providing a solution. In contrast, relaxation methods provide partial solutions which eventually (hopefully) converge to the true solution. Thus, it is customary to refer to methods like GE and LU factorization as *direct methods*, while relaxation techniques are called *indirect* or *iterative*. Indirect methods have not been an unqualified success, and they often work well only on certain classes of circuits. But they involve only matrix-vector multiplications, and so can be more efficient than direct methods, especially on very large problems, and especially on parallel computers.

In order to solve $Ax = b$, indirect methods generate a sequence of partial solutions $x^{(0)}, x^{(1)}, \ldots, x^{(k)}, \ldots$, such that $x^{(k+1)}$ can be obtained from $x^{(k)}$ with little computational effort, with the hope that the sequence converges to the true solution in a small number of steps. The starting point for the study of both Gauss-Jacobi and Gauss-Seidel is to write, for any matrix A, the expansion:

$$A = L + D + U \tag{3.186}$$

where L is *strictly* lower triangular, D is diagonal, and U is *strictly* upper triangular, i.e.,

$$L = \begin{bmatrix} 0 & 0 & 0 & \cdots & 0 \\ a_{21} & 0 & 0 & \cdots & 0 \\ a_{31} & a_{32} & 0 & \cdots & 0 \\ \vdots & \vdots & \vdots & \ddots & \vdots \\ a_{n1} & a_{n2} & a_{n3} & \cdots & 0 \end{bmatrix}, \quad D = \begin{bmatrix} a_{11} & 0 & 0 & \cdots & 0 \\ 0 & a_{22} & 0 & \cdots & 0 \\ 0 & 0 & a_{33} & \cdots & 0 \\ \vdots & \vdots & \vdots & \ddots & \vdots \\ 0 & 0 & 0 & \cdots & a_{nn} \end{bmatrix},$$

and:

$$U = \begin{bmatrix} 0 & a_{12} & a_{13} & \cdots & a_{1n} \\ 0 & 0 & a_{23} & \cdots & a_{2n} \\ 0 & 0 & 0 & \cdots & a_{3n} \\ \vdots & \vdots & \vdots & \ddots & \vdots \\ 0 & 0 & 0 & \cdots & 0 \end{bmatrix}$$

so that the system $Ax = b$ becomes:

$$(L + D + U)x = b \tag{3.187}$$

In the following, we assume that D has no zero entries, so that D^{-1} exists. Provided A is nonsingular, this condition can always be achieved by some permutation (pivoting) of the rows of A and b.

Cautionary note: These L and U matrices in (3.186) have a zero diagonal; they should not to be confused with the L and U matrices of LU factorization!

3.3.1 Gauss-Jacobi

With $(L + D + U)x = b$, we can write: $Dx = b - (L + U)x$, or:

$$x = D^{-1}b - D^{-1}(L + U)x \tag{3.188}$$

Gauss-Jacobi (GJ) starts with any $x^{(0)} \in \mathbb{R}^n$, usually $x^{(0)} = 0$, then iterates using:

$$x^{(k+1)} = D^{-1}b - D^{-1}(L + U)x^{(k)}, \quad for\ k = 0, 1, \ldots \tag{3.189}$$

$$
\begin{array}{l}
\textbf{while } (\text{not converged}) \textbf{ do} \\
\quad \textbf{for } (i = 1, \ldots, n) \textbf{ do} \\
\quad\quad x_i^{(k+1)} = b_i \\
\quad\quad \textbf{for } (j = 1, \ldots, i - 1) \textbf{ do} \\
\quad\quad\quad x_i^{(k+1)} = x_i^{(k+1)} - a_{ij} x_j^{(k)} \\
\quad\quad \textbf{for } (j = i + 1, \ldots, n) \textbf{ do} \\
\quad\quad\quad x_i^{(k+1)} = x_i^{(k+1)} - a_{ij} x_j^{(k)} \\
\quad\quad x_i^{(k+1)} = x_i^{(k+1)} / a_{ii}
\end{array}
$$

Figure 3.13: The Gauss-Jacobi algorithm.

This process can be expressed, at each iteration of k, as:

$$
x_i^{(k+1)} = \frac{1}{a_{ii}} \left(b_i - \sum_{j=1}^{i-1} a_{ij} x_j^{(k)} - \sum_{j=i+1}^{n} a_{ij} x_j^{(k)} \right), \quad \text{for } i = 1, 2, \ldots, n \quad (3.190)
$$

where, by convention, $\sum_{j=1}^{0} a_{ij} x_j^{(k)} = 0$ and $\sum_{j=n+1}^{n} a_{ij} x_j^{(k)} = 0$. Thus, the Gauss-Jacobi algorithm is as shown in Fig. 3.13; it requires $\approx n^2$ operations per iteration.

3.3.2 Gauss-Seidel

With $(L + D + U)x = b$, we can write: $(D + L)x = b - Ux$, or:

$$
x = (L + D)^{-1} b - (L + D)^{-1} U x \quad (3.191)
$$

Gauss-Seidel (GS) starts with any $x^{(0)} \in \mathbb{R}^n$, usually $x^{(0)} = 0$, then iterates using:

$$
x^{(k+1)} = (L + D)^{-1} b - (L + D)^{-1} U x^{(k)}, \quad \text{for } k = 0, 1, \ldots \quad (3.192)
$$

To see how this can be efficiently implemented, rewrite the iteration as:

$$
D x^{(k+1)} = b - L x^{(k+1)} - U x^{(k)} \quad (3.193)
$$

which, because L is lower triangular *and* has a zero diagonal, can be expressed, at each iteration of k, as:

$$
x_i^{(k+1)} = \frac{1}{a_{ii}} \left(b_i - \sum_{j=1}^{i-1} a_{ij} x_j^{(k+1)} - \sum_{j=i+1}^{n} a_{ij} x_j^{(k)} \right), \quad \text{for } i = 1, 2, \ldots, n
$$

$$(3.194)$$

where, by convention, $\sum_{j=1}^{0} a_{ij} x_j^{(k+1)} = 0$ and $\sum_{j=n+1}^{n} a_{ij} x_j^{(k)} = 0$. Thus, the Gauss-Seidel algorithm is as shown in Fig. 3.14; it requires $\approx n^2$ operations per iteration. Notice that the difference between the two methods is that Gauss-Seidel uses the latest available solution for the partial vector $x_1, x_2, \ldots, x_{i-1}$.

```
while (not converged) do
    for (i = 1, ..., n) do
        x_i^(k+1) = b_i
        for (j = 1, ..., i − 1) do
            x_i^(k+1) = x_i^(k+1) − a_ij x_j^(k+1)
        for (j = i + 1, ..., n) do
            x_i^(k+1) = x_i^(k+1) − a_ij x_j^(k)
        x_i^(k+1) = x_i^(k+1) / a_ii
```

Figure 3.14: The Gauss-Seidel algorithm.

3.3.3 Convergence

One is obviously interested in whether Gauss-Jacobi and Gauss-Seidel converge to the correct solution, and how quickly. In fact, there are four key questions for either algorithm:

1. What exactly do we mean by "convergence" for a vector sequence?
2. Is convergence guaranteed, and under what conditions?
3. If the iteration converges, does it converge to the correct solution?
4. If the iteration converges, how fast does it converge?

To give precise answers to these questions requires a brief digression, to review the notions of vector convergence and matrix eigenvalues.

Vector Convergence   To give a precise meaning to the notion of convergence of a sequence of vectors $\left\{x^{(k)}\right\}_{k=0}^{\infty} = x^{(1)}, x^{(2)}, \ldots$, we make use of vector norms.

Definition 3.2. (Vector convergence) We say that the sequence $\left\{x^{(k)}\right\}_{k=0}^{\infty}$ of real vectors in $\mathbb{R}^n$ *converges* to $x^* \in \mathbb{R}^n$ with respect to the norm $\| \cdot \|$ if:

$$\lim_{k \to \infty} \|x^{(k)} − x^*\| = 0 \tag{3.195}$$

Definition 3.3. (Norm equivalence) Two norms are said to be *equivalent* if, whenever a sequence $\left\{x^{(k)}\right\}_{k=0}^{\infty}$ converges to a vector x^* with respect to one of them, it also converges to the same x^* with respect to the other.

Theorem 3.8. *(Norm equivalence theorem) For any finite-dimensional real or complex vector space, $\mathbb{R}^n$ or $\mathbb{C}^n$, all vector norms are equivalent.*

Therefore, we use the convenient l_∞ norm to characterize vector convergence: A sequence $\left\{x^{(k)}\right\}_{k=0}^{\infty}$ converges to a vector x^* if and only if:

$$\lim_{k \to \infty} x_i^{(k)} = x_i^*, \quad \text{for all } i = 1, 2, \ldots, n \tag{3.196}$$

and we write:

$$\lim_{k \to \infty} x^{(k)} = x^* \tag{3.197}$$

Eigenvalues   Let $A \in \mathbb{R}^{n \times n}$; if there exists a non-zero $x \in \mathbb{C}^n$ and a scalar $\lambda \in \mathbb{C}$ such that:

$$Ax = \lambda x, \quad x \neq 0 \tag{3.198}$$

then λ is said to be an *eigenvalue* of A, and x is an *eigenvector* of A associated with λ. Notice that the vector 0 cannot be an eigenvector, and, if x is an eigenvector associated with λ, then αx is also an eigenvector associated with λ, where $\alpha \in \mathbb{C}$ and $\alpha \neq 0$.

The set of all eigenvalues of A, called the *spectrum of* of A, is denoted $\sigma(A)$. The spectrum is invariant under transposition: $\sigma(A) = \sigma(A^T)$. If A is triangular (or diagonal), then $\sigma(A) = \{a_{11}, a_{22}, \ldots, a_{nn}\}$. A matrix $A \in \mathbb{R}^{n \times n}$ is nonsingular if and only if $0 \notin \sigma(A)$. If A is symmetric, then all its eigenvalues are *real* and, if they are ordered as $\lambda_1 \leq \lambda_2 \leq \cdots \leq \lambda_n$, then:

$$\lambda_1 x^T x \leq x^T A x \leq \lambda_n x^T x, \quad \forall x \in \mathbb{R}^n \tag{3.199}$$

If $A^q = 0$ for some integer $q > 0$, then A is said to be *nilpotent*. If $A \in \mathbb{R}^{n \times n}$ is nilpotent, then $\exists q \leq n$ such that $A^q = 0$. It can be shown that A is nilpotent if and only if all its eigenvalues are 0. Every *strictly* triangular matrix is nilpotent.

If $A \in \mathbb{R}^{n \times n}$, then the polynomial $p(z) \triangleq \det(zI - A)$ is called the *characteristic polynomial* of A, and it has degree n. A scalar λ is an eigenvalue of A if and only if λ is a root of $p(z)$, i.e., $p(\lambda) = 0$. Therefore, a matrix $A \in \mathbb{R}^{n \times n}$ has at least one eigenvalue, and no more than n *distinct* eigenvalues. And, since $A \in \mathbb{R}^{n \times n}$ is real, then any complex eigenvalues of A always appear in conjugate pairs, so that $\lambda \in \sigma(A)$ if and only if $\lambda^* \in \sigma(A)$.

Finally, the *spectral radius* of $A \in \mathbb{R}^{n \times n}$, denoted $\rho(A)$ is defined as:

$$\rho(A) \triangleq \max\{|\lambda| : \lambda \in \sigma(A)\} \tag{3.200}$$

Notice that $\rho(A) \geq 0$, and that $\rho(A) = 0$ if and only if A is nilpotent.

Convergence   The following theorem is available for convergence of a general class of iterative methods.

Theorem 3.9. *Let $M \in \mathbb{R}^{n \times n}$ and let the sequence $\{x^{(k)}\}_{k=0}^{\infty}$ be generated according to the iteration:*

$$x^{(k+1)} = M x^{(k)} + c \tag{3.201}$$

where $x^{(0)}, c \in \mathbb{R}^n$; then this sequence converges if and only if $\rho(M) < 1$, and it converges to $x^* = \left[(I - M)^{-1} c\right]$.

Notice that, if M is nilpotent, in which case $\rho(M) = 0$, then clearly the iteration in (3.201) converges in $q \leq n$ steps. Motivated by (3.201), the Gauss-Jacobi iteration can be written as:

$$x^{(k+1)} = M_{GJ}x^{(k)} + c_{GJ} \tag{3.202}$$

where $M_{GJ} = -D^{-1}(L + U)$ and $c_{GJ} = D^{-1}b$. If $\rho(M_{GJ}) < 1$, then GJ converges to:

$$x^* = (I - M_{GJ})^{-1}c_{GJ} = \left(D^{-1}(D + L + U)\right)^{-1}D^{-1}b$$
$$= (D + L + U)^{-1}b = A^{-1}b$$

So that, if it converges, then GJ converges to the correct solution! Proceeding similarly, the Gauss-Seidel iteration can be written as:

$$x^{(k+1)} = M_{GS}x^{(k)} + c_{GS} \tag{3.203}$$

where $M_{GS} = -(L + D)^{-1}U$ and $c_{GS} = (L + D)^{-1}b$. Notice that, since U has a zero diagonal then M_{GS} has a zero first column, and therefore it is singular and has 0 as one of its eigenvalues. If $\rho(M_{GS}) < 1$, then GS converges to:

$$x^* = (I - M_{GS})^{-1}c_{GS} = \left((L + D)^{-1}(L + D + U)\right)^{-1}(L + D)^{-1}b$$
$$= (L + D + U)^{-1}b = A^{-1}b$$

So that, if it converges, then GS converges to the correct solution!

Conditions for Convergence   We can now deal with the question of *whether* Gauss-Jacobi and Gauss-Seidel are convergent. The following theorem is available.

Theorem 3.10. *If A is strictly diagonally dominant, then:*

$$\rho(M_{GJ}) < 1 \quad and \quad \rho(M_{GS}) < 1 \tag{3.204}$$

Thus, a sufficient condition for both GJ and GS to converge is that A is strictly diagonally dominant. This condition is *not* necessary. In practice, diagonal dominance of the NA matrix, even though not strict, often leads to a convergent system. With MNA, in order to deal with group 2 elements, one tries to reorder the MNA matrix so it becomes *nearly* diagonally dominant. We also have access to the following result.

Theorem 3.11. *If A is such that $M_{GJ} \geq 0$ (i.e., every entry of M_{GJ} is ≥ 0), then one and only one of the following is true:*

1. $\rho(M_{GJ}) = \rho(M_{GS}) = 0$
2. $0 < \rho(M_{GS}) < \rho(M_{GJ}) < 1$
3. $\rho(M_{GJ}) = \rho(M_{GS}) = 1$
4. $1 < \rho(M_{GJ}) < \rho(M_{GS})$

Rate of Convergence   Let $\mathcal{E}_k = \|x^{(k)} - x^*\|$ for some vector norm $\|\cdot\|$. If there exists a constant $0 < c < 1$ such that:

$$\mathcal{E}_{k+1} \leq c\mathcal{E}_k \tag{3.205}$$

for all k sufficiently large, then we say that the *convergence rate* is linear. For both GJ and GS, it can be shown that, for any $\epsilon > 0$, there exists a vector norm $\|\cdot\|$ such that:

$$\mathcal{E}_{k+1} \leq (\rho(M) + \epsilon)\,\mathcal{E}_k, \quad \forall k \tag{3.206}$$

where M is either M_{GJ} or M_{GS}. Therefore, if $\rho(M) < 1$, then (3.205) is satisfied, the convergence rate is *linear*, and the convergence constant, i.e., c in (3.205), is $\approx \rho(M)$.

Although both GJ and GS have a linear convergence rate, we can use the value of $\rho(M)$ to decide which is better. We should use the method with the smaller $\rho(M)$. A smaller $\rho(M)$ is said to give a faster convergence *speed*. The following result provides further guidance.

Theorem 3.12. *For any two matrices A and B, if* $|a_{ij}| \leq b_{ij}, \forall i, j$, *then:*

$$\rho(A) \leq \rho(B) \tag{3.207}$$

When this theorem is applicable, one can use it to decide which method would have a faster convergence speed. However, practical experience is that, if both GJ and GS will converge, then it is usually faster to use GS. Recall that GS uses more recent partial information about the solution. On the other hand, there are cases where GJ will converge while GS will not, as we illustrate with the following example.

Example   We will show a case where GJ converges but GS does not; consider the system:

$$\begin{bmatrix} 1 & 2 & -2 \\ 1 & 1 & 1 \\ 2 & 2 & 1 \end{bmatrix} \begin{bmatrix} x_1 \\ x_2 \\ x_3 \end{bmatrix} = \begin{bmatrix} 1 \\ 3 \\ 5 \end{bmatrix}, \quad \text{whose solution is } x^* = \begin{bmatrix} 1 \\ 1 \\ 1 \end{bmatrix} \tag{3.208}$$

For Gauss-Jacobi:

$$M_{GJ} = \begin{bmatrix} 0 & -2 & 2 \\ -1 & 0 & -1 \\ -2 & -2 & 0 \end{bmatrix} \tag{3.209}$$

and one finds that $\det(\lambda I - M_{GJ})^{-1} = \lambda^3$, so that M_{GJ} is nilpotent and GJ should converge in a finite $q \leq 3$ number of steps. Indeed, if $x^{(0)} = 0$, then $x^{(3)} = x^*$:

$$x^{(0)} = \begin{bmatrix} 0 \\ 0 \\ 0 \end{bmatrix}, \quad x^{(1)} = \begin{bmatrix} 1 \\ 3 \\ 5 \end{bmatrix}, \quad x^{(2)} = \begin{bmatrix} 5 \\ -3 \\ -3 \end{bmatrix}, \quad x^{(3)} = \begin{bmatrix} 1 \\ 1 \\ 1 \end{bmatrix}, \quad x^{(4)} = \begin{bmatrix} 1 \\ 1 \\ 1 \end{bmatrix}$$

For Gauss-Seidel:

$$M_{GS} = \begin{bmatrix} 0 & -2 & 2 \\ 0 & 2 & -3 \\ 0 & 0 & 2 \end{bmatrix} \tag{3.210}$$

Note that the first column is 0, as one would expect for GS, so that one of the eigenvalues is 0; the other two eigenvalues turn out to be 2 and -2. Thus, $\rho(M_{GS}) = 2 > 1$ and, in this case, GS will *diverge*:

$$x^{(0)} = \begin{bmatrix} 0 \\ 0 \\ 0 \end{bmatrix}, \quad x^{(1)} = \begin{bmatrix} 1 \\ 2 \\ -1 \end{bmatrix}, \quad x^{(2)} = \begin{bmatrix} -5 \\ 9 \\ -3 \end{bmatrix}, \quad x^{(3)} = \begin{bmatrix} -23 \\ 29 \\ -7 \end{bmatrix},$$

$$x^{(4)} = \begin{bmatrix} -71 \\ 81 \\ -15 \end{bmatrix}, \quad \cdots, \quad x^{(10)} = \begin{bmatrix} -13,823 \\ 14,337 \\ -1,023 \end{bmatrix}, \quad \cdots$$

3.4 PARTITIONING TECHNIQUES

Another method for speeding up the solution of large systems is to partition the problem or, equivalently, to partition the matrix. A large variety of partitioning approaches have been applied to circuit simulation, often in combination with a relaxation type approach. Partitioning aims to benefit from a *locality* property that circuits often have, namely that strong interactions are usually among nearby nodes. It is hoped that one can break up a circuit at a *small* number of key global boundaries, across which circuit interactions will not be strong. In practice, partitioning can become expensive, and accuracy can occasionally suffer, so these methods are not an unqualified success. We will focus on one partitioning approach, called *node tearing*.

3.4.1 Node Tearing

Node tearing may be summarized as follows:

1. Start with a connected network graph.
2. Find a set of nodes which, if removed, would cause the graph to become disconnected; these are called the *tearing set*. It is customary to include the ground node in the tearing set.

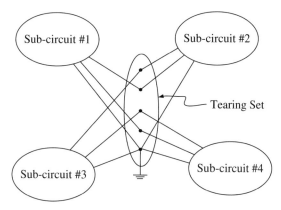

Figure 3.15: An illustration of node tearing, where a circuit is partitioned into four sub-circuits.

3. Discover the sub-circuits of the (disconnected) graph and build the system equation for each of them.
4. Combine these with the equations for the tearing set.

One hopes to solve the resulting system on a *parallel computer*, leading to improved run-time. An example of the resulting partitions is shown in Fig. 3.15.

For simplicity, we will discuss the use of node tearing in the context of the nodal analysis (NA) formulation, but this can be easily extended to MNA. The resulting system equation, which one can see by reference to Fig. 3.15, includes a system matrix which is said to be in *bordered block diagonal* (BBD) form:

$$
\begin{bmatrix}
A_1 & 0 & \cdots & 0 & B_1 \\
0 & A_2 & \cdots & 0 & B_2 \\
\vdots & \vdots & \ddots & \vdots & \vdots \\
0 & 0 & \cdots & A_m & B_m \\
C_1 & C_2 & \cdots & C_m & D_t
\end{bmatrix}
\begin{bmatrix}
x_1 \\
x_2 \\
\vdots \\
x_m \\
x_t
\end{bmatrix}
=
\begin{bmatrix}
b_1 \\
b_2 \\
\vdots \\
b_m \\
b_t
\end{bmatrix}
\tag{3.211}
$$

where A_i, B_i, C_i, and D_t are matrices such that each A_i is $n_i \times n_i$ (n_i is the number of nodes in the ith sub-circuit), each B_i is $n_i \times n_t$ (n_t is the number of nodes in the tearing set), each C_i is $n_t \times n_i$, matrix D_t is $n_t \times n_t$, and the vectors x and b are accordingly partitioned into x_i and b_i (keep in mind that these are vectors, not individual vector components). The solution plan is to solve each sub-circuit as a smaller problem, and then somehow combine the solutions. Effectively, one aims to solve the set of equations:

$$
\begin{cases}
A_i x_i + B_i x_t = b_i, & \forall i = 1, 2, \ldots, m \\
\sum_{i=1}^{m} C_i x_i + D_t x_t = b_t
\end{cases}
\tag{3.212}
$$

where the second equation will be referred to as the *tearing set equation*. The equations are not totally decoupled, of course, but the m equations for the sub-circuits can be *partially solved* in parallel, as we will see. We will study both a direct approach (using LU factorization) and an indirect approach (using GJ and GS) for solving this type of system.

3.4.2 Direct Methods

From the equation for each sub-circuit, we get:

$$x_i = A_i^{-1}(b_i - B_i x_t), \quad \forall i = 1, 2, \cdots, m \tag{3.213}$$

Substitute this into the tearing set equation, to get:

$$\left(D_t - \sum_{i=1}^{m} C_i A_i^{-1} B_i \right) x_t = \left(b_t - \sum_{i-1}^{m} C_i A_i^{-1} b_i \right) \tag{3.214}$$

which we write compactly as:

$$D_t^* x_t = b_t^* \tag{3.215}$$

where we define:

$$D_t^* \triangleq D_t - \sum_{i=1}^{m} C_i A_i^{-1} B_i \tag{3.216}$$

$$b_t^* \triangleq b_t - \sum_{i=1}^{m} C_i A_i^{-1} b_i \tag{3.217}$$

If this equation is solved for x_t, we can then find each x_i using (3.213). Obviously, this would not be done by explicit matrix inversion. Instead, the elimination of x_i from the tearing set equation is done using Gaussian elimination. Specifically, we will use the block GE approach, based on Gauss's method for LU factorization, as follows. Start by performing a partial LU, as we saw in the block GE approach, on A_1 only, leading to:

$$
\begin{bmatrix}
(L_1|U_1) & 0 & \cdots & 0 & L_1^{-1}B_1 \\
0 & A_2 & \cdots & 0 & B_2 \\
\vdots & \vdots & \ddots & \vdots & \vdots \\
0 & 0 & \cdots & A_m & B_m \\
C_1 U_1^{-1} & C_2 & \cdots & C_m & (D_t - C_1 A_1^{-1} B_1)
\end{bmatrix}
\begin{bmatrix}
x_1 \\
x_2 \\
\vdots \\
x_m \\
x_t
\end{bmatrix}
=
\begin{bmatrix}
L_1^{-1}b_1 \\
b_2 \\
\vdots \\
b_m \\
b_t - C_1 A_1^{-1} b_1
\end{bmatrix}
\tag{3.218}
$$

Notice that A_i, B_i, and b_i, for $i = 2, 3, \ldots, m$ are not modified because there is a zero matrix in the corresponding rows, in the columns below A_1. Likewise,

C_i, for $i = 2, 3, \ldots, m$ are not modified because there is a zero matrix in the corresponding columns, in the rows to the right of A_1. This is repeated for $A_2, A_3, \ldots, A_m$ and, for each A_i, the modifications are limited to B_i, C_i, b_i, b_t, and D_t, in a way which is parallelizable. The resulting system equation is:

$$
\begin{bmatrix}
(L_1|U_1) & 0 & \cdots & 0 & L_1^{-1}B_1 \\
0 & (L_2|U_2) & \cdots & 0 & L_2^{-1}B_2 \\
\vdots & \vdots & \ddots & \vdots & \vdots \\
0 & 0 & \cdots & (L_m|U_m) & L_m^{-1}B_m \\
C_1 U_1^{-1} & C_2 U_2^{-1} & \cdots & C_m U_m^{-1} & D_t^*
\end{bmatrix}
\begin{bmatrix}
x_1 \\ x_2 \\ \vdots \\ x_m \\ x_t
\end{bmatrix}
=
\begin{bmatrix}
L_1^{-1}b_1 \\ L_2^{-1}b_2 \\ \vdots \\ L_m^{-1}b_m \\ b_t^*
\end{bmatrix}
\tag{3.219}
$$

where D_t^* and b_t^* are as defined above. Thus, the net result of the modifications to D_t and b_t is to provide the values of D_t^* and b_t^* that are required to solve for x_t based on (3.215), which we do using another LU factorization.

With x_t in hand, we can solve each sub-circuit, in parallel, for its own voltages, based on the original system equation:

$$
L_i U_i x_i = b_i - B_i x_t
\tag{3.220}
$$

where, notice that, the L_i and U_i factors are already available, so that no additional factorization is required. In fact, there is no need to do forward substitution, because the final system equation is already in the form:

$$
U_i x_i = L_i^{-1} b_i - L_i^{-1} B_i x_t
\tag{3.221}
$$

so that we only need to do backward substitution. The total computational cost is clearly $\mathcal{O}\left(n_t^3 + \sum_{i=1}^{m} n_i^3\right)$, which is much less than $\mathcal{O}\left(\left(n_t + \sum_{i=1}^{m} n_i\right)^3\right)$.

3.4.3 Indirect Methods

Both Gauss-Jacobi (GJ) and Gauss-Seidel (GS) can be adapted to work with the BBD formulation. Starting with the system equations:

$$
\begin{cases}
A_i x_i + B_i x_t = b_i, & \forall i = 1, 2, \ldots, m \\
\sum_{i=1}^{m} C_i x_i + D_t x_t = b_t
\end{cases}
\tag{3.222}
$$

we transform them into the equivalent form:

$$
\begin{cases}
A_i x_i = b_i - B_i x_t, & \forall i = 1, 2, \ldots, m \\
D_t x_t = b_t - \sum_{i=1}^{m} C_i x_i
\end{cases}
\tag{3.223}
$$

and GJ and GS are then applied to these equations, in slightly different ways, as follows.

Gauss-Jacobi The GJ approach starts with an initial guess on $x_i^{(0)}$ and $x_t^{(0)}$, then applies an iteration to compute $x_i^{(k+1)}$ and $x_t^{(k+1)}$ from $x_t^{(k)}$ and $x_i^{(k)}$, as follows:

$$\begin{cases} A_i x_i^{(k+1)} = b_i - B_i x_t^{(k)}, & \forall i = 1, 2, \ldots, m \\ D_t x_t^{(k+1)} = b_t - \sum_{i=1}^{m} C_i x_i^{(k)} \end{cases} \tag{3.224}$$

This requires one to pre-compute the LU factors for A_i and D_t, which can be done in parallel. Thus, computationally, the process is straightforward. But what about convergence? In block-matrix form, the iteration is as follows:

$$
\begin{bmatrix} A_1 & 0 & \cdots & 0 & 0 \\ 0 & A_2 & \cdots & 0 & 0 \\ \vdots & \vdots & \ddots & \vdots & \vdots \\ 0 & 0 & \cdots & A_m & 0 \\ 0 & 0 & \cdots & 0 & D_t \end{bmatrix}
\begin{bmatrix} x_1^{(k+1)} \\ x_2^{(k+1)} \\ \vdots \\ x_m^{(k+1)} \\ x_t^{(k+1)} \end{bmatrix}
$$

$$
= \begin{bmatrix} b_1 \\ b_2 \\ \vdots \\ b_m \\ b_t \end{bmatrix} - \begin{bmatrix} 0 & 0 & \cdots & 0 & B_1 \\ 0 & 0 & \cdots & 0 & B_2 \\ \vdots & \vdots & \ddots & \vdots & \vdots \\ 0 & 0 & \cdots & 0 & B_m \\ C_1 & C_2 & \cdots & C_m & 0 \end{bmatrix}
\begin{bmatrix} x_1^{(k)} \\ x_2^{(k)} \\ \vdots \\ x_m^{(k)} \\ x_t^{(k)} \end{bmatrix}
$$

which, assuming the A_i's and D_t are nonsingular, can be written as:

$$
\begin{bmatrix} x_1^{(k+1)} \\ x_2^{(k+1)} \\ \vdots \\ x_m^{(k+1)} \\ x_t^{(k+1)} \end{bmatrix}
= \begin{bmatrix} A_1^{-1} b_1 \\ A_2^{-1} b_2 \\ \vdots \\ A_m^{-1} b_m \\ D_t^{-1} b_t \end{bmatrix}
- \begin{bmatrix} 0 & 0 & \cdots & 0 & A_1^{-1} B_1 \\ 0 & 0 & \cdots & 0 & A_2^{-1} B_2 \\ \vdots & \vdots & \ddots & \vdots & \vdots \\ 0 & 0 & \cdots & 0 & A_m^{-1} B_m \\ D_t^{-1} C_1 & D_t^{-1} C_2 & \cdots & D_t^{-1} C_m & 0 \end{bmatrix}
\begin{bmatrix} x_1^{(k)} \\ x_2^{(k)} \\ \vdots \\ x_m^{(k)} \\ x_t^{(k)} \end{bmatrix}
$$

The 0-diagonal matrix is certainly not diagonally dominant, but one *hopes* that the process converges anyway.

Gauss-Seidel The GS approach is slightly different: it uses the latest $x_i^{(k+1)}$ when computing $x_t^{(k+1)}$, as follows:

$$\begin{cases} A_i x_i^{(k+1)} = b_i - B_i x_t^{(k)}, & \forall i = 1, 2, \ldots, m \\ D_t x_t^{(k+1)} = b_t - \sum_{i=1}^{m} C_i x_i^{(k+1)} \end{cases} \tag{3.225}$$

As with GJ, this requires one to pre-compute the LU factors for A_i and D_t, which can be done in parallel. Again, computationally, this process is straightforward. As for convergence, note that, in block-matrix form, the iteration is as follows:

$$
\begin{bmatrix}
A_1 & 0 & \cdots & 0 & 0 \\
0 & A_2 & \cdots & 0 & 0 \\
\vdots & \vdots & \ddots & \vdots & \vdots \\
0 & 0 & \cdots & A_m & 0 \\
C_1 & C_2 & \cdots & C_m & D_t
\end{bmatrix}
\begin{bmatrix}
x_1^{(k+1)} \\
x_2^{(k+1)} \\
\vdots \\
x_m^{(k+1)} \\
x_t^{(k+1)}
\end{bmatrix}
=
\begin{bmatrix}
b_1 \\
b_2 \\
\vdots \\
b_m \\
b_t
\end{bmatrix}
-
\begin{bmatrix}
0 & 0 & \cdots & 0 & B_1 \\
0 & 0 & \cdots & 0 & B_2 \\
\vdots & \vdots & \ddots & \vdots & \vdots \\
0 & 0 & \cdots & 0 & B_m \\
0 & 0 & \cdots & 0 & 0
\end{bmatrix}
\begin{bmatrix}
x_1^{(k)} \\
x_2^{(k)} \\
\vdots \\
x_m^{(k)} \\
x_t^{(k)}
\end{bmatrix}
$$

In this case, simple symbolic inversion is not possible, but one hopes for better convergence compared to GJ, because of the use of latest $x_i^{(k+1)}$. It is anticipated that tearing techniques will become more important, with the increased use of multi-core microprocessors and parallel computers.

3.5 SPARSE MATRIX TECHNIQUES

The number of elements in a large electrical network is typically only 2–4 times the number of nodes. Given that elements usually have 2–4 terminals, then the vertex degrees in network graphs must be small. Indeed, given the preceding observations, it can be shown that the average vertex degree in a large network graph should be less than some upper bound which is in the range 6–12. Thus, even though the number of edges in a graph is, theoretically, $\mathcal{O}(n^2)$, where n is the number of nodes, in practice it is only $\mathcal{O}(n)$ for large circuit graphs. As a result, circuit matrices are highly *sparse*, i.e., most of their elements are zero, and one can benefit greatly from the use of *sparse matrix techniques*, in order to speed up circuit simulation.

With *sparse matrix techniques*, zero-valued matrix entries are not stored and, to the extent possible, not manipulated. The resulting savings can be quite significant. GE/LU, which is theoretically $\mathcal{O}(n^3)$, becomes $\approx \mathcal{O}(n^{1.5})$ in practice for sparse matrices. For highly sparse matrices, this is reported to be as low as $\approx \mathcal{O}(n^{1.1})$. Overall, modern simulators are observed to be $\mathcal{O}(n^\alpha)$, empirically, where α is in the range 1.2–1.5.

Sparse matrix techniques are more difficult to implement, but are worth the effort for large circuits. In general, this is about more than just performing a sparse GE/LU. Almost every algorithm for matrix manipulation has a specialized variant for sparse matrices. There are specialized algorithms for matrix-vector multiplication and for forward/backward substitution. We will give only a limited treatment of sparse matrix techniques, in order to motivate commonly used schemes of *storage* and *pivoting for sparsity*. We start with some general remarks about sparse matrices:

1. We earlier introduced what we called a *technical* definition of sparsity: an $n \times n$ matrix is called *sparse* if its number of non-zero entries is $\mathcal{O}(n)$,

otherwise it is said to be *dense* or *full*. However, in practice, exactly
how much of a matrix must be zero in order for it to be worthwhile to
treat it as sparse, depends on several factors, such as the type of algo-
rithm/computation to be performed, the pattern of non-zeros in the matrix,
and the architecture of the computer being used. Thus, some numerical
analysts prefer a less technical definition, according to which a matrix is
said to be *sparse* if there is a significant advantage to be had by exploiting
its zero/non-zero *structure*. Note, the term *structure* or *pattern* is used to
refer to the zero/non-zero pattern of a matrix. Simulation of large circuits
definitely benefits from the use of sparse matrix methods, and large circuit
matrices *are* definitely sparse.

2. It is often possible to gain insight into sparse matrix techniques by working
   with the graph associated with the matrix, as we will see in connection
   with pivoting. In fact, results from graph theory sometimes provide direct
   answers to questions associated with sparse matrix algorithms.

3. Exactly how one makes use of sparsity depends on the machine architecture.
   On a serial computer, one aims to reduce the number of floating point computations $(+, -, \times, /)$, while keeping the overhead in check. On a vector
   or parallel computer, there are additional opportunities, and complexities,
   to be dealt with. We will limit our focus to serial machines.

We are now ready to briefly discuss sparse matrix storage, before moving on to a
detailed study of pivoting for sparsity in GE and LU factorization. This material
is based on a number of sources, including Davis (2006), Pillage et al. (1995),
and Duff et al. (1986).

3.5.1 Sparse Matrix Storage

Historically, a number of different data structures have been used for sparse
matrix storage and manipulation. Modern techniques make use of two data structures:

1. The *triplet form* is easy to create but not the most efficient to use.
2. The *compressed-column form* is more useful and is used in most sparse
   matrix algorithms, including those in MATLAB,[3] but harder to create.

A typical flow is to specify a matrix in triplet form, and then for the tools to
convert it to compressed-column form before further processing.

Triplet Form   A triplet form data structure for an $m \times n$ matrix A, with n_z real
non-zero elements, consists of two integer arrays and one real array, as follows:

- Integer array: $r[1], r[2], \ldots, r[n_z] \in \{1, 2, \ldots, m\}$

---

[3]MATLAB is a registered trademark of The MathWorks, Inc.

- Integer array: $c[1], c[2], \ldots, c[n_z] \in \{1, 2, \ldots, n\}$
- Real array: $x[1], x[2], \ldots, x[n_z] \in \mathbb{R}$ where $x[i] \triangleq A(r[i], c[i])$.

For example, the following matrix:

$$A = \begin{bmatrix} 5.6 & 0 & 1.8 & 0 \\ 2.4 & 1.5 & 0 & -0.2 \\ 0 & 1.6 & 4.0 & 0 \\ 2.8 & -3.0 & 0 & 1.7 \end{bmatrix} \tag{3.226}$$

has the following triplet form structure:

$$r = \begin{bmatrix} 3 & 2 & 4 & 1 & 2 & 4 & 4 & 2 & 1 & 3 \end{bmatrix}$$
$$c = \begin{bmatrix} 3 & 1 & 4 & 3 & 2 & 1 & 2 & 4 & 1 & 2 \end{bmatrix}$$
$$x = \begin{bmatrix} 4.0 & 2.4 & 1.7 & 1.8 & 1.5 & 2.8 & -3.0 & -0.2 & 5.6 & 1.6 \end{bmatrix}$$

A file that describes a matrix in the triplet form would typically consist of a list of all the $(r[i], c[i], x[i])$ triplets, in any order.

Compressed-Column Form   A compressed-column form for an $m \times n$ real matrix A, which may contain up to $n_{z,max}$ non-zero entries, consists of three arrays:

- Integer array: $p[1], p[2], \ldots, p[n+1] \in \{1, 2, \ldots, n_{z,max} + 1\}$
- Integer array: $r[1], r[2], \ldots, r[n_{z,max}] \in \{1, 2, \ldots, m\}$
- Real array: $x[1], x[2], \ldots, x[n_{z,max}] \in \mathbb{R}$, where $p[1] = 1$, $p[n+1] = n_z + 1$, and where $n_z \leq n_{z,max}$ is the actual number of non-zeros, such that the row indices of the non-zero entries in column j of A are stored, in any order, in:

$$r\left[p[j]\right], \quad r\left[p[j]+1\right], \quad \ldots, \quad r\left[p[j+1]-1\right] \tag{3.227}$$

and the corresponding entry values of A are stored at:

$$x\left[p[j]\right], \quad x\left[p[j]+1\right], \quad \ldots, \quad x\left[p[j+1]-1\right] \tag{3.228}$$

and, if column j has all zero entries, then $p[j] = p[j+1]$.

For example, the matrix in (3.226) has the following compressed-column form:

$$p = \begin{bmatrix} 1 & 4 & 7 & 9 & 11 \end{bmatrix}$$
$$r = \begin{bmatrix} 1 & 2 & 4 & 2 & 3 & 4 & 1 & 3 & 2 & 4 \end{bmatrix}$$
$$x = \begin{bmatrix} 5.6 & 2.4 & 2.8 & 1.5 & 1.6 & -3.0 & 1.8 & 4.0 & -0.2 & 1.7 \end{bmatrix}$$

This representation is also referred to as *storage-by-columns*, and there are other comparable representations, such as storage-by-rows, storage-by-indices, compressed-matrix storage, compressed-diagonal storage, etc. The details are an implementation issue and will not be pursued further here.

3.5.2 Sparse GE and *LU* Factorization

We now focus on the issue of pivoting for sparsity in GE and *LU* factorization. It is worthwhile starting with the probably obvious comment that performing row and column exchanges *during* GE/*LU* is *equivalent* to first *reordering* the matrix up-front (according to the same row and column exchanges), then applying GE/*LU* without any pivoting. Discussions of pivoting are often cast in terms of up-front reordering of the matrix. Recall that matrix reordering (allowing for both row and column exchange) can be expressed using two permutation matrices P and Q as:

$$PAQ = LU \tag{3.229}$$

We commented earlier that, for solving $Ax = b$, the matrix A^{-1} is often *dense*, even when A is sparse, so that the method $x = A^{-1}b$ is unable for solving large systems. The loss of sparsity, in going from A to A^{-1} is the root of the problem. Likewise, when using LU factorization, it would be undesirable if L and U turned out to be dense, and we can quantify the significance of this, as follows. Consider an LU factorization of $Ax = b$, and define the following:

- Let $|U_k|$ be the number of non-zeros in *row* k of U.
- Let $|L_k|$ be the number of non-zeros in *column* k of L.
- Let $|L|$ and $|U|$ be the number of non-zeros in L and U, respectively.

Then, it can be shown that the cost of LU factorization, using Gauss's method (counting only floating point multiplications and divisions, as usual), is:

$$\alpha = \sum_{k=1}^{n} |U_k| (|L_k| - 1) \tag{3.230}$$

and the cost of forward substitution followed by backward substitution, assuming that b is full, is:

$$\beta = |L| + |U| - n \tag{3.231}$$

Thus, the total cost of solving the system is:

$$\alpha + \beta = (|L| - n) + \sum_{k=1}^{n} |L_k||U_k| \tag{3.232}$$

Clearly, it is in our interest to make sure that L and U are sparse!

A key advantage of LU factorization is that, in contrast to the basic method of explicitly computing A^{-1}, it can be shown that, if A is sparse, then there *exist* permutation matrices P and Q, such that $PAQ = LU$, where both L and U are sparse. This is equivalent to saying that there exists a matrix reordering (using row and column exchanges) that *preserves sparsity*. The *best* ordering may not be easy to find, as we will see, but it is possible to find *good* orderings. Thus, LU factorization is practical for large systems, *because* it offers the opportunity to preserve sparsity. In the following, we will study the impact of reordering on sparsity and consider various schemes for pivot selection for sparsity.

3.5.3 Reordering and Sparsity

The sparsity of L and U depends on the matrix ordering being used. To see this, consider this simple example of a 4×4 matrix:

$$A = \begin{bmatrix} a_{11} & a_{12} & a_{13} & a_{14} \\ a_{21} & a_{22} & 0 & 0 \\ a_{31} & 0 & a_{33} & 0 \\ a_{41} & 0 & 0 & a_{44} \end{bmatrix} \quad \text{denoted:} \quad \begin{bmatrix} \times & \times & \times & \times \\ \times & \times & & \\ \times & & \times & \\ \times & & & \times \end{bmatrix} \quad (3.233)$$

where $\times$ denotes a non-zero in this *structural representation*. Applying LU factorization on this matrix, using Gauss's algorithm for LU factorization, gives the usual auxiliary matrix $S \triangleq L + U - I$ as:

$$A = \begin{bmatrix} u_{11} & u_{12} & u_{13} & u_{14} \\ l_{21} & u_{22} & u_{23} & u_{24} \\ l_{31} & l_{32} & u_{33} & u_{34} \\ l_{41} & l_{42} & l_{43} & u_{44} \end{bmatrix} \quad (3.234)$$

where:

$$u_{11} = a_{11}, \quad u_{12} = a_{12}, \qquad\qquad u_{13} = a_{13}, \quad u_{14} = a_{14}$$
$$l_{21} = a_{21}/u_{11}, \quad u_{22} = a_{22} - l_{21}u_{12}, \quad u_{23} = 0 - l_{21}u_{13}, \quad u_{24} = 0 - l_{21}u_{14}$$
$$l_{31} = a_{31}/u_{11}, \qquad\qquad\qquad l_{32} = (0 - l_{31}u_{12})/u_{22},$$
$$u_{33} = a_{33} - l_{31}u_{13} - l_{32}u_{23}, \qquad u_{34} = 0 - l_{31}u_{14} - l_{32}u_{24}$$
$$l_{41} = a_{41}/u_{11}, \qquad\qquad\qquad l_{42} = (0 - l_{41}u_{12})/u_{22}$$
$$l_{43} = (0 - l_{41}u_{13} - l_{42}u_{23})/u_{33}, \quad u_{44} = a_{44} - l_{41}u_{14} - l_{42}u_{24} - l_{43}u_{34}$$

Barring possible cancellations of exactly equal and opposite terms, the result of each of these computations is a non-zero, so that the S matrix is a full matrix. Effectively, a sparse A matrix has been transformed thus:

$$A = \begin{bmatrix} \times & \times & \times & \times \\ \times & \times & & \\ \times & & \times & \\ \times & & & \times \end{bmatrix} \quad \rightarrow \quad S = \begin{bmatrix} \times & \times & \times & \times \\ \times & \times & \times & \times \\ \times & \times & \times & \times \\ \times & \times & \times & \times \end{bmatrix} \quad (3.235)$$

This is highly undesirable! The source of the problem is evident if we examine the very first step of Gauss's algorithm for LU factorization (equivalent to GE), which gives:

$$\begin{bmatrix} u_{11} = a_{11} & u_{12} = a_{12} & u_{13} = a_{13} & u_{14} = a_{14} \\ l_{21} = a_{21}/u_{11} & a_{22} - l_{21}u_{12} & 0 - l_{21}u_{13} & 0 - l_{21}u_{14} \\ l_{31} = a_{31}/u_{11} & 0 - l_{31}u_{12} & a_{33} - l_{31}u_{13} & 0 - l_{31}u_{14} \\ l_{41} = a_{41}/u_{11} & 0 - l_{41}u_{12} & 0 - l_{41}u_{13} & a_{44} - l_{41}u_{14} \end{bmatrix}$$

so that the matrix is already full after the very first step:

$$A = \begin{bmatrix} \times & \times & \times & \times \\ \times & \times & & \\ \times & & & \times \\ \times & & & \times \end{bmatrix} \quad \rightarrow \quad \begin{bmatrix} \times & \times & \times & \times \\ \times & \times & \times & \times \\ \times & \times & \times & \times \\ \times & \times & \times & \times \end{bmatrix} \tag{3.236}$$

Fill-Ins   The introduction of a non-zero in S where there was none in A is referred to as a *fill-in*. Early fill-ins lead to more fill-ins as the algorithm proceeds. It is in our interest to avoid fill-ins, so as to preserve sparsity! Note that, in GE or in Gauss's algorithm for LU factorization, when we are eliminating x_k, then

> *a fill-in is introduced whenever a 0 entry (of the S matrix) has an $\times$ in its row in column k <u>and</u> an $\times$ in its column in row k.*

Returning to the above example, consider a reordering of the matrix, as follows:

$$PAQ = \begin{bmatrix} a_{44} & 0 & 0 & a_{41} \\ 0 & a_{22} & 0 & a_{21} \\ 0 & 0 & a_{33} & a_{31} \\ a_{14} & a_{12} & a_{13} & a_{11} \end{bmatrix} \quad \text{denoted:} \quad \begin{bmatrix} \times & & & \times \\ & \times & & \times \\ & & \times & \times \\ \times & \times & \times & \times \end{bmatrix} \tag{3.237}$$

where rows 1 and 4 have been exchanged and columns 1 and 4 have been exchanged, using:

$$P = \begin{bmatrix} 0 & 0 & 0 & 1 \\ 0 & 1 & 0 & 0 \\ 0 & 0 & 1 & 0 \\ 1 & 0 & 0 & 0 \end{bmatrix} \quad \text{and} \quad Q = \begin{bmatrix} 0 & 0 & 0 & 1 \\ 0 & 1 & 0 & 0 \\ 0 & 0 & 1 & 0 \\ 1 & 0 & 0 & 0 \end{bmatrix} \tag{3.238}$$

In this case, LU factorization on this matrix, using Gauss's algorithm, gives the auxiliary matrix $S \triangleq L + U - I$ based on:

$$u_{11} = a_{44}, u_{12} = 0, \qquad\qquad u_{13} = 0, u_{14} = a_{41}$$
$$l_{21} = 0, u_{22} = a_{22}, \qquad\qquad u_{23} = 0, u_{24} = a_{21}$$
$$l_{31} = 0, l_{32} = 0, \qquad\qquad u_{33} = a_{33}, u_{34} = a_{31}$$
$$l_{41} = a_{14}/u_{11}, \qquad\qquad l_{42} = (a_{12} - l_{41}u_{12})/u_{22}$$
$$l_{43} = (a_{13} - l_{41}u_{13} - l_{42}u_{23})/u_{33}, \quad u_{44} = a_{11} - l_{41}u_{14} - l_{42}u_{24} - l_{43}u_{34}$$

so that S is just as sparse as the original matrix A, which we depict as:

$$PAQ = \begin{bmatrix} \times & & & \times \\ & \times & & \times \\ & & \times & \times \\ \times & \times & \times & \times \end{bmatrix} \quad \rightarrow \quad S = \begin{bmatrix} \times & & & \times \\ & \times & & \times \\ & & \times & \times \\ \times & \times & \times & \times \end{bmatrix} \quad (3.239)$$

which is a much better outcome! However, this ideal scenario cannot always be achieved and, in general, fill-ins cannot be completely avoided.

3.5.4 Pivoting for Sparsity

Based on the above test case, we are tempted to reorder (pivot) the matrix so that the earlier processed rows and columns have fewer non-zero elements. This way the reordered matrix would resemble that in (3.239), rather than that in (3.235). There is some merit in this plan, as we will see later on in connection with diagonal pivoting. However, in general, it is not sufficient; the problem is much more complex than this. There are $(n!)^2$ possible permutations of the rows and columns of a square $n \times n$ matrix, so that the search space for an optimal ordering is huge! In fact, it can be shown that the problem is $\mathcal{NP}$-hard, and we must be satisfied with using suboptimal schemes.

Typically, one computes certain metrics to help choose a pivot element, and then performs the required row and column exchanges. The metrics aim to monitor the impact on sparsity. As well, accuracy considerations must be factored in, so as to maintain algorithm stability; small-magnitude pivots must be somehow avoided. Because both row and column exchanges may be required, LU factorization is performed using Gauss's method, rather than Crout or Doolittle. Being equivalent to GE, we will refer to the solution approach as GE/LU, as we did earlier.

By reference to Gauss's algorithm in Fig. 3.8, note that, during GE/LU, as one is poised to restart the outer loop for the kth time, we have that:

1. Rows and columns $1, \ldots, (k-1)$ have been fully processed and "turned into" rows of U and columns of L, typically using in-place computation.
2. Rows and columns $k, \ldots, n$ have been only partially processed, and they form what is called the *remaining reduced sub-matrix*, which is denoted $A^{(k)}(\{k, \ldots, n\}, \{k, \ldots, n\})$ or, more succinctly, as $A^{(k)}(k:n, k:n)$.

The reader should visualize this situation and keep it in mind, as this helps understand the various pivoting metrics to be discussed.

In the next section, we will study what is probably the most commonly used pivoting scheme in the circuit simulation, due to Markowitz. The sections after that will cover some alternative schemes, leading up to some modern schemes which have probably not been fully tested yet in the circuit simulation area. Before launching on this path, we give some general remarks:

1. It is possible for a zero to be created, in the remaining reduced sub-matrix, due to exact cancellation in a subtraction during GE/LU. However, this is very rare, due to roundoff error. Most often, pivoting schemes ignore the possibility of exact cancellation. This has an impact on the fill-in count, because the fact of having one less non-zero would be missed, but this is not a significant issue. The more serious issue is that the creation of a zero may have been missed, raising the possibility that a zero pivot may be encountered at some point.

2. For efficiency reasons, some pivoting schemes are concerned only with the zero/non-zero structure of the matrix. They do not monitor element values, and they ignore the possibility of exact cancellation. Such techniques may be called *structural* or *static*. They are also often called *local* because they do not look for global optima; they take a greedy approach based on only local information.

3. Any practical pivoting scheme must, in the end, avoid using very small pivots so as to maintain accuracy, and must watch out for zero creation due to exact cancellation. Therefore, practical methods must combine pivoting for sparsity with some form of pivoting for accuracy. One way of doing this is to use threshold pivoting, as we will see. Another more recent approach, as we will also see below, is to perform column exchanges for sparsity, and then allow further row exchanges for accuracy.

4. For circuit simulation, the zero/non-zero structure of the matrix is *fixed* throughout the simulation process. The values may change, but not the structure, as we will see in the rest of this text. Therefore, one hopes to be able to perform pivoting for sparsity up-front, as a one-time cost at the beginning of the simulation, and then to only do pivoting for accuracy during the rest of the simulation, in a way that does not destroy sparsity. This is possible with some pivoting strategies, but not with others. Many existing circuit simulators choose pivots up-front with no concern to the possible change in element values during the simulation.

3.5.5 Markowitz Pivoting

Markowitz (1957) proposed the following pivoting scheme, which has proven to be quite effective, in spite of its simplicity. At the beginning of the kth outer iteration of GE/LU, let $r_i^{(k)}$ be the number of non-zero elements in row i of the remaining reduced sub-matrix $A^{(k)}(k:n, k:n)$, and let $c_j^{(k)}$ be the number of non-zero elements in column j of that same sub-matrix. Then, for every non-zero $a_{ij}^{(k)}$ in $A^{(k)}(k:n, k:n)$, compute its *Markowitz number* as:

$$\mu_{ij}^{(k)} = (r_i^{(k)} - 1)(c_j^{(k)} - 1) \tag{3.240}$$

and pick the (i, j) entry with the *smallest* Markowitz number as the pivot. Note that $\mu_{ij}^{(k)} \geq 0$ because it is only computed for non-zero elements, so that $r_i^{(k)} \geq 1$

and $c_j^{(k)} \geq 1$. In case of a tie, one option, used in SPICE, is to pick the pivot with the smaller $c_j^{(k)}$. Another option is to use the pivot of larger magnitude. As an example, consider a matrix with the following structure:

$$\begin{bmatrix} \times & & \\ \times & & \times \\ & \times & \times \end{bmatrix} \tag{3.241}$$

In the first iteration, with $k = 1$, the row and column counters, $r_i^{(1)}$ and $c_j^{(1)}$, are as shown:

$$\begin{array}{ccc} 2 & 1 & 2 \end{array}$$
$$\begin{array}{c} 1 \\ 2 \\ 2 \end{array} \begin{bmatrix} \times & & \\ \times & & \times \\ & \times & \times \end{bmatrix} \tag{3.242}$$

from which the Markowitz numbers, computed only for the non-zero entries, are as follows:

$$\begin{bmatrix} 0 & & \\ 1 & & 1 \\ & 0 & 1 \end{bmatrix} \tag{3.243}$$

We break the tie between the two 0 entries, using the lower column counter, so that element $(3, 2)$ is the chosen pivot. Using row and column exchange, this element is moved to the $(1, 1)$ position, elimination is performed using that pivot, and the matrix entries are updated. In the next iteration, the same approach is repeated for the matrix $A^{(2)}(2 : 3, 2 : 3)$.

The method is heuristic and suboptimal, but it works well in practice, which can probably be explained a few different ways:

1. It approximately minimizes the number of multiplications to be performed on this sub-matrix in this iteration, which is $r_i^{(k)}(c_j^{(k)} - 1)$.
2. It modifies the least number of coefficients in the remaining sub-matrix. Recall, during GE/LU, when eliminating x_k, a non-zero term is added to $a_{ij}^{(k)}$ in the remaining reduced sub-matrix if and only if there is an $\times$ in row i in column k and an $\times$ in column j in row k. There are exactly $(r_i^{(k)} - 1)(c_j^{(k)} - 1)$ such combinations.
3. It *approximately* minimizes the number of fill-ins in this iteration. This is *approximate* because $\mu_{ij}^{(k)}$ is the number of *potential* fill-ins, not the number of true fill-ins, that would be created. Note that $\mu_{ij}^{(k)}$ would be the number of *true* fill-ins if the remaining sub-matrix was all zero, except for the row and column of this pivot.

Because it is suboptimal, there are cases where the Markowitz scheme does not give the best ordering, and the fill-ins are not as few as could have been. There have been various attempts to improve on this scheme, notably the method by Berry (1971), which aims to count true fill-ins, but the cost of the alternative schemes has often been high and unjustifiable. Empirical studies have shown that about 5% further fill-in reduction may be obtained with such schemes, but at high computational cost.

Any practical application of the Markowitz scheme must avoid using very small-valued pivots, i.e., it must factor in accuracy concerns. One way of doing this is to use threshold pivoting, as follows. An element $a_{rs}^{(k)}$ is deemed *acceptable* as a pivot in $A^{(k)}(k : n, k : n)$ only if:

$$|a_{rs}^{(k)}| \geq \delta |a_{ij}^{(k)}|, \quad \forall i, j \in \{k, \ldots, n\} \tag{3.244}$$

where $0 < \delta < 1$ is a preset *threshold parameter*. A search for a pivot is made from among only those non-zero elements of $A^{(k)}(k : n, k : n)$ that are *acceptable*. The best value of δ is problem-specific and, in circuit simulation, a value of $\delta = 0.001$ has been found useful. Thus, this approach "interleaves" considerations of sparsity and accuracy. Notice that pivot choices made "for accuracy" impact the fill-ins and the matrix zero/non-zero structure, which affects the decisions to be made downstream "for sparsity."

Cost of Markowitz   The Markowitz numbers must be recomputed in every successive iteration of GE/LU. This can be expensive, although only integer multiplications are required for computing the Markowitz numbers. Some strategies have been proposed for *simplifying* the Markowitz scheme and thereby reducing its cost.

In certain cases, the structure of the matrix remains fixed over a number of successive instances of the problem. In circuit simulation, for example, this occurs in successive iterations of Netwon's method. For such cases, it is appealing to generate an up-front once-only reordering of the matrix, based only on its structure, and then to reuse this ordering for all future instances. The Markowitz scheme itself is unchanged, except for one minor detail. Once a pivot has been picked, we do not actually perform the variable elimination to find the fill-ins. Instead, we use the fill-in criterion given on page 114 to *predict* the changes to the matrix structure. This ignores the possibility of exact cancellation and, to emphasize this fact, the predicted fill-ins in the structure are referred to as *structural non-zeros*. More significantly, this scheme does not factor accuracy and stability into pivot selection, so that one may end up using small pivots. Thus, this scheme is of limited value in general, but may be an option when the overhead of the regular Markowitz scheme is simply unacceptable.

Another possible variation is to not compute the Markowitz numbers at all, but to make use of the (cheaper to find) $r_i^{(k)}$ and $c_j^{(k)}$. This scheme, given in Pillage et al. (1995) is as follows:

1. Find the row(s) with the smallest $r_i^{(k)}$ value(s). In general, there may be a tie, so that one finds a *set* of rows with the *same* smallest $r_i^{(k)}$ value.

2. In each row in this set, find the largest magnitude non-zero element, and designate that as a candidate pivot.

3. Among all the candidate pivots so identified, choose the one with the smallest $c_j^{(k)}$ as the pivot.

This approach is interesting in that it tries to strike a balance between sparsity and accuracy. While the row selection is done based on sparsity only, the column selection takes accuracy into account. We will see below that there are other recent approaches that work in similar ways, choosing rows for accuracy and columns for sparsity, but which are perhaps faster and more rigorous. Although, their suitability for circuit-type matrices is not yet clear.

3.5.6   Diagonal Pivoting

Diagonal pivoting is the case when pivots are chosen from among only the diagonal elements. In some cases, this can be a good idea, as we will now see. First, note that when using a diagonal pivot, one applies the *same* row and column exchanges, so that the overall matrix reordering can be expressed as:

$$PAP^T \tag{3.245}$$

which is also called a *symmetric reordering* of the matrix, because an initially symmetric matrix *remains* symmetric under diagonal pivoting. There are two cases where diagonal pivoting is guaranteed to give the best pivots, namely 1) when the matrix is diagonally dominant and 2) when it is symmetric positive definite (SPD), as follows.

If the matrix is diagonally dominant, then we are guaranteed that it has no zero elements on the diagonal, because a (non-zero) diagonally dominant matrix always has $|a_{ii}| > 0$. Furthermore, according to Duff et al. (1986), if A is diagonally dominant, then GE without any pivoting is guaranteed to be automatically stable, because in this case it can be shown that:

$$\max_{i,j,k} |a_{ij}^{(k)}| \leq 2 \max_{i,j} |a_{ij}| \tag{3.246}$$

so that, using the diagonal pivots, no significant element growth would occur.

On the other hand, if the matrix is SPD, then again we are guaranteed that the diagonal has no zero elements, because an SPD matrix always has $a_{ii} > 0$. We also have, according to Duff et al. (1986), that if A is SPD, then GE without any pivoting is automatically stable, because it can be shown that:

$$\max_{i,j,k} |a_{ij}^{(k)}| \leq \max_{i,j} |a_{ij}| \tag{3.247}$$

so that, again, no significant element growth would take place.

These results mean that one does not need to worry about small pivots, and can use GE/LU, using the existing diagonal, *without any stability monitoring*. Of course, it remains to choose the best ordering of the diagonal elements so as to preserve sparsity. The problem is much simplified, however, because the number of potential pivots is now much smaller, and because *the ordering can be chosen based only on sparsity considerations*.

But, is diagonal pivoting applicable to circuits? Recall that, for linear resistive networks, with the consistency requirements, with no controlled sources and no voltage sources, the MNA matrix is symmetric positive definite (SPD) *and* diagonally dominant, so that diagonal pivoting is applicable. In the general case, where the MNA matrix is not necessarily SPD or diagonally dominant, diagonal pivoting has been used in circuit simulation, because it is often the case that the MNA matrix is *nearly symmetrical*.

In the SPD case in particular, some very efficient methods for pivot selection have been developed, as we will see in detail in the following sections. The availability of such efficient algorithms is quite appealing, and has led to the following strategy for the case when a circuit matrix is not quite SPD. As we will see below, the algorithms for the SPD case actually work with only a structural description of the matrix. Therefore, when A is not quite SPD, we can consider the symmetric structural pattern of (the symmetric) $A + A^T$, which is very easy to generate, and we can use it for diagonal pivoting, using the algorithms for the SPD case, which ignore accuracy, and then use that same ordering for the A matrix. When this is done for a matrix that is not truly SPD or diagonally dominant it becomes prudent to monitor stability, and one may want to use threshold pivoting.

However, practical experience with circuit simulation has shown that this simplistic approach (based on $A + A^T$) is not very good. It is often the case that, for circuits with controlled sources, the largest element in a row or column may not be on the diagonal. Typical SPICE implementations use an up-front, one-time only, Markowitz-based ordering, with a "preference" for diagonal pivots. This can create problems for MNA, where dealing with voltage sources and inductors leads to zero diagonal elements in their equations. However, in such cases, the MNA equations are typically pre-processed before solving them to avoid this, by either row exchange or row additions, as described in Pillage et al. (1995). In any case, as we will see below, there are other ways of using the solution for the SPD special case to reorder general matrices, although is not clear how well these methods are suited for circuit-type matrices. We will describe these techniques after some detailed coverage of the SPD case.

3.5.7   The Symmetric (SPD) Case

Many techniques have been developed for the special case of a symmetric positive definite (SPD) matrix, using diagonal pivoting. As we saw above, a key advantage in this case is that stability is guaranteed, so that one need not worry about small or zero pivots. This simplifies pivoting for sparsity, because there is no interference from the accuracy concerns, and very efficient techniques become possible. In this case, it is known that the diagonal entry with the smallest Markowitz number has the smallest Markowitz number in the whole sub-matrix:

$$\min_i \mu_{ii}^{(k)} = \min_{ij} \mu_{ij}^{(k)} \tag{3.248}$$

Due to symmetry, $r_i^{(k)} = c_i^{(k)}$, and $\mu_{ii}^{(k)} = (r_i^{(k)} - 1)^2 = (c_i^{(k)} - 1)^2$, so that minimizing $\mu_{ii}^{(k)}$ becomes equivalent to minimizing $r_i^{(k)}$ (equivalently, $c_i^{(k)}$). Furthermore, $r_i^{(k)} = c_i^{(k)}$ is the vertex *degree* for node i, in a specially constructed graph associated with the matrix zero/non-zero structure. Thus, looking for a diagonal element with the smallest Markowitz number becomes equivalent to finding the *minimum degree* vertex in the graph. The specialization of the Markowitz method to the SPD case, by Tinney and Walker (1967), is thus called the *minimum degree (MD) algorithm*. We will examine this in detail.

Matrix Graphs   For a general square $n \times n$ matrix A, we can construct a *directed* graph that reflects the zero/non-zero pattern of A, as follows. The set of nodes is $V \triangleq \{1, 2, \dots, n\}$, corresponding to the matrix rows (equivalently, columns). The set of edges is the set of *ordered* pairs, $E \triangleq \{(i, j) : a_{ij} \neq 0, i \neq j\}$. Notice that the graph has no self-loops and that it depends only on the zero/non-zero structure of the matrix. Sparsity of the matrix may be studied by examining the graph.

For a *structurally symmetric* matrix A, we redefine the graph, so that it becomes *undirected*, by simply redefining the set of edges, E, so that we create an undirected edge between nodes i and j if $a_{ij} \neq 0$, with $i \neq j$, and, as before, E is the set of all such edges in the graph. Here, too, the graph has no self-loops and depends only on the zero/non-zero structure of the matrix. We are interested primarily in the undirected (symmetric matrix) case.

In this case, it is clear that the vertex degree $d(i)$ is easily related to the Markowitz numbers:

$$d(i) = r_i - 1 = c_i - 1 \quad \text{and} \quad \mu_{ii} = d(i)^2 \tag{3.249}$$

and that $\mu_{kk} = \min_i \mu_{ii}$ if and only if $d(k) = \min_i d(i)$. Thus, again, finding a pivot element with the smallest μ_{ii} can be done by finding a minimum degree vertex in the graph.

Vertex Elimination   There is also a corresponding graph transformation for the GE/LU step of variable elimination, called *vertex elimination*, defined as follows:

1. Remove vertex i.
2. If any two of its neighbors are not already connected by an edge, then introduce a new edge between them.

As a result, the neighbors of a vertex that has been eliminated form a *clique* (a complete subgraph). Any edges that are created as a result of vertex elimination correspond to fill-ins in the matrix resulting from elimination of that variable. In fact, every such edge corresponds to *two* fill-ins in the matrix, at symmetric locations.

Diagonal pivot selection and variable elimination, in the outer loop of GE/LU on a SPD matrix, can be "simulated" on the graph by repeatedly finding a minimum degree vertex in the graph, and eliminating that vertex, until all vertices have been eliminated. Using Markowitz numbers to implement such an algorithm is the Tinney and Walker *minimum degree (MD) algorithm*. A tie-breaking strategy is required, and there is empirical data that shows that the chosen strategy has a significant impact on the results. However, there is no well-defined general tie-breaking strategy.

Minimum Degree The minimum degree (MD) reordering strategy has been very successful in practice, in spite of its simplicity. It is quite fast, given good data structures and a good implementation. When the graph is a tree, it can be shown that MD gives the optimal result, with zero fill-ins. But, in general, it remains sub-optimal and heuristic. Despite its apparent simplicity, however, MD hides much complexity below the surface, arising from two complications:

Storage requirements: A straightforward implementation would have us explicitly create new edges corresponding to fill-ins, but this requires excessive storage. One way around this, using so-called *quotient graphs*, leads to an algorithm that performs edge/fill-in creation *implicitly*.

Redundant computations: Two equations that have identical non-zero patterns correspond to graph nodes that are neighbors and also have the same neighbors. Such nodes are called *indistinguishable*. Significant savings in computation are achieved by grouping indistinguishable nodes together and performing so-called *mass elimination* on them.

With such optimizations, the complexity of MD becomes $\mathcal{O}(|V|^2 E)$ but, for sparse matrices, it is much faster than this worst-case bound. The earliest *efficient* implementation of MD was given in the multiple minimum degree (MMD) algorithm, by George and Liu (1989). More recently, the state-of-the-art is a more efficient approach, called the average minimum degree (AMD) algorithm, by Amestoy et al. (1996). AMD uses an approximate way to compute the vertex degrees, so that the theoretical complexity becomes $\mathcal{O}(|V||E|)$. On large problems, it is reported that AMD can be 10–100 times faster than MMD, with fewer fill-ins, but this is not always guaranteed.

3.5.8 Extension to the Non-SPD Case

As a result of an MD algorithm, we get an up-front symmetric reordering of the whole SPD matrix, i.e., we have the permutation matrix P to form:

$$PAP^T \tag{3.250}$$

Being independent of any accuracy concerns, this is then factorized as is, using a Cholesky decomposition, because of the SPD property, giving:

$$PAP^T = LL^T \tag{3.251}$$

If the structure of the matrix is fixed but its values change, while remaining SPD, we can refactor the matrix using the same ordering P. Ideally, one would like to have similar capabilities for the general non-SPD case; this would certainly be useful for circuit simulation. There is a way to do this, due to George and Ng (1985), based on the following result.

Theorem 3.13. *Let B be a nonsingular, possibly non-symmetric, $n \times n$ matrix. It is known that B can be efficiently permuted so that it has no zero diagonal elements, and so we readily assume that $b_{ii} \neq 0, \forall i$. Because $B^T B$ is SPD, let L_c be its Cholesky factor:*

$$B^T B = L_c L_c^T \tag{3.252}$$

Now, let PB be a row permutation of B resulting from the use of standard Gaussian elimination with partial pivoting (GEPP) that gives:

$$PB = LU \tag{3.253}$$

Then, it can be shown that the non-zero pattern of $L_c + L_c^T$ includes the non-zero pattern of $L + U$, so that if element (i, j) in $L_c + L_c^T$ is 0, then element (i, j) in $L + U$ is also 0. In other words, L and U are at least as sparse as L_c.

This key result can be used to formulate an algorithm for reordering and factorization of a general non-SPD matrix A, as follows.

Let A be a nonsingular, possibly non-symmetric, $n \times n$ matrix. Use an MD algorithm to efficiently find a symmetric reordering Q^T for the SPD matrix $A^T A$, aimed at a *sparse* Cholesky decomposition:

$$Q^T \left(A^T A \right) Q = L_c L_c^T \tag{3.254}$$

but there is no need to actually perform this factorization. Let N be a permutation matrix so that $B = NAQ$ has a zero-free diagonal by row permutation, so that L_c is also the Cholesky factor of $B^T B$:

$$B^T B = \left(Q^T A^T N^T \right) (NAQ) = Q^T \left(A^T A \right) Q = L_c L_c^T \tag{3.255}$$

Now, let MB be a row permutation of B resulting from the use of standard Gaussian elimination with partial pivoting (GEPP) that gives:

$$MB = LU \tag{3.256}$$

or:

$$PAQ = LU \tag{3.257}$$

where $P \triangleq MN$. Then, using the above result, the non-zero pattern of $L_c + L_c^T$ *includes* the non-zero pattern of $L + U$, so that L and U are *sparse*. The resulting strategy in practice becomes as follows:

1. Use an MD algorithm to find a *sparsity-preserving* symmetric reordering Q^T for $A^T A$, so that the Cholesky factor L_c is *sparse*.
2. Apply Q as a column permutation on A, then apply standard Gaussian elimination with partial pivoting (GEPP) on AQ, leading to $PAQ = LU$. Note, GEPP will avoid zero pivots, so that N is implicit in $P = MN$. The use of GEPP provides an *accuracy-preserving* row-permutation.
3. Write the final factorization of A as $PAQ = LU$.

Irrespective of what row permutation M is used by GEPP (for accuracy), we are guaranteed that L and U are sparse, at least as sparse as L_c.

COLMMD and COLAMD   Thus, the above approach combines a choice of columns for sparsity with a choice of rows for accuracy. The column permutation guarantees a certain level of sparsity, irrespective of what row permutation may be applied for accuracy. These choices are not interleaved, as was the case earlier, because the column permutation is chosen up-front, irrespective of any row choices. This effectively *decouples* the sparsity and accuracy concerns, so that accuracy decisions do not impact sparsity decisions. Circuit simulation, with its fixed matrix structure, should benefit greatly from this type of approach. The "catch," historically, has been that the cost of explicitly forming $A^T A$ is high, because $A^T A$ can be dense even when A is sparse.

This disadvantage was lifted in the algorithm COLMMD, by Gilbert et al. (1992). The structure of A was used to *implicitly* deduce the structure of $A^T A$ and provide the required column permutation. This was further improved in the algorithm COLAMD, by Davis et al. (2004). Both COLMMD and COLAMD are available for MATLAB.

Because $A^T A$ is typically denser than A, it is not clear if L_c is actually as sparse as one would like L and U to be. Indeed, one would expect L_c to be about as sparse as $A^T A$, because Q^T is *sparsity-preserving*, but we would like L and U to be as sparse as A. Thus, the L_c bound on fill-ins may be too pessimistic. Nevertheless, the speed of MD makes this approach worth considering. It is not known if any recent circuit simulation systems have made use of the COLMMD and COLAMD techniques. There is some evidence in the literature that the L_c bound on fill-ins may indeed be pessimistic for circuit-type matrices.

KLU   Finally, there are variations on the above themes. For example, the recent KLU package, by Davis and Stanley (2004), first puts the matrix into a block-triangular form (BTF), then performs factorization on the blocks. KLU uses AMD for each block, because they are so nearly symmetrical, and COLAMD turns out to be pessimistic for these cases. The KLU package was tailored for circuit-type matrices, with the observation that these matrices have certain key properties. They are reported to have a zero-free or nearly zero-free diagonal, to be roughly structurally symmetric but with unsymmetric values, to be permutable to block triangular form (BTF), to be highly sparse but that the blocks in the BTF can be dense, and it is reported that they *can* be ordered so that the LU factors remain sparse. KLU is claimed to achieve 1000 times speed-up over the original Markowitz-based scheme (Kundert's Sparse 1.3) used in early SPICE implementations. It is not clear how it compares to more recent circuit simulation implementations. Additional information is available on the KLU package at the web site: www.cise.ufl.edu/research/sparse/klu.

Notes   For a reference on the theory and application of matrices, see Horn and Johnson (1985). General material on direct methods for solving linear systems is available in the following sources. In Chua and Lin (1975), see sections 4.3–4.6. In Ruehli (1986), see chapter 6. In Vlach and Singhal (1994), see sections 2.4–2.8 and 20.1. In Pillage et al. (1995), see sections 3.4–3.5 and chapter 7. In Duff et al. (1986), see chapters 4 and 7, and appendix A. In Golub and Van Loan (1989), see sections 2.1–2.4, 2.7, and 3.1–3.5. In Higham (2002), see sections 1.1–1.6, 2.1–2.4, 7.1, and chapters 12 and 15. And in Davis (2006), see chapter 7. For more on indirect methods, see Ruehli (1987) and Saad (2003).

Sparse matrix solvers continue to be a lively research topic. For more focused study of this area, see the following. For the original Markowitz scheme, see Markowitz (1957). For a review of MD algorithms, see George and Liu (1989). For the COLMMD algorithm, see Gilbert et al. (1992). For the AMD algorithm, see Amestoy et al. (1996). For the COLAMD algorithm, see Davis et al. (2004). Finally, for the KLU package, see Davis and Stanley (2004) and the news item in Sipics (2007), as well as the web site given above.

Problems

3.1. Show that the computational cost of the basic Gaussian elimination (GE) algorithm, counting only multiplications and divisions, is $n^3/3 + n^2 - n/3$.

3.2. Revise Crout's algorithm to give the $PA = LDU$ decomposition.

3.3. Give a detailed listing of the row variant of Crout's algorithm, and find an expression for its computational complexity.

3.4. Derive a Gauss's algorithm for LU factorization that provides $u_{ii} = 1$.

3.5. If $A = CC^T$, where $C \in \mathbb{R}^{n \times n}$ is nonsingular, show that A is SPD.

3.6. Derive the Cholesky algorithm.

3.7. Derive a Gauss-Cholesky algorithm for SPD matrices.

3.8. If A is a nonsingular $n \times n$ matrix, and x is an n-vector, prove that:

$$\min_{x \neq 0} \frac{\|Ax\|}{\|x\|} = \frac{1}{\|A^{-1}\|}$$

where matrix norms are induced norms.

3.9. Prove that the condition number of a matrix is at least 1.

3.10. Prove that, for any induced norm, $\|AB\| \leq \|A\|\|B\|$, where A and B are matrices.

3.11. Prove that, if $\|\cdot\|$ is a vector norm on $\mathbb{R}^n$, then there exists a $\gamma > 0$, $\gamma \in \mathbb{R}$, such that, $\forall x \in \mathbb{R}^n$, $\|x\| \leq \gamma \|x\|_\infty$.

3.12. Prove that, for any $A \in \mathbb{R}^{n \times n}$ and any two permutation matrices $P, Q \in \mathbb{R}^{n \times n}$, we have:

$$\kappa_p(PAQ) = \kappa_p(A)$$

so that the condition number is pivoting-invariant.

3.13. Give an algorithm that accepts a matrix description in compressed-column form and produces its transpose, also in compressed-column form.

3.14. Write an algorithm for converting a matrix from a triplet form data structure to a compressed-column form.

3.15. Write the Gauss-Jacobi algorithm in a form that uses the compressed-column form for a sparse matrix.

3.16. (Computer Project) Based on the code developed previously in problem 2.10, write a C or C++ program that solves *any linear resistive circuit with no controlled sources* using MNA. Your linear solver should use Gauss's method for LU-factorization, using partial (row) pivoting, and using an in-place computation so that L and U simply over-write the system matrix. Partial pivoting (using row-exchanges) should be performed to find the best pivot for accuracy, and not only to avoid a zero pivot. You can ignore issues of sparsity. Your implementation should be general, in the sense that it should accept any linear circuit description consisting of any combination of linear resistors and independent voltage and current sources. Use your code to solve the test circuit given in problem 2.10, where the $10\,\Omega$ and $50\,\Omega$ resistors are required to be in group 2. The correct solution is $V(4) = 1.9888\,\text{V}$, $V(8) = 1\,\text{V}$, $V(3) = 2.00879\,\text{V}$, $V(2) = 1.80879\,\text{V}$, $V(6) = 1.98814\,\text{V}$, $V(5) = 2\,\text{V}$, $V(1) = 1.88527\,\text{V}$, $V(7) = 3.98814\,\text{V}$, $I(R8) = 198.88\,\text{mA}$, $I(R3) = 3.82\,\text{mA}$, $I(V3) = 0\,\text{A}$, $I(V2) = -199.88\,\text{mA}$, and $I(V1) = -198.88\,\text{mA}$.

CHAPTER 4

Solution of Nonlinear Algebraic Circuit Equations

In the presence of nonlinear elements, the network equations can be formulated as a system of nonlinear equations. Solving such systems is not trivial and, in fact, is much harder than solving systems of linear equations. As we will see, the practical approach for solving nonlinear equations is to repeatedly *linearize* them and solve the resulting *linear* systems. In general, nonlinear systems of equations can have a unique solution, no solution, multiple solutions, or an infinity of solutions. Practical methods for solving nonlinear systems can only hope to provide the *approximate* value of "a solution," if at all; they never provide closed-form solutions, only numerical and approximate ones. We will study the formulation of nonlinear network equations, the general solution methods, and their application to circuit simulation.

4.1 NONLINEAR NETWORK EQUATIONS

The need to solve a system of nonlinear equations arises in several ways as part of circuit simulation. For one thing, it comes up under DC Analysis, for finding either the quiescent steady state ($t = \infty$) solution under DC inputs, the initial ($t = 0$) solution required to initiate Transient Analysis, or for finding the DC transfer characteristic, by means of a DC-sweep. As well, the need to solve nonlinear equations arises throughout Transient Analysis, as the circuit response at every time-point is solved given the circuit response at the previous time point(s). Considering the *formulation* of the nonlinear network equations, it should be clear that KCL and KVL remain as *linear* relationships. Nonlinearity is due only to the *nonlinear element equations*. In the following, we will study nonlinear elements, nonlinear equation formulation, and the preparation required for DC Analysis.

Circuit Simulation, by Farid N. Najm
Copyright © 2010 John Wiley & Sons, Inc.

4.1.1 Nonlinear Elements

Nonlinear elements come in three varieties:

1. Nonlinear resistors, both voltage-controlled, with an element equation $i = f(v)$, and current-controlled, with $v = f(i)$, where i and v are the current in the element and the voltage across it.
2. Nonlinear capacitors, $C(v)$, and nonlinear inductors, $L(i)$, where v is the voltage across the capacitor and i the current in the inductor.
3. Nonlinear controlled sources, with the element equation $i = f(x)$ or $v = f(x)$ where, in general, x is the vector of variables in the MNA system.

In connection with nonlinear controlled sources, and as a notational convention, we will *always* assume the following. Given an element equation in the form $y = f(x)$, where x is the MNA variable vector, and suppose that y is itself an MNA variable, which it can sometimes be. Then, we will implicitly assume that $f(\cdot)$ does not depend on y via x. In other words, $f(x)$ does not depend on the component of the vector x that corresponds to y. This is required in order to ensure that the y value in the element equation can always be *explicitly* evaluated, given x, by a single evaluation of the function $f(\cdot)$.

We will postpone the coverage of nonlinear capacitors and inductors until later, when we cover the simulation of dynamic circuits. Thus, for now, we will study nonlinear equation formulation excluding dynamic elements; the equations will be only algebraic, not differential. Hence the title of this chapter.

It is often the case that simple nonlinear controlled sources can be replaced by a sub-circuit consisting of a nonlinear resistor and a couple of *linear* controlled sources. For example, a nonlinear VCVS implementing $v_e = f(v_x)$ can be replaced by the sub-circuit shown in Fig. 4.1. In such cases, it is possible to formulate the network equations in such a way that the *only* nonlinear elements are nonlinear resistors. However, more complex controlled sources cannot be so simplified. On the other hand, it is always possible to represent a nonlinear resistor by a nonlinear controlled source. Thus, in all cases, one can formulate the network equations in such a way that the *only* nonlinear elements are nonlinear controlled sources. Using controlled sources is a more versatile approach in practice, and is often the approach used in commercial simulators.

Therefore, in general, and this is the approach we will adopt, it suffices to consider that we have only two types of nonlinear resistive elements:

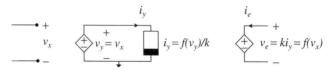

Figure 4.1: An equivalent circuit for a nonlinear VCVS, $v_e = f(v_x)$, using linear controlled sources.

1. Those whose current can be written as an explicit function of other variables, i.e., $i = f(x)$, where x is the vector of MNA variables. We will refer to these elements as simply *controlled current sources* (CCS).

2. Those whose voltage can be written as an explicit function of other variables, i.e., $v = f(x)$, where x is the vector of MNA variables. We will refer to these elements as simply *controlled voltage sources* (CVS).

We are now ready to consider equation formulation.

4.1.2   Nonlinear MNA Formulation

We will revisit the MNA equations and see how they are modified by the introduction of nonlinear resistive elements. Recall that, in formulating the MNA equations, we had divided the elements into two groups, and correspondingly divided the element currents (i_1, i_2) and branch voltages (u_1, u_2). As well, KCL was partitioned:

$$A_1 i_1 + A_2 i_2 = 0 \tag{4.1}$$

and KVL was broken up:

$$u_1 = A_1^T v \quad \text{and} \quad u_2 = A_2^T v \tag{4.2}$$

where v is the vector of nodal voltages.

In the *linear* case, recall that elements in group 1 were such that their currents can be expressed as explicit functions of other variables. Thus, they could only be (linear) resistors, independent current sources, VCCS, or CCCS. To this list, we now add *nonlinear controlled current sources* $i = f(v, i_2)$. As a result, the branch equations for group 1 elements can be expressed in the following matrix form:

$$i_1 + Z_{12} i_2 = Y_{11} u_1 + Y_{12} u_2 + \alpha(v, i_2) + s_1 \tag{4.3}$$

where $\alpha(v, i_2)$ is a vector function of the MNA variables, in which $\alpha_j(v, i_2)$ is either 0 or is the nonlinear function corresponding to one nonlinear CCS. Writing KCL and plugging i_1 from the above, along with KVL, leads to the top part of the MNA equations, as:

$$\left(A_1 Y_{11} A_1^T + A_1 Y_{12} A_2^T \right) v + (A_2 - A_1 Z_{12}) i_2 + A_1 \alpha(v, i_2) = -A_1 s_1 \tag{4.4}$$

As for nonlinear voltage sources, they belong in group 2, whose branch equations are thereby augmented to become in this form:

$$\beta(v, i_2) + Z_{22} i_2 = Y_{21} u_1 + Y_{22} u_2 + s_2 \tag{4.5}$$

where $\beta(v, i_2)$ is a vector function of the MNA variables, in which $\beta_j(v, i_2)$ is either 0 or is the nonlinear function corresponding to one nonlinear CVS.

In general, in order to also allow the inclusion of nonlinear controlled current sources in group 2, this can be written in the more general form:

$$\beta(v, i_2) + Z_{22}i_2 = Y_{21}u_1 + Y_{22}u_2 + \gamma(v, i_2) + s_2 \tag{4.6}$$

where $\gamma(v, i_2)$ is a vector function of the MNA variables, in which $\gamma_j(v, i_2)$ is either 0 or is the nonlinear function corresponding to one nonlinear CCS. Note that, if an entry of $\beta(v, i_2)$ is not identically 0, then the corresponding entry of $\gamma(v, i_2)$ is identically zero, and vice-versa. Thus, the bottom part of the MNA equations becomes:

$$-\left(Y_{21}A_1^T + Y_{22}A_2^T\right)v + Z_{22}i_2 + \beta(v, i_2) - \gamma(v, i_2) = s_2 \tag{4.7}$$

and the final nonlinear MNA system (excluding dynamic elements, for now) becomes:

$$\begin{bmatrix} \left(A_1Y_{11}A_1^T + A_1Y_{12}A_2^T\right) & \left(A_2 - A_1Z_{12}\right) \\ -\left(Y_{21}A_1^T + Y_{22}A_2^T\right) & Z_{22} \end{bmatrix}\begin{bmatrix} v \\ i_2 \end{bmatrix} + \begin{bmatrix} A_1\alpha(v, i_2) \\ \beta(v, i_2) - \gamma(v, i_2) \end{bmatrix}$$
$$= \begin{bmatrix} -A_1s_1 \\ s_2 \end{bmatrix} \tag{4.8}$$

Thus, the MNA equations have been modified by the *addition* of nonlinear terms to the equations. For the top part, the KCL at each node is augmented by adding, with appropriate signs, all the nonlinear $\alpha(v, i_2)$'s for nonlinear group 1 elements incident on that node. For the bottom part, each nonlinear group 2 element contributes a new equation with a nonlinear term, as in $\beta(v, i_2)$ or $\gamma(v, i_2)$. Similar development can be done for the sparse tableau approach, but we will focus on MNA.

We now introduce some further notation, for clarity of presentation. Let $x = \begin{bmatrix} v \\ i_2 \end{bmatrix}$ denote the combined solution vector. Let G denote the system matrix that multiplies x. Let $s = \begin{bmatrix} -A_1s_1 \\ s_2 \end{bmatrix}$ denote the right-hand side (RHS) vector. Therefore, the MNA system can be written more compactly as:

$$Gx + \begin{bmatrix} A_1\alpha(v, i_2) \\ \beta(v, i_2) - \gamma(v, i_2) \end{bmatrix} = s \tag{4.9}$$

Then, because A_1 is an incidence matrix, notice that each entry of the nonlinear vector function $\begin{bmatrix} A_1\alpha(v, i_2) \\ \beta(v, i_2) - \gamma(v, i_2) \end{bmatrix}$ is either identically 0, a single nonlinear function (with coefficient ± 1) corresponding to a single CCS or CVS element in group 2, or a linear combination (with coefficients ± 1) of one or more nonlinear functions corresponding to CCS elements in group 1. Then, to simplify further, let $g(x)$ be a vector whose every entry $g_i(x)$ is the nonlinear function corresponding to a single nonlinear element. Note, the number of entries in $g(x)$ is equal to the total number of nonlinear elements in the network. Then, the MNA system can be more compactly expressed as:

$$Gx + Hg(x) = s \tag{4.10}$$

where H is a matrix whose entries are either 0 or ± 1. Finally, let:

$$f(x) = Gx + Hg(x) - s \tag{4.11}$$

so that:

$$f(x) = 0 \tag{4.12}$$

is the nonlinear system of equations to be solved.

Example We will formulate the MNA equations for the circuit in Fig. 4.2, in which the nonlinear elements are as follows:

- For diode D_1: $i_D = I_s \left(e^{u_D/\eta V_T} - 1 \right)$, with $u_D = v_1 - v_2$.
- For resistor R_2: $u_R = i_2^3 (i_2 - 1)$, with $u_R = v_3 - 0 = v_3$.

As for element classification, group 1 consists of R_1, R_3, R_4, and D_1, while group 2 consists of R_2 and V_s. To write the top part of the MNA system, we write the KCL at every node:

At node 1: $$I_s \left(e^{(v_1-v_2)/\eta V_T} - 1 \right) + \frac{v_1 - v_3}{1\Omega} + i_s = 0$$

At node 2: $$-I_s \left(e^{(v_1-v_2)/\eta V_T} - 1 \right) + \frac{v_2}{1\Omega} + \frac{v_2 - v_3}{1\Omega} = 0$$

At node 3: $$\frac{v_3 - v_1}{1\Omega} + \frac{v_3 - v_2}{1\Omega} + i_2 = 0$$

To write the bottom part of the MNA system, we write the group 2 element equations:

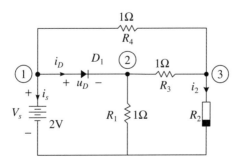

Figure 4.2: A nonlinear circuit used to illustrate equation formulation.

- For R_2: $v_3 - i_2^3 (i_2 - 1) = 0$.
- For V_s: $v_1 - 2V = 0$.

This leads to the nonlinear MNA equations:

$$
\begin{bmatrix}
1 & 0 & -1 & 0 & 1 \\
0 & 2 & -1 & 0 & 0 \\
-1 & -1 & 2 & 1 & 0 \\
0 & 0 & 1 & 0 & 0 \\
1 & 0 & 0 & 0 & 0
\end{bmatrix}
\begin{bmatrix}
v_1 \\ v_2 \\ v_3 \\ i_2 \\ i_s
\end{bmatrix}
+
\begin{bmatrix}
I_s \left(e^{(v_1-v_2)/\eta V_T} - 1\right) \\
-I_s \left(e^{(v_1-v_2)/\eta V_T} - 1\right) \\
0 \\
-i_2^3 (i_2 - 1) \\
0
\end{bmatrix}
=
\begin{bmatrix}
0 \\ 0 \\ 0 \\ 0 \\ 2
\end{bmatrix}
\qquad (4.13)
$$

which can be written as:

$$
\begin{bmatrix}
1 & 0 & -1 & 0 & 0 \\
0 & 2 & -1 & 0 & 0 \\
-1 & -1 & 2 & 1 & 0 \\
0 & 0 & 1 & 0 & 0 \\
1 & 0 & 0 & 0 & 0
\end{bmatrix}
\begin{bmatrix}
v_1 \\ v_2 \\ v_3 \\ i_2 \\ i_s
\end{bmatrix}
+
\begin{bmatrix}
1 & 0 \\
-1 & 0 \\
0 & 0 \\
0 & -1 \\
0 & 0
\end{bmatrix}
\begin{bmatrix}
I_s \left(e^{(v_1-v_2)/\eta V_T} - 1\right) \\
i_2^3 (i_2 - 1)
\end{bmatrix}
=
\begin{bmatrix}
0 \\ 0 \\ 0 \\ 0 \\ 2
\end{bmatrix}
$$

which is in the standard form:

$$
Gx + Hg(x) = s \qquad (4.14)
$$

with:

$$
G =
\begin{bmatrix}
1 & 0 & -1 & 0 & 0 \\
0 & 2 & -1 & 0 & 0 \\
-1 & -1 & 2 & 1 & 0 \\
0 & 0 & 1 & 0 & 0 \\
1 & 0 & 0 & 0 & 0
\end{bmatrix},
\qquad
H =
\begin{bmatrix}
1 & 0 \\
-1 & 0 \\
0 & 0 \\
0 & -1 \\
0 & 0
\end{bmatrix},
\qquad
s =
\begin{bmatrix}
0 \\ 0 \\ 0 \\ 0 \\ 2
\end{bmatrix}
$$

and:

$$
x =
\begin{bmatrix}
v_1 \\ v_2 \\ v_3 \\ i_2 \\ i_s
\end{bmatrix},
\qquad
g(x) =
\begin{bmatrix}
I_s \left(e^{(v_1-v_2)/\eta V_T} - 1\right) \\
i_2^3 (i_2 - 1)
\end{bmatrix}
$$

As was the case with linear circuits, the above matrix equation can be set up by inspection of the circuit, as the circuit description file is parsed.

4.1.3   Preparing for a DC Analysis

In general, one starts with a network that includes dynamic elements. When intending to run any type of DC Analysis, and before being able to set up the MNA equations seen above, the dynamic elements must be removed, somehow. Thus, one must "prepare" the circuit for running DC Analysis.

To see how this should be done, note that, as we will see later on, the general dynamic nonlinear MNA system is of this form:

$$Gx(t) + Hg(x(t)) + D(x)x'(t) = s(t) \qquad (4.15)$$

where, $D(x)$ is a matrix of element stamps of, possibly nonlinear, capacitors and inductors. At a DC steady state, with $x'(t) = 0$, this reduces to the familiar nonlinear system:

$$Gx + Hg(x) = s \qquad (4.16)$$

In order to arrive at a circuit that corresponds to this DC system, it would seem that simply disabling all the dynamic elements should be sufficient. To disable a capacitor, we replace it by an open circuit because, at DC, $dv/dt = 0$, so that $i = C\frac{dv}{dt} = 0$, an open circuit. To disable an inductor we replace it by a short circuit because, at DC, $di/dt = 0$, so that $v = L\frac{di}{dt} = 0$, a short circuit. Note that a short circuit can be represented in MNA by a $0\,\text{V}$ voltage source.

However, this can be problematic because, as a result, some nodes can be left isolated, which may create a singular or ill-conditioned Jacobian (the meaning and significance of this will become clear in the following sections). To overcome this, it is common to attach a large resistor, e.g., $100 \times 10^6\,\Omega$, from every node to ground, or across certain pn-junctions. With all the dynamic elements disabled and such resistors added, the network is typically solvable with little or no numerical problems. Such resistance simulates the presence of leakage current in circuits, thereby injecting some often needed realism into the circuit model. Thus, it is not to be "sneered at" as a remedy. Numerically, this helps ensure that no nodes are isolated, which helps provide a Jacobian that is nonsingular and often more diagonally dominant.

4.2   SOLUTION TECHNIQUES

Solving nonlinear equations is a classical problem with a very rich history, dating back to the ancient Babylonians who used an iterative method to find square roots. More recently, the best known and most influential approach is called *Newton's method* and has its roots in the work of Isaac Newton in the late 1600s. Another British mathematician, Joseph Raphson, later published a simplified description of the method and, as a result, the method is also sometimes referred to as the *Newton-Raphson method*. We will refer to it by its more common name, as simply *Newton's method*. This method is the basis for a large class of modern techniques that are at the heart of all practical nonlinear solvers, including quasi-Newton methods, modified-Newton methods, inexact-Newton methods, Newton-Krylov methods, etc. We will see that Newton's method can be derived as a special case of a certain *fixed point method*. We will return to Newton's method, and fixed point, after covering some preliminary introductory material.

The material in this section is based on a number of sources, including the texts by Kelley (1995), Burden and Faires (2005), Bartle (1976), Dennis and Schnabel (1996), Chua and Lin (1975), and Press et al. (2007).

4.2.1 Iterative Methods and Convergence

As mentioned earlier, practical methods for solving a general $f(x) = 0$ can, at best, provide an approximate value of "a" solution. They cannot, a priori, answer questions relating to *existence* or *uniqueness* of the solutions. An approximate solution is either found within the allotted time, or not. Practical approaches typically consist of an *iterative method* that generates a sequence $\{x^{(k)}\}_{k=0}^{\infty}$ of *candidate solutions*. The iterative method starts with an initial candidate solution $x^{(0)}$ and stops when the sequence has *converged* to a solution, which we denote as x^*. We earlier defined the notion of convergence of vector sequences, in connection with indirect (iterative) methods for solving linear systems, and we saw that a vector sequence $\{x^{(k)}\}_{k=0}^{\infty}$ converges to a vector x^* if and only if $\lim_{k \to \infty} x_i^{(k)} = x_i^*$, for all i. We also studied the notion of convergence rate, but we now discuss it in some more detail.

Rate of Convergence   Of crucial importance, in practice, is the question of how *fast* a sequence converges; we seek iterative methods that produce sequences that converge quickly. Thus, we are interested in the *rate of convergence*.

Suppose the sequence $\{x^{(k)}\}_{k=0}^{\infty}$ converges to x^*, with $x^{(k)} \neq x^*$ for all k. We say that this sequence *converges q-linearly* to x^* if there exists a number $c \in (0, 1)$ such that:

$$\lim_{k \to \infty} \frac{\|x^{(k+1)} - x^*\|}{\|x^{(k)} - x^*\|} = c \tag{4.17}$$

and the number c is called the *asymptotic error constant*. An alternate, equivalent, definition of q-linear convergence is to say that there exists a $\sigma \in (0, 1)$ such that:

$$\|x^{(k+1)} - x^*\| \leq \sigma \|x^{(k)} - x^*\| \tag{4.18}$$

for all k sufficiently large, and σ is called the *q-factor*. If (4.17) holds with $c = 0$, then we say the sequence *converges q-superlinearly*. If the sequence converges, but if the convergence is neither q-linear nor q-superlinear, then it is said to *converge q-sublinearly*.

When the convergence is superlinear, then we can further distinguish between different rates of convergence using the notion of *convergence order*, as follows.

We say that the sequence *converges q-superlinearly with order* α, if there exist $\alpha > 1$ and $c > 0$ such that:

$$\lim_{k \to \infty} \frac{\|x^{(k+1)} - x^*\|}{\|x^{(k)} - x^*\|^{\alpha}} = c \tag{4.19}$$

or, equivalently, if there exists a $\sigma > 0$ such that:

$$\|x^{(k+1)} - x^*\| \leq \sigma \|x^{(k)} - x^*\|^{\alpha} \tag{4.20}$$

for all k sufficiently large. Typically, a sequence with a higher convergence order converges faster. Convergence with order 2 is called *q-quadratic convergence*, and convergence with order 3 is called *q-cubic convergence*. One can describe a q-linearly convergent sequence as having order 1.

Some remarks are noteworthy at this point:

- The asymptotic constants, c or σ, do affect the speed of convergence, but are not as important as the order.
- A q-superlinearly convergent sequence is also q-linearly convergent with q-factor σ, for *any* $\sigma \in (0, 1)$.
- When comparing iterative methods, one needs to study not only the order of convergence but also the cost of each iteration.
- With a q-quadratic method, the number of correct significant digits is approximately *doubled* with every iteration.

The "q" prefix stands for "quotient," because of the ratio used to define convergence in (4.17) and (4.19). This is to differentiate this type of convergence from the following more general notion of "r-convergence", in which "r" stands for "root": If $\epsilon_k \to 0$ is q-<type> convergent and if $\|x^{(k)} - x^*\| \leq \epsilon_k, \forall k$, then $\{x^{(k)}\}_{k=0}^{\infty}$ is said to be r-<type> convergent. Here, q-<type> can denote any of the above, q-linear, q-superlinear, q-quadratic, etc. This "r" notion of convergence is more useful for certain algorithms. Finally, the above convergence notions can be further extended, as follows. If, we replace $x^{(k+1)}$ by $x^{(k+j)}$ throughout, then the convergence is said to be *j-step q-linear, j-step q-superlinear, j-step q-quadratic*, etc.

Examples   This sequence converges q-sublinearly to 0:

$$\left\{ \frac{1}{k+1} \right\}_{k=0}^{\infty} = 1, \frac{1}{2}, \frac{1}{3}, \frac{1}{4}, \frac{1}{5}, \cdots \tag{4.21}$$

This sequence converges q-linearly to 0, with asymptotic constant 1/2:

$$\left\{ \frac{1}{2^k} \right\}_{k=0}^{\infty} = 1, \frac{1}{2}, \frac{1}{4}, \frac{1}{8}, \frac{1}{16}, \cdots \tag{4.22}$$

Table 4.1: A comparison of linear and quadratic convergence rates.

k	$1/2^k$	$1/2^{2^k}$
1	0.500×10^{-0}	0.250×10^{-0}
2	0.250×10^{-0}	0.625×10^{-1}
3	0.125×10^{-0}	0.391×10^{-2}
4	0.625×10^{-1}	0.153×10^{-4}
5	0.312×10^{-1}	0.233×10^{-9}
6	0.156×10^{-1}	0.542×10^{-19}
7	0.781×10^{-2}	0.294×10^{-38}

This sequence converges q-quadratically to 0:

$$\left\{ \frac{1}{2^{2^k}} \right\}_{k=0}^{\infty} = \frac{1}{2}, \frac{1}{4}, \frac{1}{16}, \frac{1}{256}, \frac{1}{65536}, \cdots \tag{4.23}$$

It is instructive to compare q-linear convergence, as in the case of (4.22), with q-quadratic convergence, as in the case of (4.23), and the results are shown in Table 4.1. From this, it is clear that quadratic convergence is indeed *much* faster than linear convergence!

4.2.2   Introduction to Newton's Method

We are now ready to start looking at solution methods. Recall, our problem is that, with $x \in \mathbb{R}^n$ and $f : \mathbb{R}^n \to \mathbb{R}^n$, we are interested to find a solution of:

$$f(x) = 0 \tag{4.24}$$

The most widely used solution method for this is Newton's method, whose description can be very simple and brief, as shown in Fig. 4.3, where $J(x)$ is the *Jacobian* matrix of $f(\cdot)$, evaluated at x, i.e.:

$$J(x)_{ij} \triangleq \frac{\partial f_i}{\partial x_j}(x) \tag{4.25}$$

The convergence theory for Newton's method is *local*, i.e., it depends on the initial candidate solution $x^{(0)}$ being close enough to the true solution x^*. Specifically, three conditions are required, which are *sufficient* for convergence. First, the function $f(x)$ must have a solution x^*, and $f(x)$ must be continuously differentiable near x^*. Second, the Jacobian $J(x)$ must be *Lipschitz continuous* near x^*, i.e., there must exist a $\gamma > 0$ such that, for all x, y sufficiently near x^*, we have:

$$\| J(x) - J(y) \| \leq \gamma \| x - y \| \tag{4.26}$$

> **Input:** function $f(\cdot)$ and initial guess x
> **Output:** a value of x for which $f(x) \approx 0$
>   **while** $(f(x) \not\approx 0)$ **do**
>   Solve $J(x)\delta = -f(x)$
> $x = x + \delta$

Figure 4.3: Newton's method.

and γ is called the *Lipschitz constant*. Third, the Jacobian, evaluated at the true solution, $J(x^*)$, must be nonsingular. Given these three conditions, then, if $x^{(0)}$ is close enough to x^*, it can be shown that $J(x^{(k)})$ is nonsingular, $\forall k$, and the algorithm converges q-quadratically to x^*.

We could stop here! However, in order to get a full appreciation of this method, and its links to other methods, we will examine the various ways in which it may be derived, leading up to it in small steps. For example, it is possible to derive Newton's method from either a Taylor series viewpoint or as a special case of a fixed point method.

Plan of Work   In the remainder of this introductory section, we will review some concepts from calculus, leading up to a study of the question of Lipschitz continuity of the Jacobian. In the next section, we then consider solutions for the one-dimensional case, so as to convey the geometric intuition behind Newton's method. Several root-finding methods in one-dimension will be noted, and Newton's method singled out. As well, we will illuminate the links between Newton's method and the fixed point method, and present some convergence results. We will then discuss the multidimensional case, illustrating two ways in which the method can be derived, and give its convergence theory. Finally, we will give a brief discussion of quasi-Newton methods.

Lipschitz Continuity   A function f, defined from $D \subset \mathbb{R}^p$ to $\mathbb{R}^q$, is *continuous* in D if and only if, for any $\epsilon > 0$ and $u \in D$, there is a $\delta(\epsilon, u) > 0$ such that if $x \in D$ and $\|x - u\| \leq \delta$, then $\|f(x) - f(u)\| \leq \epsilon$. Notice that δ depends on both ϵ and u; a stronger continuity condition can be defined for the case when δ does not depend on u, as follows. A function f, defined on $D \subset \mathbb{R}^p$ to $\mathbb{R}^q$, is said to be *uniformly continuous* in D if, for any $\epsilon > 0$, there is a $\delta(\epsilon) > 0$ such that if $x, u \in D$ and $\|x - u\| \leq \delta$, then $\|f(x) - f(u)\| \leq \epsilon$. Clearly, if a function is uniformly continuous then it is continuous, but the converse is not true.

A set $D \subset \mathbb{R}^p$ is said to be *open* if, for any $x \in D$, there exists $r > 0$, such that $S \triangleq \{y \in \mathbb{R}^p : \|x - y\| < r\}$ is a subset of D. A set $D \subset \mathbb{R}^p$ is said to be *closed* if its complement in $\mathbb{R}^p$ is open. A set $D \subset \mathbb{R}^p$ is said to be *bounded* if there exist real numbers a_i, b_i, for $i = 1, 2, \ldots, p$, such that $D \subset \{x \in \mathbb{R}^p : a_i \leq x_i \leq b_i, \forall i = 1, 2, \ldots, p\}$. One can show that, if f is continuous on a *closed* and *bounded* $K \subset \mathbb{R}^p$, then f is uniformly continuous on K.

A stronger condition still is the *Lipschitz condition*, as follows. If f has domain $D \subset \mathbb{R}^p$ and range in $\mathbb{R}^q$, then we say that f is *Lipschitz continuous* in D if there exists a constant $L > 0$ such that:

$$\|f(x) - f(y)\| \le L\|x - y\| \tag{4.27}$$

for all x, y in D, and L is called the *Lipschitz constant*. If a function is Lipschitz continuous then it is uniformly continuous, but the converse is not true. Any linear function $f : \mathbb{R}^p \to \mathbb{R}^q$ is Lipschitz continuous. The Lipschitz condition is not sufficient for differentiability. For example, the function:

$$f(x) = \begin{cases} x \sin(1/x), & x \ne 0, \\ 0, & x = 0. \end{cases} \tag{4.28}$$

is Lipschitz continuous but not differentiable at $x = 0$. Thus, loosely speaking, Lipschitz continuity constitutes a little more than continuity, but a little less than differentiability.

Sufficient Conditions   What are the sufficient conditions in order for a function to be Lipschitz? This question is obviously relevant to the use of Newton's method. We now consider some answers. In the one-dimensional case, let f be continuous on $[a, b] \subset \mathbb{R}$ and its derivative $f'(x)$ exist in (a, b). If $f'(x)$ is bounded on (a, b), i.e., if $\exists M \in \mathbb{R}$ such that $|f'(x)| \le M$, for all $a < x < b$, then f is Lipschitz continuous on $[a, b]$.

The multidimensional case requires some further preliminary definitions. A function $f : D \subset \mathbb{R}^n \to \mathbb{R}$ is said to be *continuously differentiable* at $x \in D$, if $(\partial f(x)/\partial x_i)$ exist *and* are continuous at x, $\forall i = 1, 2, \ldots, n$. It is said to be continuously differentiable in an open set $K \subset D$, if it is continuously differentiable at every $x \in K$. A function $f : D \subset \mathbb{R}^n \to \mathbb{R}^m$ is said to be *continuously differentiable* at $x \in D$, if each of its component functions f_i is continuously differentiable at x. It is said to be continuously differentiable in an open set $K \subset D$, if it is continuously differentiable at every $x \in K$. In this case, its Jacobian matrix $J(x)$ exists and is continuous in D. Recall, a subset D of $\mathbb{R}^n$ is said to be *convex* if, whenever $x, y \in D$, and given a real number $t \in (0, 1)$, then $z = (1 - t)x + ty \in D$.

With this, we can now give the sufficient condition in the multidimensional case. Let $D \subset \mathbb{R}^p$ be *open* and *convex*, and let $f : D \to \mathbb{R}^q$ be continuously differentiable in D. Let the Jacobian of f have a bounded norm in D, i.e., $\|J(x)\| \le M, \forall x \in D$, for some $M > 0$, and based on some induced matrix norm. Then f is Lipschitz continuous in D, with the Lipschitz constant M.

The Jacobian Is Lipschitz   To understand the conditions under which Newton's method is convergent, it is important to study when the Jacobian is Lipschitz continuous. The Jacobian is a matrix-valued or simply a "matrix function," not a vector-valued, function, so that some further definitions are needed. Using an induced matrix norm, we say that a matrix function $J(x)$, defined from $D \subset \mathbb{R}^n$ to $\mathbb{R}^{m \times n}$, is Lipschitz continuous in D if $\exists M > 0$ such that:

$$\forall x, y \in D, \qquad \|J(x) - J(y)\| \leq M \|x - y\| \tag{4.29}$$

We say a function $J : D \subset \mathbb{R}^n \to \mathbb{R}^{m \times n}$ is *continuously differentiable* at $x \in D$, if each of its components $J_{ij}(x)$ is continuously differentiable at x. It is said to be continuously differentiable in an open set $K \subset D$, if it is continuously differentiable at every $x \in K$. Note, if $f(x)$ has a Jacobian $J(x)$, then saying that $J(x)$ is continuously differentiable means that all its $\partial^2 f_k / \partial x_i \partial x_j$ entries exist *and* are continuous.

Using the preceding results, one can prove the following:

Theorem 4.1. *Let $D \subset \mathbb{R}^n$ be open and convex, and let $f : D \to \mathbb{R}^m$ have a Jacobian $J(x)$ which is continuously differentiable in D. Suppose also that all the 2nd derivatives, $\partial^2 f_k / \partial x_i \partial x_j$ are bounded in D, i.e., $\exists M > 0$, such that $\forall i, j, k$, and $\forall x \in D$, $|\partial^2 f_k(x)/\partial x_i \partial x_j| \leq M$. Then $J(x)$ is Lipschitz continuous in D, with constant $\gamma = n^3 M$, so that:*

$$\forall x, y \in D, \qquad \|J(x) - J(y)\| \leq \gamma \|x - y\| \tag{4.30}$$

Thus, a sufficient condition for Lipschitz continuity of the Jacobian is that all the 2nd derivatives of f must be continuous and bounded. With this, we are now ready to return to the study of Newton's method, and we start with the one-dimensional case.

4.2.3 The One-Dimensional Case

In the one-dimensional case, $f : \mathbb{R} \to \mathbb{R}$, and we are interested to solve the following scalar equation, which is basically the classical *root finding* problem:

$$f(x) = 0 \tag{4.31}$$

The apparent similarity between this problem and the multidimensional case is misleading—it is *much* harder to solve the multidimensional case! The key difference is that, in one dimension, it is possible to *bracket* or "trap" a root between two values of x and then to "hunt it down." A root is said to be *bracketed* in (a, b) if $f(a)$ and $f(b)$ have opposite signs. In multiple dimensions, you can never be sure that the root is there at all, until you have found it.

There are many methods for finding a root of a nonlinear function, often differentiated by whether the derivatives are available/required or not. Given a *continuous* function, and supposing that the derivatives are *not* available, then, as described in Press et al. (2007), the following methods are available:

- The *bisection* method reduces by 1/2 the size of the bracketing interval in every iteration; it is guaranteed to find a root, but is not very fast.

- The *secant* method and the *false position* (or *regula falsi*) method can be faster than bisection, both requiring smooth functions near the root. Regula falsi maintains a bracketed root, while the secant method does not, although secant is generally faster.
- *Brent's method* (c. 1973) is well known and superior to the above; it combines bisection and interpolation to find a bracketed root.
- *Ridder's method* (c. 1979) is a close competitor, is easier to code, and also is guaranteed to converge; it is a powerful variant on regula falsi.

Of the above methods, secant is the only one that does not require, and does not maintain, a bracket around the root. However, it is quite appealing because it extends to multiple dimensions.

The above techniques may be described as *derivative-free* methods. When derivatives *are* available, the best known method is Newton's method, mainly due to its fast q-quadratic convergence. However, even if derivatives are available, they can often be expensive to compute and, in practice, one is often interested in cheaper alternatives. The secant method may be viewed as a means to approximate the derivative, and there is one other method that tries to do the same, called the *Newton-chord method*. Like secant, the Newton-chord method does not require a bracketed root and, like secant, it can be extended to multiple dimensions.

Thus, we will describe only three one-dimensional methods: Newton's method, the secant method, and the Newton-chord method. We will then see that both Newton's method and the Newton-chord method are special cases of the general fixed point method.

Newton's Method   Since we are working in a single dimension, we will simplify the notation and denote successive candidate solutions using subscripts: x_0, x_1, x_2, etc. Let x_k be the *current* candidate solution, and suppose that x_k is sufficiently close to the true solution x^*. In a small neighborhood around x_k that includes x^*, we can approximate the function $f(x)$ by a *local linearized model*. Newton's method uses the tangent to the curve $y = f(x)$, at x_k, as the local linearized model around x_k. The secant and chord methods use different linearized models. Strictly speaking, these models are not linear, but *affine*. To construct a local linearized model, consider a Taylor series expansion of $f(x)$ around x_k:

$$f(x) = f(x_k) + (x - x_k)f'(x_k) + \frac{(x - x_k)^2}{2}f''(\xi) \qquad (4.32)$$

for some ξ between x and x_k, i.e., either $\xi \in (x, x_k)$ or $\xi \in (x_k, x)$. In a small neighborhood around x_k, we expect the third term above to be very small, due to the square, and we form our local linearized model as:

$$M_k(x) = f(x_k) + (x - x_k)f'(x_k) \qquad (4.33)$$

To the extent that this affine model is a good approximation to $f(x)$ in the neighborhood, we may solve $M_k(x) = 0$ instead of solving $f(x) = 0$. Setting $M_k(x) = 0$ leads to:

$$x = x_k - \frac{f(x_k)}{f'(x_k)} \tag{4.34}$$

assuming that $f'(x_k) \neq 0$. We do not expect that this value of x is equal to the true solution x^*, because the affine model is only an approximation. However, we would *hope* that this is a *better* candidate solution than the one we had previously, and this leads to Newton's method as the iteration:

$$x_{k+1} = x_k - \frac{f(x_k)}{f'(x_k)} \tag{4.35}$$

By repeatedly applying the above as a correction to the candidate solution, we hope to iteratively move closer to the true solution x^*. Graphically, the progress of the algorithm is depicted in Fig. 4.4.

Example   It is instructive to study how Newton's method applies in simple cases, such as in the example given in Dennis and Schnabel (1996), in which:

$$f(x) = x^2 - a \tag{4.36}$$

where $a > 0$ and the solution is $x^* = \sqrt{a}$, starting with some $x_0 \neq 0$. The Newton iteration in this case becomes:

$$x_{k+1} = x_k - \frac{x_k^2 - a}{2x_k} = \frac{x_k^2 + a}{2x_k} \tag{4.37}$$

$$= \frac{\left(x_k - \sqrt{a}\right)^2}{2x_k} + \sqrt{a} \tag{4.38}$$

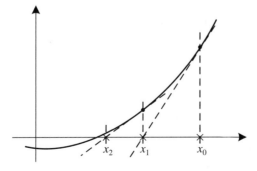

Figure 4.4: A graphical illustration of Newton's method in one dimension.

so that:

$$x_{k+1} - \sqrt{a} = \frac{\left(x_k - \sqrt{a}\right)^2}{2x_k} \tag{4.39}$$

and we can write an error update equation as:

$$\left|x_{k+1} - \sqrt{a}\right| = \frac{\left|x_k - \sqrt{a}\right|^2}{|2x_k|} \tag{4.40}$$

which, with $\mathcal{E}_k \triangleq |x_k - x^*|$, gives:

$$\mathcal{E}_{k+1} = \frac{\mathcal{E}_k^2}{|2x_k|} \tag{4.41}$$

It is evident from this that the error does not necessarily decrease in every iteration—it depends on the value of $\mathcal{E}_k/|2x_k|$—so that convergence is not guaranteed. However, if we are close enough to $x^* = \sqrt{a}$, in this case if $|x_k - \sqrt{a}| < |2x_k|$, then $\mathcal{E}_{k+1} < \mathcal{E}_k$, and the algorithm will converge to the true solution.

Convergence This is the convergence result for the one-dimensional Newton's method:

Theorem 4.2. *Let $f : \mathbb{R} \to \mathbb{R}$ be* continuously differentiable *(i.e., f' exists and both f and f' are continuous) in an open interval $\mathcal{D}$ that contains a solution x^*. Let f' be* Lipschitz continuous *in $\mathcal{D}$, and let there exist a $\rho > 0$ such that $|f'(x)| \geq \rho$ for every $x \in \mathcal{D}$ (in particular, note that $f'(x^*) \neq 0$). Then, if x_0 is sufficiently close to x^*, it follows that $f'(x_k) \neq 0, \forall k$, and the sequence $\{x_k\}_{k=0}^{\infty}$ exists, and it converges q-quadratically to x^*.*

It is interesting to note that, if $f'(x^*) = 0$, then it can be shown that Newton's method converges, but only q-linearly. The requirement, in the theorem, that f' is Lipschitz continuous can be achieved by ensuring that $f''(x)$ exists and is bounded in $\mathcal{D}$, i.e., $\exists M > 0 : |f''(x)| \leq M, \forall x \in \mathcal{D}$. Another form of the theorem can be given, in which the Lipschitz condition on f' is dropped, and replaced by the requirement that $f'''(x)$ exist in $\mathcal{D}$. These conditions, that either $f''(x)$ is bounded or that $f'''(x)$ exists, become desirable requirements for device modeling for circuit simulation. These conditions are not *necessary*; they are part of the *sufficient* conditions for quadratic convergence.

Although the above results are reassuring, in practice we can never be sure that we are starting close enough to a root of $f(x)$. This is particularly problematic during DC Analysis, where one typically does not have a very reliable initial solution. However, it is less of a problem during Transient Analysis because, with the use of a small enough time step, we have a good initial solution, namely the solution at the previous time point. In general, far from a root, one can get

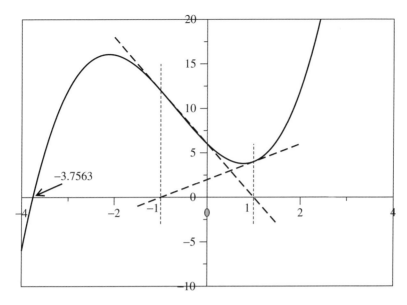

Figure 4.5: Oscillation in Newton's method.

grossly inaccurate, meaningless corrections, and the sequence of candidate solutions can diverge! Thus, Newton's method is useful for its fast *local* convergence, but must be combined with other strategies to give a reliable *global* method.

One example of a problem that can arise in connection with Newton's method is shown in Fig. 4.5. Here, we are solving the equation $f(x) = 0$, where $f(x) = x^3 + 2x^2 - 5x + 6$, and where the exact solution is $x^* \approx -3.7563$. If we start with an initial solution of $x_0 = -1$, the tangent to the curve at that point would lead us to the next solution $x_1 = 1$, and the computations are simple enough (mostly integer, $x_1 = -1 + (12/6)$) that no roundoff error would be expected. In the next step, the tangent to the curve at the point $x_1 = 1$ would return us to the previous point, $x_2 = -1$, again with no roundoff (using $x_2 = 1 - (4/2)$). The result is an endless *oscillation* between the two points -1 and 1. Realistically, depending on exactly how the division is implemented, one may get some amount of roundoff error, but it may take a very long time to get out of the oscillation, if at all. Indeed, a c implementation of Newton's method applied to this problem, using double precision arithmetic, was still oscillating between ± 1.000000 after 10^6 iterations! On the other hand, if we happen to reach the point $x_k \approx 0.9257$, for some k value, then it is clear from Fig. 4.6 that we would have another problem. The tangent at that point would lead us to $x_{k+1} \approx -2.1196$, which happens to be the point at which the curve has a maximum and where the slope of the tangent is zero, and the next point is either at infinity or is a very large value. Either way, we would *diverge*, and we may or may not be able to return from that excursion to the neighborhood of the true solution.

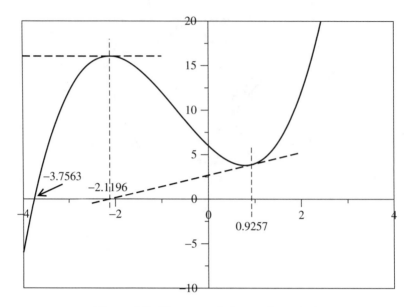

Figure 4.6: Divergence in Newton's method.

The Secant Method   As mentioned earlier, one drawback of Newton's method is the need to compute the derivatives $f'(x_k)$. To avoid this, notice that:

$$f'(x_k) = \lim_{x \to x_k} \frac{f(x) - f(x_k)}{x - x_k} \tag{4.42}$$

so that, if x_{k-1} is close to x_k, we can write:

$$f'(x_k) \approx \frac{f(x_{k-1}) - f(x_k)}{x_{k-1} - x_k} \tag{4.43}$$

Using this as an approximation to $f'(x_k)$ in Newton's method, we get:

$$x_{k+1} = x_k - \frac{x_k - x_{k-1}}{f(x_k) - f(x_{k-1})} f(x_k) \tag{4.44}$$

This is called the *secant method*, and is illustrated in Fig. 4.7. Under fairly general conditions, the secant method is q-superlinearly convergent, and its order of convergence is equal to the so-called *golden ratio*:

$$\phi = \left(1 + \sqrt{5}\right)/2 = 1.6180339887\ldots \tag{4.45}$$

provided that we start close enough to the true solution x^*. It can also be shown to be 2-step q-quadratic, so that $\mathcal{E}_{k+1} \le c\, \mathcal{E}_{k-1}^2$, where, as before, $\mathcal{E}_k \triangleq |x_k - x^*|$. Secant has the dangerous property that, when $|x_k - x_{k-1}|$ is very small, the finite

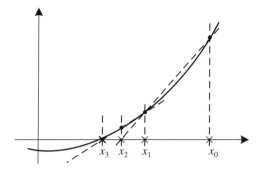

Figure 4.7: A graphical illustration of the secant method.

difference approximation to $f'(x_k)$ suffers severe roundoff errors. However, its extension to multiple dimensions leads to some very powerful quasi-Newton methods, such as Broyden's method.

The Newton-Chord Method   Another way to approximate the derivative is to simply compute $f'(x_0)$ and use that as the approximation to $f'(x_k)$, for all subsequent k values. With this, Newton's method becomes:

$$x_{k+1} = x_k - \frac{f(x_k)}{f'(x_0)} \tag{4.46}$$

This is called the *Newton-chord method* (or, simply, the *chord method*) and is shown in Fig. 4.8. Under fairly general conditions, and if we start near enough to x^*, this method is q-linearly convergent.

Fixed Point   We now consider the very important fixed point method. If $g : \mathbb{R} \to \mathbb{R}$, then a point $x^* \in \mathbb{R}$ is said to be a *fixed point* for $g(x)$ if:

$$g(x^*) = x^* \tag{4.47}$$

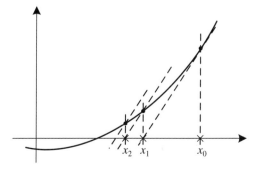

Figure 4.8: A graphical illustration of the Newton-chord method.

There is an intimate relationship between root finding for nonlinear equations and finding a fixed point of a function. Indeed, it is easy to see that x^* is a solution of $f(x) = 0$ if and only if it is a fixed point of:

$$g(x) = x - \phi f(x) \qquad (4.48)$$

where $\phi \neq 0$ is some real number. Even though our main interest is in root finding, the fixed point form of the problem is easier to analyze. In fact, the fixed point problem often provides insight for developing powerful root-finding techniques. As to the existence and uniqueness of fixed points, we have the following result.

Theorem 4.3. *If $g : \mathbb{R} \to \mathbb{R}$ is a continuous function on its domain $[a, b]$ and if its range is a subset of $[a, b]$, then g has at least one fixed point in $[a, b]$. If, in addition, $g'(x)$ exists on (a, b) and there exists $k < 1$ such that $|g'(x)| \leq k$, for all $x \in (a, b)$, then the fixed point in $[a, b]$ is unique.*

Thus, if the graph of the function lies in a square $a \leq x \leq b, a \leq y \leq b$, and if its slope is bounded and less than 1, then the function has a unique fixed point where its graph meets the diagonal line $y = x$, as shown in Fig. 4.9.

Fixed Point Method   In order to find a fixed point of a function, we can consider the iterative method shown in Fig. 4.10. This is called a *fixed point method*; it operates as illustrated in Fig. 4.11 and it generates the sequence $\{x_k\}$ according to:

$$x_{k+1} = g(x_k) \qquad (4.49)$$

In general, the fixed point method may converge to a unique solution, may diverge, or may converge to a non-unique solution. The method is guaranteed to converge in certain cases, as given in the following theorem.

Theorem 4.4. *Let $g : \mathbb{R} \to \mathbb{R}$ have domain $[a, b]$ and range in $[a, b]$, and suppose that g' exists on (a, b) and a constant $k < 1$ exists such that:*

$$|g'(x)| \leq k, \quad \forall x \in (a, b) \qquad (4.50)$$

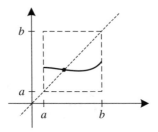

Figure 4.9: Existence of a fixed point [after Burden and Faires (2005)].

> **Input:** function $g(\cdot)$ and initial guess x
> **Output:** a value of x for which $g(x) \approx x$
>   **while** $(g(x) \not\approx x)$ **do**
> $x = g(x)$

Figure 4.10: The fixed point algorithm.

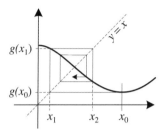

Figure 4.11: Illustration of progress of the fixed point method [after Chua and Lin (1975)].

Then, for any $x_0 \in [a, b]$, the sequence generated by $x_{k+1} = g(x_k)$ converges to the unique fixed point x^* in $[a, b]$. Furthermore, if $g'(x^*) \neq 0$, then the sequence is q-linearly convergent.

Quadratic Fixed-Point Method   Under certain conditions, fixed point becomes q-quadratically convergent, as follows.

Theorem 4.5. *Let $g : \mathbb{R} \to \mathbb{R}$ have a fixed point x^* and suppose that g' exists and is Lipschitz continuous on an open interval containing x^*, with $g'(x^*) = 0$. Then, if x_0 is close enough to x^*, the fixed point method is q-quadratically convergent to x^*.*

Quadratic convergence being so attractive, we should try to construct a q-quadratic fixed point problem $g(x)$ for a given root finding problem $f(x)$. Suppose we choose $g(x) = x - \phi(x)f(x)$ and, inspired by the above theorem, let us require that $g'(x^*) = 0$, where $f(x^*) = 0$, then:

$$0 = g'(x^*) = 1 - \phi'(x^*)f(x^*) - \phi(x^*)f'(x^*) = 1 - \phi(x^*)f'(x^*) \qquad (4.51)$$

so that we need to ensure that $\phi(x^*) = 1/f'(x^*)$, assuming $f'(x^*) \neq 0$. One way to do this is to choose $\phi(x) = 1/f'(x)$, for all x, so that the resulting fixed point method would be:

$$x_{k+1} = g(x_k) = x_k - \frac{f(x_k)}{f'(x_k)} \qquad (4.52)$$

which, of course, is simply Newton's method. Notice that, if we choose $\phi(x) = 1/f'(x_0) = constant$, then we would get the Newton-chord method which, as we know, is q-linearly convergent.

In spite of the fact that Newton's method is faster than a general fixed point method, it is not always the winner. Consider the function $f(x) = x + 2x^{1/3} - 4$, whose solution is ≈ 1.6410, and suppose we start with $x_0 = 27$. You would find that a fixed point method on $g(x) = x - f(x)$ converges in about 20 iterations, while Newton's method on $f(x)$ converges to 0. If you start Newton's method with $x_0 = 0.001$, it would converge to the correct solution in about 7 iterations.

4.2.4 The Multidimensional Case

In multiple dimensions, we can again derive Newton's method starting from either a Taylor series based argument, or as a special case of fixed point. In a small neighborhood around $x^{(k)}$ that includes x^*, we can approximate the function $f(x)$ by a *local linearized model*. Newton's method uses the *tangent hyperplane* to the surface $y = f(x)$, at $x^{(k)}$, as the local linearized model around $x^{(k)}$. Secant and Chord use different linearized models. Strictly speaking, these models are not linear, but *affine*.

To construct a local linearized model, consider a Taylor series expansion of $f(x)$ around x_k, then, for each $i = 1, 2, \ldots, n$ we get:

$$
f_i(x) = f_i(x^{(k)}) + \left. \frac{\partial f_i}{\partial x_1} \right|_{x^{(k)}} \left(x_1 - x_1^{(k)} \right)
$$

$$
+ \left. \frac{\partial f_i}{\partial x_2} \right|_{x^{(k)}} \left(x_2 - x_2^{(k)} \right) + \cdots + \left. \frac{\partial f_i}{\partial x_n} \right|_{x^{(k)}} \left(x_n - x_n^{(k)} \right)
$$

$$
+ \text{higher-order terms involving} \left(x_i - x_i^{(k)} \right)^m
$$

where $m \geq 2$. In a small neighborhood around $x^{(k)}$, we would expect the higher-order terms to be small and, therefore, we form our local linearized model as:

$$
M_k(x) = f(x^{(k)}) +
\begin{bmatrix}
\dfrac{\partial f_1}{\partial x_1} & \dfrac{\partial f_1}{\partial x_2} & \cdots & \dfrac{\partial f_1}{\partial x_n} \\
\dfrac{\partial f_2}{\partial x_1} & \dfrac{\partial f_2}{\partial x_2} & \cdots & \dfrac{\partial f_2}{\partial x_n} \\
\vdots & \vdots & & \vdots \\
\dfrac{\partial f_n}{\partial x_1} & \dfrac{\partial f_n}{\partial x_2} & \cdots & \dfrac{\partial f_n}{\partial x_n}
\end{bmatrix}
(x - x^{(k)})
\tag{4.53}
$$

or:

$$
M_k(x) = J(x^{(k)})(x - x^{(k)}) + f(x^{(k)})
\tag{4.54}
$$

To the extent that this affine model is a good approximation to $f(x)$ in the neighborhood, we may solve $M_k(x) = 0$ instead of solving $f(x) = 0$. Setting $M_k(x) = 0$ leads to:

$$
x = x^{(k)} - J(x^{(k)})^{-1} f(x^{(k)})
\tag{4.55}
$$

We do not expect that this value of x is equal to the true solution x^*, because the affine model is of course approximate. However, we would *hope* that this is a *better* candidate solution than the one we had previously, and this leads to:

$$x^{(k+1)} = x^{(k)} - J(x^{(k)})^{-1} f(x^{(k)}) \qquad (4.56)$$

which is the familiar form of Newton's method. By repeatedly applying the above "correction" to the candidate solution, we hope to iteratively move closer to the true solution x^*.

The main convergence result for Newton's method, is as follows:

Theorem 4.6. *Let $f : \mathbb{R}^n \to \mathbb{R}^n$ be continuously differentiable in an open convex set $\mathcal{D} \subset \mathbb{R}^n$ that contains a solution x^*. Let $J(x)$ be Lipschitz continuous in $\mathcal{D}$, and let $J(x^*)$ be nonsingular. Then, if x_0 is sufficiently close to x^*, it follows that $J(x^{(k)})^{-1}$ exists $\forall k$, so the sequence $\{x^{(k)}\}_{k=0}^{\infty}$ exists, and it converges q-quadratically to x^*.*

Note that, if f is itself an affine function, then Newton's method solves the problem exactly in one iteration. If a component f_i of f is affine, then $f_i(x^{(k)}) = 0$ for all $k \geq 1$, irrespective of the initial candidate solution $x^{(0)}$. We will next study the links between Newton's method and the fixed point problem in multiple dimensions.

Fixed Point in Multiple Dimensions   Let $g : \mathbb{R}^n \to \mathbb{R}^n$, then a point $x^* \in \mathbb{R}^n$ is said to be a *fixed point* of $g(x)$ if:

$$g(x^*) = x^* \qquad (4.57)$$

As we saw in the scalar case, we will examine the relationship between root finding and fixed point in multiple dimensions. Notice that x^* is a solution of $f(x) = 0$ if and only if it is a fixed point of:

$$g(x) = x - \Phi f(x) \qquad (4.58)$$

where Φ is some nonsingular matrix.

Theorem 4.7. *Let $\mathcal{D} = \{x \in \mathbb{R}^n | x_i \in [a_i, b_i], \forall i = 1, 2, \ldots, n\}$, for some collection of constants a_i and b_i, and let $g : \mathbb{R}^n \to \mathbb{R}^n$ be continuous on $\mathcal{D}$. If $g(x) \in \mathcal{D}$ whenever $x \in \mathcal{D}$, then g has a fixed point in $\mathcal{D}$. Furthermore, suppose, for all $i, j \in \{1, 2, \ldots, n\}$, that $\partial g_i / \partial x_j$ exists and is continuous in $\mathcal{D}$, and that a constant $L < 1$ exists with:*

$$\left| \frac{\partial g_i(x)}{\partial x_j} \right| \leq \frac{L}{n}, \quad \text{whenever } x \in \mathcal{D} \qquad (4.59)$$

Then, the sequence generated by the fixed point method $x^{(k+1)} = g(x^{(k)})$, starting from any $x^{(0)} \in \mathcal{D}$, converges to the unique *fixed point $x^* \in \mathcal{D}$, and:*

$$\left\| x^{(k)} - x^* \right\|_\infty \leq \frac{L^k}{1-L} \left\| x^{(1)} - x^{(0)} \right\|_\infty \tag{4.60}$$

This result is a special case of the *contraction mapping theorem*, given below, and shows that in general the fixed point method has q-linear convergence.

Contraction   If a function $g : \mathbb{R}^n \to \mathbb{R}^n$ is Lipschitz continuous on a domain $\mathcal{D} \subset \mathbb{R}^n$, with a Lipschitz constant $L < 1$, then g is called a *contraction*. Among other things, this means that, starting from any $x^{(0)} \in \mathcal{D}$, the step sizes taken by a fixed point method decrease by at least L at each iteration:

$$\left\| x^{(k+1)} - x^{(k)} \right\| = \left\| g(x^{(k)}) - g(x^{(k-1)}) \right\| \leq L \left\| x^{(k)} - x^{(k-1)} \right\| \tag{4.61}$$

so that:

$$\left\| x^{(k+1)} - x^{(k)} \right\| < \left\| x^{(k)} - x^{(k-1)} \right\| \tag{4.62}$$

Theorem 4.8. *(Contraction Mapping Theorem) Let $g : \mathcal{D} \to \mathcal{D}$, where $\mathcal{D}$ is a closed subset of $\mathbb{R}^n$, and let g be a contraction on $\mathcal{D}$ with the constant $L < 1$, then the contraction g has a unique fixed point $x^* \in \mathcal{D}$ and, for any $x^{(0)} \in \mathcal{D}$, the sequence $\{x^{(k)}\}$ generated by $x^{(k+1)} = g(x^{(k)})$ remains in $\mathcal{D}$ and converges q-linearly to x^*, with constant L, and:*

$$\left\| x^{(k)} - x^* \right\| \leq \frac{L^k}{1-L} \left\| x^{(1)} - x^{(0)} \right\| \tag{4.63}$$

Quadratic Fixed-Point Method   It remains to explore what conditions lead to a fixed point method which is q-quadratically convergent, and how that relates to Newton's method.

Theorem 4.9. *Let $g : \mathbb{R}^n \to \mathbb{R}^n$ have a fixed point at $x^* \in \mathbb{R}^n$. Suppose there exists $\delta > 0$ such that $\partial^2 g_i / (\partial x_j \partial x_k)$ exists and is continuous on the open set $\mathcal{D} = \{x \in \mathbb{R}^n | \, \|x - x^*\| < \delta\}$, $\forall i, j, k \in \{1, 2, \dots, n\}$. Suppose there exists $M > 0$ such that, $\forall x \in \mathcal{D}$, $|\partial^2 g_i(x)/(\partial x_j \partial x_k)| \leq M$, for all $i, j, k \in \{1, 2, \dots, n\}$. Suppose $\partial g_i(x^*)/\partial x_k = 0$, for all $i, k \in \{1, 2, \dots, n\}$. Then, if $x^{(0)}$ is chosen sufficiently close to x^*, the sequence generated by the fixed point method $x^{(k+1)} = g(x^{(k)})$ converges q-quadratically to x^*.*

Thus, if the second derivatives are all continuous and bounded around x^* and the first derivatives all vanish at x^*, then fixed point is q-quadratic. Guided by this, we can design a q-quadratic fixed point problem to solve the root finding problem $f(x) = 0$, as follows. Suppose we choose $g(x) = x - \Phi(x)f(x)$, where $\Phi(x^*)$ is a nonsingular matrix, so that x^* is a zero of $f(x)$ if and only if it is a fixed point of $g(x)$. Inspired by the above theorem, if we set $\partial g_i(x^*)/\partial x_k = 0$, for all $i, k \in \{1, 2, \ldots, n\}$, then it is easy to show that $\Phi(x^*) J(x^*) = I$ and, therefore:

$$\Phi(x^*) = J(x^*)^{-1} \tag{4.64}$$

Thus, a suitable choice for $\Phi(x)$ would be $\Phi(x) = J(x)^{-1}$, for all x, so that the fixed point method becomes:

$$x^{(k+1)} = g(x^{(k)}) = x^{(k)} - J(x^{(k)})^{-1} f(x^{(k)}) \tag{4.65}$$

which, of course, is simply Newton's method.

Termination   When does one terminate the Newton iterations? It is tempting to terminate the method when the corrections $\delta \triangleq \| x^{(k+1)} - x^{(k)} \|$ become small enough, perhaps smaller than some built-in threshold. But how small is small enough? There is no perfect answer to this question; the answer depends on the particular problem instance. It is good practice to check that the *relative* magnitude $\delta / \| x_{nom} \|$ is small, where x_{nom} is some *nominal* value of x that, ideally, should be representative of "typical" x values in the particular problem instance. Typically, one makes an a priori choice of a threshold value of *relative tolerance*, $\tau_{rel} > 0$, say $\tau_{rel} = 0.1\%$, and terminates the algorithm when $(\delta / \| x_{nom} \|) \leq \tau_{rel}$. It remains to consider what the value of x_{nom} should be and how it should be derived from the problem instance.

One approach, and this is often used, is to set $x_{nom} = x^{(0)}$, so that Newton's method is terminated when $x^{(0)}$ has been "improved" to a high enough degree. Thus, the criterion becomes to check if $(\delta / \| x^{(0)} \|) \leq \tau_{rel}$. But, if $\| x^{(0)} \|$ is very small, then it may be overkill to aim for $\delta \leq \tau_{rel} \| x^{(0)} \|$, so this is often augmented with an *absolute tolerance* threshold, $\tau_{abs} > 0$, and the approach becomes to check if:

$$\| x^{(k+1)} - x^{(k)} \| \leq \tau_{rel} \| x^{(0)} \| + \tau_{abs} \tag{4.66}$$

Furthermore, one should watch out for the following. It is known that, if $J(x^{(k)})$ becomes nearly singular, then the corrections $\delta \triangleq \| x^{(k+1)} - x^{(k)} \|$ approach zero. In this case if one is only monitoring step sizes, then one may falsely conclude that convergence has been achieved. In one dimension, this is the case when the slope of the curve becomes very large so that the tangent is nearly vertical. Thus, one should also check the *residual*, $\| f(x^{(k)}) - 0 \|$, such as by using the additional termination condition:

$$\| f(x^{(k)}) \| \leq \tau_{rel} \| f(x^{(0)}) \| + \tau_{abs} \tag{4.67}$$

Remarks   The above results, relating to the convergence of Newton's method, have some implications for device modeling. Recall that one of the sufficient conditions for convergence of Newton's method is that the Jacobian must be Lipschitz continuous. In turn, as we saw earlier, a sufficient condition for this is that the 2nd partial derivatives $\partial^2 f_k / \partial x_i \partial x_j$ must all be continuous and bounded.

This provides a desirable smoothness property for nonlinear device models in circuit simulation: the 2nd derivatives of the device equations must be continuous and bounded. This can be a significant complication for device modeling, especially at the boundaries between different device operating modes. Nevertheless, this is highly desirable in order to improve the chances of convergence of Newton's method.

Finally, it is useful to reiterate that Newton's method works quite well on *nearly linear* problems, because it finds the zero of a nonsingular affine function in a single iteration.

4.2.5 Quasi-Newton Methods

There are two problem areas with Newton's method that motivate the study of alternative so-called *quasi-Newton methods*, namely:

1. *Locality:*   Newton's method works well if $x^{(0)}$ is close enough to x^* and if $J(x^*)$ is nonsingular. As a result, it is a good *local method*, and should be augmented with a *global strategy* to ensure convergence in the general case.
2. *The Jacobian:*   The Jacobian may not be analytically available, or may be expensive to compute. The Jacobian may also become singular or ill-conditioned, so that finding the correction cannot be done reliably.

To address these difficulties, quasi-Newton methods typically include a *global strategy*, i.e., some means to improve the convergence prospects when far away from the solution, as well as some scheme to *overcome the expense of Jacobian evaluations*. This remains an important research topic, although much has advanced in the field in the last 50 years.

Global Strategies   The vector $s^{(k)} \triangleq x^{(k+1)} - x^{(k)}$ is referred to as the *step* taken in every iteration, and the *full Newton step* is denoted:

$$s_N^{(k)} = -J(x^{(k)})^{-1} f(x^{(k)}) \tag{4.68}$$

A good component of any global strategy is some means to take steps, in the Newton direction, that are shorter than the full Newton step. This is done when far away from the solution, because otherwise one can over-shoot the solution region altogether, or run into overflow problems. When we get closer to the true solution, the full Newton steps are definitely taken, so as to benefit from the quadratic convergence rate. One may refer to these schemes as *stepping strategies*.

In the general literature, outside any specific problem domain, a commonly used, often successful, stepping strategy is to take steps that ensure that the norm $\|f(x)\|_2$ is reduced with every step (iteration). Specifically, a scalar objective function $\psi(x)$ is monitored to ensure that it is reduced as a result of every step, where:

$$\psi(x) = \frac{1}{2} \left\| f(x) \right\|_2^2 = \frac{1}{2} f(x)^T f(x) \tag{4.69}$$

where the $1/2$ is there for algebraic convenience. Clearly $\psi(x) \geq 0$, $\forall x$, and, at the solution x^*, we have $\psi(x^*) = 0$, so that:

$$f(x) = 0 \iff \psi(x) = 0 \tag{4.70}$$

If $J(x)$ is nonsingular, then one can show that the Newton direction is a *descent direction* for $\psi(x)$ but, nevertheless, too large a step size *can* cause $\psi(x)$ to increase. To take shorter steps, there are rules, such as those by Armijo, Goldstein, and Wolfe, that guarantee a decrease of $\psi(x)$ in every step. For additional details, the reader should consult Dennis and Schnabel (1996).

For a given problem domain, such as circuit simulation in our case, better domain-specific techniques for taking shorter steps become possible. Later on, we will see examples of such strategies, which are called *damping* in the circuit simulation area.

Efficient Jacobian Replacements In order to avoid the expensive Jacobian evaluations in practice, one can consider several modifications of Newton's method, such as the following:

- Use a finite-difference approximation to approximate the Jacobian. If the finite-difference step size is properly chosen, this method can be q-quadratically convergent. But this can be expensive in terms of function evaluations and may not be worth the effort.
- Use a Newton-Chord method. Use a fixed matrix, possibly $J(x^{(0)})$, as a fixed approximation for the Jacobian at all future steps. This method is q-linearly convergent.
- Use Broyden's method. This is a generalization/extension of the secant method to the multidimensional case. The data, from all the steps taken so far, is used to construct an approximation to the Jacobian at the present point. This method is q-superlinearly convergent and is also $2n$-step q-quadratic.

Traditionally, these methods are not used in circuit simulation, because the Jacobian can be constructed by inspection, and is not as hard to build as it can be in other disciplines. The true Jacobian, used in Newton's method, offers the benefit of quadratic convergence. Nevertheless, due to the cost of having to refactor the Jacobian in every iteration, modern "fast SPICE" simulators are using some alternative approaches. This is an advanced topic and will not be discussed further here.

4.3   APPLICATION TO CIRCUIT SIMULATION

Armed with problem-specific knowledge, one can often improve on the generic and general formulation of Newton's and quasi-Newton methods. In applying Netwon's and quasi-Newton methods to circuit simulation, we will study three areas of improvement:

1. The Jacobian can be constructed by inspection, by linearizing each nonlinear element around the present operating point.
2. Linearized macromodels can be developed for complex devices with fixed model topology, which helps reduce the size of the MNA system.
3. Stepping strategies, to implement a quasi-Newton approach, can be developed based on knowledge of specific semiconductor device operation.

Such techniques are implemented in all modern circuit simulators.

4.3.1   Linearization and Companion Models

The winning approach to solving nonlinear circuits is to not solve their nonlinear equations directly but, instead, to linearize the circuits, build their corresponding linear equations, solve them, and repeat until convergence. We will see that this approach is the direct result of applying Newton's method to the nonlinear MNA equations of the circuit.

Affine Approximation   We start with some useful definitions and terminology in the one-dimensional case. If $g : \mathbb{R} \to \mathbb{R}$ is differentiable at $x_0 \in \mathbb{R}$, then we define an approximation, which we call the *affine approximation* of g at x_0, as the affine function:

$$\hat{g}_{x_0}(x) \triangleq g'(x_0)(x - x_0) + g(x_0) \qquad (4.71)$$

where we use the familiar notation:

$$g'(x_0) \triangleq \frac{dg}{dx}(x_0) = \frac{dg}{dx}\bigg|_{x_0} \qquad (4.72)$$

Notice that $\hat{g}_{x_0}(x)$ is the unique function that satisfies the two conditions:

$$\hat{g}_{x_0}(x_0) = g(x_0) \qquad \text{and} \qquad \hat{g}'_{x_0}(x) = g'(x_0), \forall x \qquad (4.73)$$

Geometrically, $y = \hat{g}_{x_0}(x)$ is the equation of the tangent line to the graph of the function $y = g(x)$ at x_0. In a neighborhood around x_0, $\hat{g}_{x_0}(x)$ is a useful linear approximation of the function $g(x)$; it is a *linearization* of $g(x)$. The affine approximation is motivated by a Taylor series expansion, in the same way as we saw earlier in connection with Newton's method.

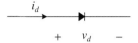

Figure 4.12: A diode.

Companion Model   Let $g(x)$ be the nonlinear function corresponding to a certain nonlinear element e, which can be either a CCS or a CVS, biased at some x_0. We define the *companion model* of e, at x_0, as an equivalent (linear) circuit whose terminal response (either current or voltage) is given by $\hat{g}_{x_0}(x)$. Effectively, the affine approximation is the element equation of the companion model, viewed as a composite element. Note that the companion model is not unique; several circuit realizations can have the same terminal characteristics $\hat{g}_{x_0}(x)$.

For example, consider a diode (effectively, a nonlinear resistor, captured as a CCS), as shown in Fig. 4.12, that has the element equation:

$$i_d = g(v_d) = I_{sat}\left[e^{v_d/\eta V_T} - 1\right] \qquad (4.74)$$

Assuming the diode is biased at $v_d^{(k)}$ and $i_d^{(k)}$, define:

$$G_{eq}^{(k)} \triangleq g'\left(v_d^{(k)}\right) = \left.\frac{di_d}{dv_d}\right|_{v_d^{(k)}} = \frac{I_{sat}}{\eta V_T}e^{v_d^{(k)}/\eta V_T} \qquad (4.75)$$

so that the affine approximation at $v_d^{(k)}$ is:

$$i = \hat{g}_{v_d^{(k)}}(v) = G_{eq}^{(k)}\left(v - v_d^{(k)}\right) + i_d^{(k)}$$

$$= G_{eq}^{(k)}v + I_{eq}^{(k)}$$

where:

$$I_{eq}^{(k)} \triangleq i_d^{(k)} - G_{eq}^{(k)}v_d^{(k)} = I_{sat}\left[e^{v_d^{(k)}/\eta V_T} - 1\right] - \frac{I_{sat}}{\eta V_T}v_d^{(k)}e^{v_d^{(k)}/\eta V_T} \qquad (4.76)$$

Then, the companion model of the diode is as shown in Fig. 4.13.

Affine Approximation in Multiple Dimensions   It will also be useful to define the notion of an affine approximation in multiple dimensions, as follows. If $g : \mathbb{R}^n \to \mathbb{R}^m$ has partial derivatives $\partial g_i/\partial x_j$ at $x^{(k)} \in \mathbb{R}^n$, $\forall i, j$, we define the *affine approximation* of g at $x^{(k)}$, for $x \in \mathbb{R}^n$, as the affine function:

$$\hat{g}_{x^{(k)}}(x) \triangleq J_g(x^{(k)})\left(x - x^{(k)}\right) + g(x^{(k)}) \qquad (4.77)$$

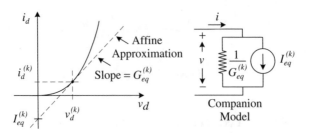

Figure 4.13: Companion model for a diode.

where $J_g(x)$ is the $m \times n$ Jacobian matrix of g:

$$
J_g(x) = \begin{bmatrix}
\dfrac{\partial g_1}{\partial x_1} & \dfrac{\partial g_1}{\partial x_2} & \cdots & \dfrac{\partial g_1}{\partial x_n} \\[2ex]
\dfrac{\partial g_2}{\partial x_1} & \dfrac{\partial g_2}{\partial x_2} & \cdots & \dfrac{\partial g_2}{\partial x_n} \\[2ex]
\vdots & \vdots & & \vdots \\[2ex]
\dfrac{\partial g_m}{\partial x_1} & \dfrac{\partial g_m}{\partial x_2} & \cdots & \dfrac{\partial g_m}{\partial x_n}
\end{bmatrix}
\tag{4.78}
$$

Notice that $\hat{g}_{x^{(k)}}(x)$ is the unique function that satisfies the two conditions:

$$
\hat{g}_{x^{(k)}}(x^{(k)}) = g(x^{(k)}) \quad \text{and} \quad J_{\hat{g}_{x^{(k)}}}(x) = J_g(x^{(k)}), \forall x
\tag{4.79}
$$

If the partial derivatives are continuous at $x^{(k)}$, then $y = \hat{g}_{x^{(k)}}(x)$ is the equation of the tangent hyperplane to the surface $y = g(x)$ at $x^{(k)}$. In this case, $\hat{g}_{x^{(k)}}(\cdot)$ is called the (multidimensional) *derivative* of g, also called the *Fréchet derivative* of g, or the *differential* of g, at $x^{(k)}$. In a neighborhood around $x^{(k)}$, $\hat{g}_{x^{(k)}}(x)$ is a useful linear approximation of the function $g(x)$; it is a *linearization* of $g(x)$. The affine approximation is motivated by a Taylor series expansion, in the same way as we saw earlier in connection with Newton's method.

4.3.2   Some Test Cases

We will consider some instructive test cases that will lead us towards a general solution approach.

Example   Consider the diode circuit shown in Fig. 4.14, where the diode element equation is:

$$
i_d = g(v_d) = I_{sat}\left[e^{v_d/\eta V_T} - 1\right]
\tag{4.80}
$$

Combining KCL, KVL, and the diode element equation leads to:

$$
\frac{v_d}{R_S} + g(v_d) = I_S
\tag{4.81}
$$

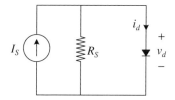

Figure 4.14: A simple diode circuit.

If we let:

$$f(v_d) = \frac{v_d}{R_S} + g(v_d) - I_S \tag{4.82}$$

this leads to a 1-dimensional nonlinear equation to be solved:

$$f(v_d) = 0 \tag{4.83}$$

In order to find the value of v_d that solves the system, we use Newton's method and apply the iteration:

$$v_d^{(k+1)} = v_d^{(k)} - \frac{f(v_d^{(k)})}{f'(v_d^{(k)})} \tag{4.84}$$

where:

$$f(v_d^{(k)}) = \frac{v_d^{(k)}}{R_S} + g(v_d^{(k)}) - I_S \tag{4.85}$$

$$f'(v_d^{(k)}) = \frac{1}{R_S} + g'(v_d^{(k)}) \tag{4.86}$$

This means that we seek a $v_d^{(k+1)}$ such that:

$$f'(v_d^{(k)}) \left(v_d^{(k+1)} - v_d^{(k)} \right) + f(v_d^{(k)}) = 0 \tag{4.87}$$

so that we are effectively solving for a zero of the affine approximation to the circuit function f at $v_d^{(k)}$. This is to be expected, because this is how we derived Newton's method in the first place.

Now, using the affine approximation of the diode at $v_d^{(k)}$, we have:

$$f'(v_d^{(k)}) = \frac{1}{R_S} + G_{eq}^{(k)} \tag{4.88}$$

Thus, the k-th Newton iteration can be written as:

$$\left(\frac{1}{R_S} + G_{eq}^{(k)} \right) \left(v_d^{(k+1)} - v_d^{(k)} \right) + \frac{v_d^{(k)}}{R_S} + g(v_d^{(k)}) - I_S = 0 \tag{4.89}$$

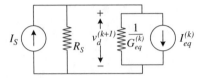

Figure 4.15: A linear circuit replacement that provides the solution for the k-th Newton iteration.

or:

$$\frac{v_d^{(k+1)}}{R_S} + \left[G_{eq}^{(k)} \left(v_d^{(k+1)} - v_d^{(k)} \right) + g(v_d^{(k)}) \right] = I_S \qquad (4.90)$$

which, using the affine approximation of the diode, can be written as:

$$\frac{v_d^{(k+1)}}{R_S} + \hat{g}_{v_d^{(k)}} \left(v_d^{(k+1)} \right) = I_S \qquad (4.91)$$

Comparing this to (4.81), it is clear that the solution of the k-th Newton iteration can be achieved by the following scheme:

1. Replace the diode by its companion model corresponding to $v_d^{(k)}$.
2. Solve the resulting *linear* circuit to find $v_d^{(k+1)}$.

Thus, $v_d^{(k+1)}$ can be obtained as the solution to the *linear* circuit shown in Fig. 4.15. Solving the linear equation resulting from the k-th Newton iteration is *equivalent* to solving this *linear* circuit.

Remarks   The companion model is a *linear circuit*, corresponding to the diode operating point $(v_d^{(k)}, i_d^{(k)})$, which is valid in the k-th Newton iteration. This companion model is an *equivalent circuit*, but only for the purpose of solving the k-th Newton iteration. The companion model represents a *linearization* of the diode equation at the operating point $(v_d^{(k)}, i_d^{(k)})$. Solving the circuit by repeated linearization is not an ad hoc method; we will see that it is valid in general, and not only in this simple case. This, in fact, becomes the *general* strategy for solving nonlinear circuits, to repeatedly linearize all nonlinear elements and solve the resulting linear circuit, until (Newton) convergence has been achieved.

Example   As another example, consider the diode circuit, shown in Fig. 4.16. The MNA equations for this circuit are:

$$\begin{bmatrix} \dfrac{1}{R_S} & 0 \\ 0 & \dfrac{1}{R} \end{bmatrix} \begin{bmatrix} v_1 \\ v_2 \end{bmatrix} + \begin{bmatrix} +1 \\ -1 \end{bmatrix} g(v_1, v_2) = \begin{bmatrix} I_S \\ 0 \end{bmatrix} \qquad (4.92)$$

where:

$$g(v_1, v_2) = i_d = I_{sat} \left[e^{(v_1 - v_2)/\eta V_T} - 1 \right] \qquad (4.93)$$

We introduce vector notation and write the above system equation as:

$$Gv + Hg(v) = s \qquad (4.94)$$

where:

$$v = \begin{bmatrix} v_1 \\ v_2 \end{bmatrix}, \qquad G = \begin{bmatrix} \dfrac{1}{R_S} & 0 \\ 0 & \dfrac{1}{R} \end{bmatrix}, \qquad H = \begin{bmatrix} +1 \\ -1 \end{bmatrix}, \qquad s = \begin{bmatrix} I_S \\ 0 \end{bmatrix}$$

We want to solve the 2-dimensional system $f(v) = 0$, where:

$$f(v) = Gv + Hg(v) - s \qquad (4.95)$$

The k-th Newton iteration consists of finding:

$$v^{(k+1)} = v^{(k)} - J_f(v^{(k)})^{-1} f(v^{(k)}) \qquad (4.96)$$

where J_f is the Jacobian of f, and this leads to:

$$J_f(v^{(k)}) \left(v^{(k+1)} - v^{(k)} \right) + f(v^{(k)}) = 0 \qquad (4.97)$$

so that we seek a zero of the affine approximation to $f(\cdot)$ at $v^{(k)}$. The Jacobian matrix, $J_f(v^{(k)})$ is given by:

$$J_f(v^{(k)}) = \begin{bmatrix} \left.\dfrac{\partial f_1}{\partial v_1}\right|_{v^{(k)}} & \left.\dfrac{\partial f_1}{\partial v_2}\right|_{v^{(k)}} \\ \left.\dfrac{\partial f_2}{\partial v_1}\right|_{v^{(k)}} & \left.\dfrac{\partial f_2}{\partial v_2}\right|_{v^{(k)}} \end{bmatrix} \qquad (4.98)$$

where:

$$\left.\frac{\partial f_1}{\partial v_1}\right|_{v^{(k)}} = \frac{1}{R_S} + \frac{I_{sat}}{\eta V_T} e^{(v_1^{(k)} - v_2^{(k)})/\eta V_T} \qquad (4.99)$$

$$\left.\frac{\partial f_1}{\partial v_2}\right|_{v^{(k)}} = -\frac{I_{sat}}{\eta V_T} e^{(v_1^{(k)} - v_2^{(k)})/\eta V_T} \qquad (4.100)$$

$$\left.\frac{\partial f_2}{\partial v_1}\right|_{v^{(k)}} = -\frac{I_{sat}}{\eta V_T} e^{(v_1^{(k)} - v_2^{(k)})/\eta V_T} \qquad (4.101)$$

$$\left.\frac{\partial f_2}{\partial v_2}\right|_{v^{(k)}} = \frac{1}{R} + \frac{I_{sat}}{\eta V_T} e^{(v_1^{(k)} - v_2^{(k)})/\eta V_T} \qquad (4.102)$$

Recall the affine approximation to the diode, where:

$$G_{eq}^{(k)} \triangleq \left.\frac{d i_d}{d v_d}\right|_{v_d^{(k)}} = \frac{I_{sat}}{\eta V_T} e^{v_d^{(k)}/\eta V_T} = \frac{I_{sat}}{\eta V_T} e^{(v_1^{(k)} - v_2^{(k)})/\eta V_T} \qquad (4.103)$$

so that:

$$J_f(v^{(k)}) = \begin{bmatrix} \frac{1}{R_S} + G_{eq}^{(k)} & -G_{eq}^{(k)} \\ -G_{eq}^{(k)} & \frac{1}{R} + G_{eq}^{(k)} \end{bmatrix} \tag{4.104}$$

$$= G + \begin{bmatrix} +1 \\ -1 \end{bmatrix} \begin{bmatrix} +G_{eq}^{(k)} & -G_{eq}^{(k)} \end{bmatrix} = G + H \begin{bmatrix} +G_{eq}^{(k)} & -G_{eq}^{(k)} \end{bmatrix} \tag{4.105}$$

Thus, the k-th Newton iteration becomes:

$$\left(G + H \begin{bmatrix} +G_{eq}^{(k)} & -G_{eq}^{(k)} \end{bmatrix} \right) \left(v^{(k+1)} - v^{(k)} \right) + G v^{(k)} + H g(v^{(k)}) - s = 0 \tag{4.106}$$

or:

$$G v^{(k+1)} + H \left\{ \begin{bmatrix} +G_{eq}^{(k)} & -G_{eq}^{(k)} \end{bmatrix} \left(v^{(k+1)} - v^{(k)} \right) + g(v^{(k)}) \right\} = s \tag{4.107}$$

Now, notice that the affine approximation to $g(v)$ at $v^{(k)}$ is:

$$\hat{g}_{v^{(k)}}(v) = J_g(v^{(k)}) \left(v - v^{(k)} \right) + g(v^{(k)}) \tag{4.108}$$

where:

$$J_g(v^{(k)}) = \begin{bmatrix} \dfrac{\partial g}{\partial v_1}\Big|_{v^{(k)}} & \dfrac{\partial g}{\partial v_2}\Big|_{v^{(k)}} \end{bmatrix} \tag{4.109}$$

and:

$$\frac{\partial g}{\partial v_1}\Big|_{v^{(k)}} = +\frac{I_{sat}}{\eta V_T} e^{(v_1^{(k)} - v_2^{(k)})/\eta V_T} = +G_{eq}^{(k)} \tag{4.110}$$

$$\frac{\partial g}{\partial v_2}\Big|_{v^{(k)}} = -\frac{I_{sat}}{\eta V_T} e^{(v_1^{(k)} - v_2^{(k)})/\eta V_T} = -G_{eq}^{(k)} \tag{4.111}$$

so that:

$$\hat{g}_{v^{(k)}}(v^{(k+1)}) = \begin{bmatrix} +G_{eq}^{(k)} & -G_{eq}^{(k)} \end{bmatrix} \left(v^{(k+1)} - v^{(k)} \right) + g(v^{(k)}) \tag{4.112}$$

and the k-th Newton iteration becomes:

$$G v^{(k+1)} + H \hat{g}_{v^{(k)}}(v^{(k+1)}) = s \tag{4.113}$$

Comparing this expression to (4.94), it is clear that, again, the k-th Newton iteration may be solved by replacing the diode by its companion model and solving the resulting linear circuit. It will be instructive to actually do this for this simple circuit and see that the final MNA system is the same; we do this next.

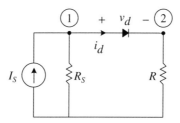

Figure 4.16: Another diode circuit.

Using the form $J_f(v^{(k)})v^{(k+1)} = J_f(v^{(k)})v^{(k)} - f(v^{(k)})$, the Newton iteration can be written as:

$$
\begin{bmatrix} \dfrac{1}{R_S} + G_{eq}^{(k)} & -G_{eq}^{(k)} \\ -G_{eq}^{(k)} & \dfrac{1}{R} + G_{eq}^{(k)} \end{bmatrix} \begin{bmatrix} v_1^{(k+1)} \\ v_2^{(k+1)} \end{bmatrix} = \begin{bmatrix} \dfrac{1}{R_S} + G_{eq}^{(k)} & -G_{eq}^{(k)} \\ -G_{eq}^{(k)} & \dfrac{1}{R} + G_{eq}^{(k)} \end{bmatrix} \begin{bmatrix} v_1^{(k)} \\ v_2^{(k)} \end{bmatrix}
$$
$$
- \begin{bmatrix} \dfrac{v_1^{(k)}}{R_S} + i_d^{(k)} - I_S \\ \dfrac{v_2^{(k)}}{R} - i_d^{(k)} \end{bmatrix}
$$

Then, recall from the diode companion model that:

$$
I_{eq}^{(k)} = i_d^{(k)} - G_{eq}^{(k)} v_d^{(k)} = i_d^{(k)} - G_{eq}^{(k)}\left(v_1^{(k)} - v_2^{(k)}\right)
$$

so that the Newton iteration becomes:

$$
\begin{bmatrix} \dfrac{1}{R_S} + G_{eq}^{(k)} & -G_{eq}^{(k)} \\ -G_{eq}^{(k)} & \dfrac{1}{R} + G_{eq}^{(k)} \end{bmatrix} \begin{bmatrix} v_1^{(k+1)} \\ v_2^{(k+1)} \end{bmatrix} = \begin{bmatrix} I_S - I_{eq}^{(k)} \\ I_{eq}^{(k)} \end{bmatrix} \tag{4.114}
$$

Now, replace the diode by its companion model, to get the circuit shown in Fig. 4.17. The MNA equations for this circuit are:

$$
\begin{bmatrix} \dfrac{1}{R_S} + G_{eq}^{(k)} & -G_{eq}^{(k)} \\ -G_{eq}^{(k)} & \dfrac{1}{R} + G_{eq}^{(k)} \end{bmatrix} \begin{bmatrix} v_1^{(k+1)} \\ v_2^{(k+1)} \end{bmatrix} = \begin{bmatrix} I_S - I_{eq}^{(k)} \\ I_{eq}^{(k)} \end{bmatrix} \tag{4.115}
$$

which are the *same* as those of the k-th Newton iteration (4.114).

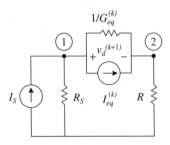

Figure 4.17: A linearization of the circuit in Fig. 4.16, using the diode companion model.

Element Stamps   Finally, we see clearly that the diode has an *element stamp* in the above equation, consisting of:

$$\begin{bmatrix} G_{eq}^{(k)} & -G_{eq}^{(k)} \\ -G_{eq}^{(k)} & G_{eq}^{(k)} \end{bmatrix} \begin{bmatrix} v_1^{(k+1)} \\ v_2^{(k+1)} \end{bmatrix} = \begin{bmatrix} -I_{eq}^{(k)} \\ I_{eq}^{(k)} \end{bmatrix} \tag{4.116}$$

In general, this suggests that we should be able to directly build the equation for the k-th Newton iteration by combining element stamps of the linear elements and element stamps of the linearized elements (companion models). We will see below that this procedure is quite general and applies to all nonlinear circuits and elements.

4.3.3 Generalization

We now consider the general nonlinear MNA equations, to be solved using Newton's method; recall that the MNA system was put in the form:

$$Gx + Hg(x) = s \tag{4.117}$$

where G and H are constant matrices, obtained by inspection of the circuit. We want to solve the system $f(x) = 0$, where:

$$f(x) = Gx + Hg(x) - s \tag{4.118}$$

which we do using Newton's method, based on the iteration:

$$x^{(k+1)} = x^{(k)} - J_f(x^{(k)})^{-1} f(x^{(k)}) \tag{4.119}$$

where J_f is the Jacobian of f, and this leads to:

$$J_f(x^{(k)}) \left(x^{(k+1)} - x^{(k)} \right) + f(x^{(k)}) = 0 \tag{4.120}$$

which shows that we seek a zero of the affine approximation of f at $x^{(k)}$. Because G and H are constant matrices, then the Jacobian of f is:

$$J_f(x) = G + H J_g(x) \tag{4.121}$$

where J_g is the Jacobian of g, so that the k-th Newton iteration becomes:

$$\left(G + HJ_g(x^{(k)})\right)\left(x^{(k+1)} - x^{(k)}\right) + Gx^{(k)} + Hg(x^{(k)}) - s = 0 \tag{4.122}$$

or:

$$Gx^{(k+1)} + H\left\{J_g(x^{(k)})\left(x^{(k+1)} - x^{(k)}\right) + g(x^{(k)})\right\} = s \tag{4.123}$$

but, the term inside the brace ({ }) is the affine approximation of g, so that this becomes:

$$Gx^{(k+1)} + H\hat{g}_{x^{(k)}}(x^{(k+1)}) = s \tag{4.124}$$

which, compared with (4.117), shows that the k-th Newton iteration can be solved if we replace g in the original system by its affine approximation.

The affine approximation of the multidimensional g is achieved if each of its entries is replaced by its affine approximation. And this, in turn, is achieved if we replace every nonlinear element in the original circuit by its companion model. This essentially proves that solving any nonlinear circuit can be done by repeatedly replacing all nonlinear elements by their companion models at $x^{(k)}$ and solving the resulting linear circuit for $x^{(k+1)}$, until convergence.

Assembling the MNA System from Element Stamps

In practice, one does not actually replace any elements in the circuit graph; instead, the required element stamps are entered directly into the linearized system equation. Recall, in the k-th Newton iteration, we want to solve the system:

$$J_f(x^{(k)})\left(x^{(k+1)} - x^{(k)}\right) = -f(x^{(k)}) \tag{4.125}$$

or:

$$J_f(x^{(k)})x^{(k+1)} = J_f(x^{(k)})x^{(k)} - f(x^{(k)}) \tag{4.126}$$

which, using $J_f(x^{(k)}) = [G + HJ_g(x^{(k)})]$ gives:

$$J_f(x^{(k)})x^{(k+1)} = HJ_g(x^{(k)})x^{(k)} - Hg(x^{(k)}) + s \tag{4.127}$$

or:

$$J_f(x^{(k)})x^{(k+1)} = s^{(k)} \tag{4.128}$$

as *the* linear MNA system to be solved. A few remarks are in order in relation to this system:

1. Were it not for the nonlinear elements, the above equation would reduce to the linear case $Gx^{(k+1)} = s$.
2. The matrix $J_f(x^{(k)}) = [G + HJ_g(x^{(k)})]$ on the left-hand side is the key system matrix that we need to build, using element stamps. This matrix is commonly referred to as "the Jacobian" of the system, but we have taken care here to distinguish it from the Jacobian of $g(x)$.

3. The right-hand side (RHS) vector $s^{(k)} \triangleq \left[H J_g(x^{(k)}) x^{(k)} - H g(x^{(k)}) + s \right]$ is another key vector to be built from element stamps.

4. Linear elements have their usual element stamps, and they contribute to the Jacobian $[G + H J_g(x^{(k)})]$ via G, and to the RHS vector via s.

5. Nonlinear elements contribute element stamps to $[G + H J_g(x^{(k)})]$ via $J_g(x^{(k)})$, and to the RHS vector via $J_g(x^{(k)}) x^{(k)} - g(x^{(k)})$.

6. Once the above linear system has been assembled, standard methods (such as GE, LU factorization, etc.) can be used to solve it.

Finally, in order to check whether to terminate the Newton iterations, as we saw earlier, one would need to compute the residual $\| f(x^{(k)}) \|$, from $f(x^{(k)}) = G x^{(k)} + H g(x^{(k)}) - s$. This means that the matrices G and H, and the nonlinear vector $g(x^{(k)})$, must be maintained and evaluated as part of the simulator implementation.

General Companion Models   In general, any two-terminal nonlinear resistive element with a (single, for now) controlling variable, be it a CCS or CVS:

$$ i = g(v) \qquad \text{or} \qquad v = g(i) \qquad (4.129) $$

has an affine approximation in either admittance form (for a CCS) or impedance form (for a CVS):

$$ i = G_{eq}^{(k)} v + I_{eq}^{(k)} \qquad \text{or} \qquad v = R_{eq}^{(k)} i + V_{eq}^{(k)} \qquad (4.130) $$

where, for a CCS, making use of $i^{(k)} = g(v^{(k)})$:

$$ G_{eq}^{(k)} = g'(v^{(k)}) \qquad \text{and} \qquad I_{eq}^{(k)} = g(v^{(k)}) - G_{eq}^{(k)} v^{(k)} \qquad (4.131) $$

and, for a CVS, making use of $v^{(k)} = g(i^{(k)})$:

$$ R_{eq}^{(k)} = g'(i^{(k)}) \qquad \text{and} \qquad V_{eq}^{(k)} = g(i^{(k)}) - R_{eq}^{(k)} i^{(k)} \qquad (4.132) $$

and a corresponding companion model in either a Norton source form or a Thévenin source form. If it has several controlling variables, $x_1, \ldots, x_m$, a CCS $i = g(x)$ can be linearized as:

$$ i = \sum_{j=1}^{m} \frac{\partial g}{\partial x_j}\bigg|_{x^{(k)}} (x_j - x_j^{(k)}) + i^{(k)} \qquad (4.133) $$

$$ = \sum_{j=1}^{m} \frac{\partial g}{\partial x_j}\bigg|_{x^{(k)}} x_j + \left(g(x^{(k)}) - \sum_{j=1}^{m} \frac{\partial g}{\partial x_j}\bigg|_{x^{(k)}} x_j^{(k)} \right) \qquad (4.134) $$

and has a companion model consisting of $m + 1$ current sources in parallel, one of which is independent while the others are each a linear CCS. If it has several controlling variables, $x_1, \ldots, x_m$, a CVS $v = g(x)$ can be linearized as:

$$v = \sum_{j=1}^{m} \frac{\partial g}{\partial x_j}\bigg|_{x^{(k)}} (x_j - x_j^{(k)}) + v^{(k)} \tag{4.135}$$

$$= \sum_{j=1}^{m} \frac{\partial g}{\partial x_j}\bigg|_{x^{(k)}} x_j + \left(g(x^{(k)}) - \sum_{j=1}^{m} \frac{\partial g}{\partial x_j}\bigg|_{x^{(k)}} x_j^{(k)} \right) \tag{4.136}$$

and has a companion model consisting of $m + 1$ voltage sources in series, one of which is independent while the others are each a linear CVS.

Complex Element Stamps For a complex device, such as a transistor or a diode, before an element stamp can be generated, one must have a model in hand, from which to generate the companion model. This is part of device modeling and is separate from simulation *per se*. In general, a complex device can have multiple companion models, depending on the mode of operation. In order to generate the element stamps, one has two options:

1. For each companion model, enter the element stamps due to each of its elements, individually, into the system Jacobian and the RHS vector. For example, for a diode, enter the stamp due to the resistor, then enter the stamp due to the current source.
2. Pre-build the element stamp for the companion model based on a study of its collective contributions to the system Jacobian and the RHS vector. For example, based on what we saw earlier, the complete element stamp for a diode in group 1 is as follows:

$$
\begin{array}{ccccc}
 & v^+ & v^- & | & \text{RHS} \\
 & \vdots & \vdots & | & \vdots \\
n^+ & \cdots \ +G_{eq}^{(k)} \ \cdots & -G_{eq}^{(k)} \ \cdots & | & -I_{eq}^{(k)} \\
 & \vdots & \vdots & | & \vdots \\
n^- & \cdots \ -G_{eq}^{(k)} \ \cdots & +G_{eq}^{(k)} \ \cdots & | & +I_{eq}^{(k)} \\
 & \vdots & \vdots & | & \vdots
\end{array}
\tag{4.137}
$$

For complex nonlinear elements and because the companion model topology is known a priori and fixed, the second option can be faster.

4.3.4 Considerations for Multiterminal Elements

A multiterminal element (MTE) is a device with more than two terminals. At first glance, it may seem that MTEs do not fit within the MNA framework that we have described so far; it would seem there are many questions. For example, how would we determine whether an element is in group 1 or group 2, and how would we assign its terminal equations in the MNA system? However, with only a mild requirement on the *form* in which their terminal characteristics are specified, MTEs can fit quite easily in MNA, as follows.

Assumption 4.1. We assume that the terminals of any MTE are partitioned up-front into two groups, group $\mathcal{I}$ and group $\mathcal{V}$, as shown in Fig. 4.18, such that, if we let vector x consist of the voltage signals at group $\mathcal{I}$ terminals and the current signals into group $\mathcal{V}$ terminals, and let vector y consist of the current signals into group $\mathcal{I}$ terminals and the voltage signals at group $\mathcal{V}$ terminals, then the MTE terminal characteristics can be expressed in the form $y = h(x)$, where $h(\ )$ is some general, possibly nonlinear, vector function.

Most MTEs of interest, e.g., semiconductor devices, can be modeled so that all their terminals are in group $\mathcal{I}$, and they are specified using $i = h(v)$.

Given this general assumption, it is clear that any terminal of an MTE may be viewed as connected internally to a *two terminal*, possibly nonlinear, CCS or CVS, $y_i = h_i(x)$, to ground, as shown in Fig. 4.19. As a result, MTEs bring no new complications and they fit quite naturally into the MNA framework we have developed. Indeed, each $h_i(x)$ terminal function of the MTE becomes a distinct component of the MNA $g(x)$ vector, as part of the nonlinear MNA system:

$$Gx + Hg(x) = s \tag{4.138}$$

For each CCS or CVS corresponding to a terminal of the MTE, we classify it as group 1 or group 2, in the usual way, and proceed with MNA.

Some remarks may be useful at this point. Recall that all nonlinear elements, be they two-terminal elements or MTEs, contribute to the MNA nonlinear vector

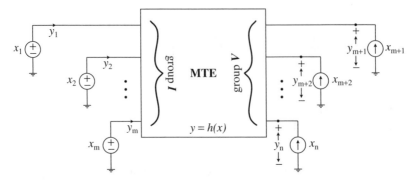

Figure 4.18: A general multiterminal element.

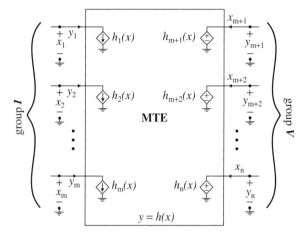

Figure 4.19: An equivalent circuit for a general multiterminal element.

function $g(x)$. This $g(x)$ need not be available as a closed-form algebraic expression. Instead, we only require that $g(x)$ be a well defined function that may be evaluated in some way, such as by means of a software function. As well, $g'(x)$ must be available, so that we can evaluate the companion model. For some elements, $g_i(x)$ or $h(x)$ may be specified by means of an equivalent circuit of linear and/or nonlinear elements. The equivalent circuit might consist of a few two-terminal elements.

4.3.5 Multivariable Differentiation

Fluency in multivariable calculus is useful in order to appreciate some subtle aspects of multiterminal affine approximations and companion models. Thus, we will briefly review differentiation in the multivariable case before proceeding further. The material in this section is based mainly on Bartle (1976) and the user should consult that source for additional background.

Several complications and interesting new features arise in connection with differentiation of functions in $\mathbb{R}^p$, where $p > 1$. This happens mainly because, in multidimensional space, it is possible to approach a point $c \in \mathbb{R}^p$ from "many directions." Recall that, in one dimension, the derivative of a function $f : \mathbb{R} \to \mathbb{R}$ at a point $c \in \mathbb{R}$ is defined as the number $L \in \mathbb{R}$, such that:

$$L = \lim_{x \to c} \frac{f(x) - f(c)}{x - c} \tag{4.139}$$

when the limit exists, and we write $f'(c) = L$. Equivalently, one can show that the derivative could have been defined as the number $L \in \mathbb{R}$, such that:

$$\lim_{x \to c} \frac{\left| f(x) - \left[L(x - c) + f(c) \right] \right|}{|x - c|} = 0 \tag{4.140}$$

This alternate definition makes precise the sense in which we can approximate the values of a function by its affine approximation, defined earlier. It is this alternate approach to the derivative that provides the insight for defining derivatives in multiple dimensions, as we will see. But, first, we will review the generalization of the notion of a *partial derivative*.

Partial Derivative   A point c is said to be an *interior point* of a set D if c belongs to an open set that is a subset of D. Thus, if $c \in \mathbb{R}^p$ is an interior point of $D \subset \mathbb{R}^p$, then, for any $u \in \mathbb{R}^p$, the point $c + tu$ belongs to D for sufficiently small $|t|$, with $t \in \mathbb{R}$.

If $f : \mathbb{R}^p \to \mathbb{R}$, then we are familiar with the notion of a *partial derivative* of $f(x)$ with respect to one of its *variables*, $\partial f / \partial x_i$. We now define the (more general) notion of a *partial derivative* of a function with respect to a *vector*, or in a certain *direction*. Let $f : D \subset \mathbb{R}^p \to \mathbb{R}^q$, and let c be an interior point of D, $t \in \mathbb{R}$, and u be any point in $\mathbb{R}^p$. The *partial derivative of f at c with respect to u* is the vector:

$$L_u = \lim_{t \to 0} \frac{f(c + tu) - f(c)}{t} \tag{4.141}$$

when the limit exists, and we write $f_u(c) = L_u \in \mathbb{R}^q$. Notice that this partial derivative is a *vector*, not a scalar (unless if $q = 1$). If u is a unit vector in $\mathbb{R}^p$, i.e., if $\|u\| = 1$, then $f_u(c)$ is often called the *directional derivative of f at c in the direction u*. If f has range in $\mathbb{R}$, i.e., $q = 1$, and if $e_i \in \mathbb{R}^p$ is such that its i-th entry is 1, while all others are zero, then f_{e_i} is the same as the familiar $\partial f / \partial x_i$. In general, $\partial f_i / \partial x_j$ is the directional derivative of the component function f_i in the direction of a unit vector aligned with the j-th coordinate axes.

We are now ready to define the notion of *derivative* in multidimensions. Let $f : D \subset \mathbb{R}^p \to \mathbb{R}^q$, let c be an interior point of D, and let $x \in D$, then the *derivative of f at c* is the $q \times p$ real matrix L, if it exists, such that:

$$\lim_{x \to c} \frac{\| f(x) - [L(x - c) + f(c)] \|}{\|x - c\|} = 0 \tag{4.142}$$

and we write $f'(c) = L$ and say that f is *differentiable* at c. Notice that this derivative is a *matrix*, not a scalar.

When it exists, this derivative is *unique*. Crucially, x can approach c from "any direction," and the same unique matrix L must satisfy the above limit in all cases; if not, then the derivative at c does not exist. There are other ways to define this same derivative that are more technical and do not rely on the notion of "approach from any direction." However, for our purposes, the above definition is sufficient and attractive, because it highlights the link to the notion of affine approximation. This definition of derivative is called the *Frechét derivative*, also called the *differential*, of f at c, and there are other alternative definitions.

If f is differentiable at c, then one can show that f is continuous at c. Furthermore, if $f : D \subset \mathbb{R}^p \to \mathbb{R}^q$ is differentiable at $c \in D$, then for any vector

$u \in \mathbb{R}^p$, the partial derivative of f at c with respect to u exists and is given by:

$$f_u(c) = f'(c)u \qquad (4.143)$$

(recall, $f'(c)$ is a matrix). Thus, the directional derivative, at c in the direction of u, may be obtained by multiplying the derivative (a matrix) by the unit vector u. The above definition begs the question of whether the matrix $f'(c)$ is simply the Jacobian of f at c. We will now deal with this question.

Partial Derivatives and the Jacobian   If $f(x)$ is differentiable at c, then all the partial derivatives $\partial f_i / \partial x_j$ exist and $f'(c)$ is simply the Jacobian matrix of f evaluated at c, i.e., $J_f(c)$. The converse of this is *not* true; existence of the Jacobian J_f at a point is not sufficient to conclude that f is differentiable at that point! However, it can be shown that if all the partial derivatives *exist* in a neighborhood of c <u>and</u> are *continuous* at c, then f is differentiable at c.

To illustrate these points, consider the function, given in Bartle (1976):

$$f(x, y) = \begin{cases} \dfrac{xy^2}{x^2 + y^2}, & \text{for } (x, y) \neq (0, 0), \\ 0, & \text{for } (x, y) = (0, 0). \end{cases} \qquad (4.144)$$

for which it can be shown, using the definition, that the partial derivative of f at $(0, 0)$ with respect to any vector $u \neq 0$, is given by:

$$f_u(0, 0) = \frac{u_1 u_2^2}{u_1^2 + u_2^2} \qquad (4.145)$$

In particular, if we let $e_1 = \begin{bmatrix} 1 & 0 \end{bmatrix}^T$ and $e_2 = \begin{bmatrix} 0 & 1 \end{bmatrix}^T$, then:

$$\frac{\partial f}{\partial x}(0, 0) = f_{e_1}(0, 0) = 0 \qquad \text{and} \qquad \frac{\partial f}{\partial y}(0, 0) = f_{e_2}(0, 0) = 0$$

thus, all partial derivatives, and the Jacobian, exist at $(0, 0)$; the function is also continuous at $(0, 0)$. However, the function is *not* differentiable at $(0, 0)$, and this is easily proven by contradiction because, if it is differentiable, it would mean that, for any $u \in \mathbb{R}^2$:

$$f_u(0, 0) = f'(0, 0)u = J_f(0, 0)u$$

$$= \frac{\partial f}{\partial x}(0, 0) \cdot u_1 + \frac{\partial f}{\partial y}(0, 0) \cdot u_2 = 0$$

which contradicts the above (4.145). The "problem" may be traced to the fact that the partial derivatives are not continuous at the origin, because, away from the origin:

$$\frac{\partial f}{\partial x}(x, y) = \frac{y^2 \left(y^2 - x^2\right)}{\left(x^2 + y^2\right)^2} \qquad (4.146)$$

so that, for any point along the y-axis, we have that $\partial f / \partial x = 1$, while, at the origin, we already know that $\partial f / \partial x = 0$.

But, what is the significance of the Jacobian and the affine approximation when the function is not differentiable? The answer is, not much. If the partial derivatives exist, but the function is not differentiable, then the Jacobian is simply a matrix of the partial derivatives at that point. The affine approximation *can* be built from the Jacobian, and it *does* represent a hyperplane, but *not* the tangent hyperplane at that point. Indeed, the tangent hyperplane to the function at that point does not exist when the function is not differentiable.

When the tangent hyperplane does not exist, the affine approximation to the function is no longer a unique hyperplane under coordinate transformations. To see this, consider that, for the above function, the affine approximation at the origin is:

$$\hat{f}(x, y) = J_f(0, 0) \left(\begin{bmatrix} x \\ y \end{bmatrix} - \begin{bmatrix} 0 \\ 0 \end{bmatrix} \right) + f(0, 0) \tag{4.147}$$

$$= \begin{bmatrix} 0 & 0 \end{bmatrix} \begin{bmatrix} x \\ y \end{bmatrix} = 0 \tag{4.148}$$

which is the $z = 0$ plane in the 3-dimensional xyz-space. Then, let us introduce a coordinate transformation by rotating the (x, y) axes by $+45°$, so that the new axes are in the directions of the unit vectors:

$$u = \begin{bmatrix} 1/\sqrt{2} \\ 1/\sqrt{2} \end{bmatrix} \quad \text{and} \quad v = \begin{bmatrix} -1/\sqrt{2} \\ 1/\sqrt{2} \end{bmatrix} \tag{4.149}$$

If we denote the coordinates in the new system by $(\tilde{x}, \tilde{y})$, then:

$$\begin{bmatrix} x \\ y \end{bmatrix} = \frac{1}{\sqrt{2}} \begin{bmatrix} +1 & -1 \\ +1 & +1 \end{bmatrix} \begin{bmatrix} \tilde{x} \\ \tilde{y} \end{bmatrix} \tag{4.150}$$

If $\tilde{f}_u(0, 0)$ is the partial derivative in the new system, then:

$$\tilde{f}_{\begin{bmatrix} 1 \\ 0 \end{bmatrix}}(0, 0) = f_u(0, 0) = \frac{1}{2\sqrt{2}} \tag{4.151}$$

and

$$\tilde{f}_{\begin{bmatrix} 0 \\ 1 \end{bmatrix}}(0, 0) = f_v(0, 0) = -\frac{1}{2\sqrt{2}} \tag{4.152}$$

so that the affine approximation in the new system becomes:

$$\hat{f}(\tilde{x}, \tilde{y}) = \begin{bmatrix} \frac{1}{2\sqrt{2}} & -\frac{1}{2\sqrt{2}} \end{bmatrix} \begin{bmatrix} \tilde{x} \\ \tilde{y} \end{bmatrix} \tag{4.153}$$

which is different from the $z = 0$ plane. Thus, where f is not differentiable, the affine approximation depends on the coordinate system, and does not correspond to a unique tangent hyperplane. This point is relevant to linearization of multiterminal elements.

4.3.6 Linearization of Multiterminal Elements

We now return to multiterminal elements (MTE), where there are three issues to discuss: affine approximations, companion models, and element stamps. As we saw earlier, once the element equation is available in the form $y = h(x)$, the affine approximation for it is well defined, provided the Jacobian $J_h(x)$ exists. There are at least two ways to obtain the affine approximation:

1. Perform a Taylor series expansion on the element equation $y = h(x)$ and retain only the first order terms and the constant.
2. Using an equivalent circuit of the MTE, linearize each element in that circuit and solve for the affine terminal characteristics $\hat{h}_i(x)$.

Notice that the existence of a tangent hyperplane, i.e., of the derivative of $h(x)$, does not seem to be required for the above procedure to be applicable; it is sufficient that the partials $\partial h_i/\partial x_j$ exist. However, there are two good reasons to insist on existence of the derivative:

1. If the derivative exists, then the affine approximation becomes unique, independent of exactly *how* it is obtained. One can obtain the affine approximation either by using a Taylor series expansion on $h(x)$, or by linearizing the elements of an equivalent circuit.
2. In order to guarantee local convergence of Newton's method, we need $\partial^2 h_k/\partial x_i \partial x_j$ to exist and be continuous. This, in turn, requires the existence and continuity of $\partial h_i/\partial x_j$ and, therefore, the existence of the tangent hyperplane and the derivative.

Therefore, to ensure existence of the derivative, it is generally required that the element equation $y = h(x)$ have continuous 1st partial derivatives.

Uniqueness of the Affine Approximation   Given that the derivative of $h(\cdot)$ exists, what is it that guarantees that the affine approximation, obtained by the two methods given above, is the same? The answer is: the linearity of the KCL and KVL equations, as follows.

Consider an MTE with a specified element equation $y = h(x)$ and an equivalent circuit consisting of a number of possibly nonlinear two-terminal elements. Construct a circuit around the MTE, in which an independent current source is applied to every terminal in group $\mathcal{V}$, and an independent voltage source is applied to every terminal in group $\mathcal{I}$. Then, the MNA equations for this circuit would effectively provide a mapping from x to y, equivalent to $h(\cdot)$.

Linearizing the MNA equations would yield the same affine approximation as linearizing $h(\cdot)$. But we already know that, due to linearity of KCL and KVL, linearizing the MNA equations is equivalent to linearizing its nonlinear vector $g(x)$. Thus, linearizing the individual elements inside the equivalent circuit is equivalent to inearizing its terminal equations.

Companion Model and Element Stamp   The companion model of an MTE, being a circuit realization of the affine approximation, is well defined and can be obtained in at least two ways:

1. If the affine approximation equations are available in algebraic form, "reverse engineer" a circuit that would realize those equations.
2. If the affine approximation is available as a result of a linearized equivalent circuit, then that circuit is itself the companion model.

Before we can generate the element stamp, we must classify the terminal signals of the MTE as contributing either to the top or bottom parts of the MNA system. Recall, each terminal signal $y_i = h_i(x)$ is either a CCS or a CVS from that terminal to ground. A CVS must belong to group 2, while a CCS may be in either group. Thus, the MTE terminals themselves are assigned as either group 1 (contributing to top part of MNA system) or group 2 (bottom part). Generation of the element stamp can then proceed, in at least two ways:

1. If the affine approximation is available as an equation, read off the stamps from the equation, depending on the terminal group assignments.
2. Using the companion model, use the stamp of each element in it to cumulatively discover the MTE element stamp, using group assignments.

We will illustrate this process in connection with BJTs and MOSFETs.

A BJT Example   Consider an npn BJT with its Ebers-Moll model, shown in Fig. 4.20, with the model equations:

$$i_e = -I_{es}\left(e^{v_{be}/V_{Te}} - 1\right) + a_R I_{cs}\left(e^{v_{bc}/V_{Tc}} - 1\right) \tag{4.154}$$

$$i_c = a_F I_{es}\left(e^{v_{be}/V_{Te}} - 1\right) - I_{cs}\left(e^{v_{bc}/V_{Tc}} - 1\right) \tag{4.155}$$

$$i_b = -(i_e + i_c) \tag{4.156}$$

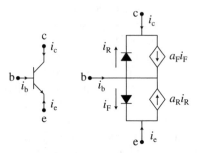

Figure 4.20: The Ebers-Moll model for an npn BJT.

These equations specify the function $y = h(x)$ that relates the MTE terminal currents (vector y) to the terminal voltages (vector x):

$$y = \begin{bmatrix} i_e \\ i_c \\ i_b \end{bmatrix} \qquad x = \begin{bmatrix} v_e \\ v_c \\ v_b \end{bmatrix} \qquad (4.157)$$

using $v_{be} = v_b - v_e$ and $v_{bc} = v_b - v_c$. Alternatively, because the equations are in terms of v_{be} and v_{bc} only, we can define:

$$y = \begin{bmatrix} i_e \\ i_c \\ i_b \end{bmatrix} \qquad x = \begin{bmatrix} v_{be} \\ v_{bc} \end{bmatrix} \qquad y = h(x) \qquad (4.158)$$

We will proceed with this more compact notation, so that the function $h(\cdot)$ is explicitly specified by the above Ebers-Moll equations.

To find the affine approximation, we will linearize the equations $h(x)$ directly, employing the notation introduced in Pillage et al. (1995), and we start by finding the Jacobian $J_h(x)$:

$$\frac{\partial i_e}{\partial v_{be}} = -\frac{I_{es}}{V_{Te}} e^{v_{be}/V_{Te}} \triangleq -g_{ee} \qquad (4.159)$$

$$\frac{\partial i_e}{\partial v_{bc}} = a_R \frac{I_{cs}}{V_{Tc}} e^{v_{bc}/V_{Tc}} \triangleq g_{ec} \qquad (4.160)$$

$$\frac{\partial i_c}{\partial v_{be}} = a_F \frac{I_{es}}{V_{Te}} e^{v_{be}/V_{Te}} \triangleq g_{ce} \qquad (4.161)$$

$$\frac{\partial i_c}{\partial v_{bc}} = -\frac{I_{cs}}{V_{Tc}} e^{v_{bc}/V_{Tc}} \triangleq -g_{cc} \qquad (4.162)$$

$$\frac{\partial i_b}{\partial v_{be}} = g_{ee} - g_{ce} \qquad (4.163)$$

$$\frac{\partial i_b}{\partial v_{bc}} = g_{cc} - g_{ec} \qquad (4.164)$$

Thus, the affine approximation becomes:

$$\begin{bmatrix} \hat{i}_e \\ \hat{i}_c \\ \hat{i}_b \end{bmatrix} = \begin{bmatrix} -g_{ee}^{(k)} & g_{ec}^{(k)} \\ g_{ce}^{(k)} & -g_{cc}^{(k)} \\ g_{ee}^{(k)} - g_{ce}^{(k)} & g_{cc}^{(k)} - g_{ec}^{(k)} \end{bmatrix} \left(\begin{bmatrix} \hat{v}_{be} \\ \hat{v}_{bc} \end{bmatrix} - \begin{bmatrix} v_{be}^{(k)} \\ v_{bc}^{(k)} \end{bmatrix} \right) + \begin{bmatrix} i_e^{(k)} \\ i_c^{(k)} \\ i_b^{(k)} \end{bmatrix}$$

The affine approximation for each terminal signal, $y_i = \hat{h}_i(x)$, may now be written as:

$$\hat{i}_e = -g_{ee}^{(k)} \hat{v}_{be} + g_{ec}^{(k)} \hat{v}_{bc} + I_e^{(k)} \qquad (4.165)$$

$$\hat{i}_c = g_{ce}^{(k)} \hat{v}_{be} - g_{cc}^{(k)} \hat{v}_{bc} + I_c^{(k)} \tag{4.166}$$

$$\hat{i}_b = -(\hat{i}_e + \hat{i}_c) \tag{4.167}$$

where:

$$I_e^{(k)} \triangleq i_e^{(k)} + g_{ee}^{(k)} v_{be}^{(k)} - g_{ec}^{(k)} v_{bc}^{(k)} \tag{4.168}$$

$$I_c^{(k)} \triangleq i_c^{(k)} - g_{ce}^{(k)} v_{be}^{(k)} + g_{cc}^{(k)} v_{bc}^{(k)} \tag{4.169}$$

and where:

$$i_e^{(k)} = -I_{es} \left(e^{v_{be}^{(k)}/V_{Te}} - 1 \right) + a_R I_{cs} \left(e^{v_{bc}^{(k)}/V_{Tc}} - 1 \right) \tag{4.170}$$

$$i_c^{(k)} = a_F I_{es} \left(e^{v_{be}^{(k)}/V_{Te}} - 1 \right) - I_{cs} \left(e^{v_{bc}^{(k)}/V_{Tc}} - 1 \right) \tag{4.171}$$

This leads to the companion model shown in Fig. 4.21, and the BJT element stamp, obtained by combining the stamps of the individual elements in the companion model is as follows:

	e	c	b		RHS
	$\vdots$	$\vdots$	$\vdots$	|	$\vdots$
$e \cdots$	$g_{ee}^{(k)}$	$-g_{ec}^{(k)}$	$(g_{ec}^{(k)} - g_{ee}^{(k)})$	|	$-I_e^{(k)}$
$c \cdots$	$-g_{ce}^{(k)}$	$g_{cc}^{(k)}$	$(g_{ce}^{(k)} - g_{cc}^{(k)})$	|	$-I_c^{(k)}$
$b \cdots$	$(g_{ce}^{(k)} - g_{ee}^{(k)})$	$(g_{ec}^{(k)} - g_{cc}^{(k)})$	$(g_{cc}^{(k)} + g_{ee}^{(k)} - g_{ce}^{(k)} - g_{ec}^{(k)})$	|	$I_e^{(k)} + I_c^{(k)}$

Note that we have assumed that all three terminals of this BJT have been classified in group 1 of the MNA system.

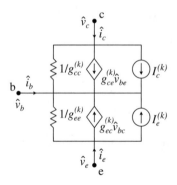

Figure 4.21: A companion model for the npn BJT.

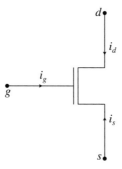

Figure 4.22: An n-channel MOSFET.

A MOSFET Example   Consider an n-channel MOSFET, as shown in Fig. 4.22, with the simple DC model:

$$
i_d = \begin{cases}
0, & \text{if } v_{gs} \leq V_t \text{ (cut-off)}, \\
\beta\left[(v_{gs} - V_t)v_{ds} - \tfrac{1}{2}v_{ds}^2\right], & \text{if } 0 \leq v_{ds} \leq v_{gs} - V_t \text{ (linear)}, \\
\tfrac{\beta}{2}(v_{gs} - V_t)^2(1 + \lambda v_{ds}), & \text{if } 0 \leq v_{gs} - V_t \leq v_{ds} \text{ (saturation)}.
\end{cases}
\tag{4.172}
$$

with:

$$
i_s = -i_d \quad \text{and} \quad i_g = 0 \tag{4.173}
$$

These equations specify the element characteristics $y = h(x)$, where in this case:

$$
y = \begin{bmatrix} i_d \\ i_s \\ i_g \end{bmatrix} \qquad x = \begin{bmatrix} v_{ds} \\ v_{gs} \end{bmatrix} \tag{4.174}
$$

To build the Jacobian $J_h(x)$, we find:

$$
G_{ds} \triangleq \frac{\partial i_d}{\partial v_{ds}} = \begin{cases}
0, & \text{if } v_{gs} \leq V_t \text{ (cut-off)}, \\
\beta\left[v_{gs} - V_t - v_{ds}\right], & \text{if } 0 \leq v_{ds} \leq v_{gs} - V_t \text{ (linear)}, \\
\frac{\beta}{2}\lambda(v_{gs} - V_t)^2, & \text{if } 0 \leq v_{gs} - V_t \leq v_{ds} \text{ (saturation)}.
\end{cases}
$$

$$
g_m \triangleq \frac{\partial i_d}{\partial v_{gs}} = \begin{cases}
0, & \text{if } v_{gs} \leq V_t \text{ (cut-off)}, \\
\beta v_{ds}, & \text{if } 0 \leq v_{ds} \leq v_{gs} - V_t \text{ (linear)}, \\
\beta(v_{gs} - V_t)(1 + \lambda v_{ds}), & \text{if } 0 \leq v_{gs} - V_t \leq v_{ds} \text{ (saturation)}.
\end{cases}
$$

where $G_{ds} > 0$ is the small-signal drain-to-source conductance and $g_m > 0$ is the small-signal transconductance. Thus, the affine approximation becomes:

$$
\begin{bmatrix} \hat{i}_d \\ \hat{i}_s \\ \hat{i}_g \end{bmatrix} = \begin{bmatrix} G_{ds}^{(k)} & g_m^{(k)} \\ -G_{ds}^{(k)} & -g_m^{(k)} \\ 0 & 0 \end{bmatrix} \left(\begin{bmatrix} \hat{v}_{ds} \\ \hat{v}_{gs} \end{bmatrix} - \begin{bmatrix} v_{ds}^{(k)} \\ v_{gs}^{(k)} \end{bmatrix} \right) + \begin{bmatrix} i_d^{(k)} \\ i_s^{(k)} \\ i_g^{(k)} \end{bmatrix}
$$

The affine approximation for each terminal signal, $y_i = \hat{h}_i(x)$, is:

$$\hat{i}_d = G_{ds}^{(k)} \hat{v}_{ds} + g_m^{(k)} \hat{v}_{gs} + I_{eq}^{(k)} \tag{4.175}$$

$$\hat{i}_s = -\hat{i}_d \tag{4.176}$$

$$\hat{i}_g = 0 \tag{4.177}$$

where:

$$I_{eq}^{(k)} \triangleq i_d^{(k)} - G_{ds}^{(k)} v_{ds}^{(k)} - g_m^{(k)} v_{gs}^{(k)} \tag{4.178}$$

and where:

$$i_d^{(k)} = \begin{cases} 0, & \text{if } v_{gs}^{(k)} \leq V_t, \\ \beta \left[(v_{gs}^{(k)} - V_t) v_{ds}^{(k)} - \frac{1}{2} v_{ds}^{(k)2} \right], & \text{if } 0 \leq v_{ds}^{(k)} \leq v_{gs}^{(k)} - V_t, \\ \dfrac{\beta}{2} (v_{gs}^{(k)} - V_t)^2 (1 + \lambda v_{ds}^{(k)}), & \text{if } 0 \leq v_{gs}^{(k)} - V_t \leq v_{ds}^{(k)}. \end{cases} \tag{4.179}$$

From this, the companion model is as shown in Fig. 4.23 and, assuming all terminals are in group 1, the element stamp is:

	d	s	g		RHS
	$\vdots$	$\vdots$	$\vdots$	$\vert$	$\vdots$
$d \cdots$	$G_{ds}^{(k)}$	$-\left(G_{ds}^{(k)} + g_m^{(k)}\right)$	$g_m^{(k)}$	$\vert$	$-I_{eq}^{(k)}$
$s \cdots$	$-G_{ds}^{(k)}$	$\left(G_{ds}^{(k)} + g_m^{(k)}\right)$	$-g_m^{(k)}$	$\vert$	$I_{eq}^{(k)}$
$g \cdots$	0	0	0	$\vert$	0

4.3.7   Elements with Internal Nodes

In general, multiterminal elements can have complex companion models, possibly including several *internal nodes*. The examples we have seen so far have only featured internal elements connected between *terminal nodes*, but no internal nodes.

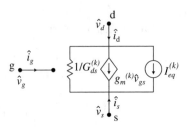

Figure 4.23: A simple companion model for the n-channel MOSFET.

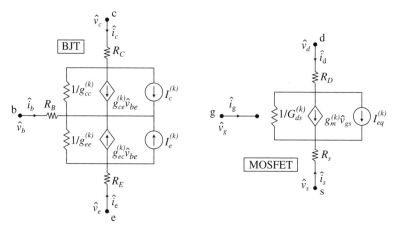

Figure 4.24: More detailed companion models for the BJT and the MOSFET, including series resistance and internal nodes.

However, even for these simple models, the mere addition of series terminal resistance to the model leads to several internal nodes, as shown in Fig. 4.24.

In principle, there is no "problem" with having such internal nodes; they can simply become additional nodes in the MNA system. However, this would drastically increase the number of MNA variables and, therefore, the size of the MNA system. We can reap *significant* savings if we are able to somehow avoid having to explicitly account for these nodes as additional MNA variables. Luckily, there is a way to do this, by benefiting from the fact that the companion model *topology* is *fixed* for any given device, as follows.

Because the companion model topology is fixed, we can perform *symbolic node elimination* on all internal nodes of the model. We demonstrate this for the simple case of a diode with series resistance, which can be modeled as a resistor in series with an ideal diode, as shown in Fig. 4.25, where we refer to a diode without any series resistance as an *ideal* diode. With v_d as the internal (ideal) diode voltage, the element equation is the (system of) two equations:

$$i = I_{sat}\left[e^{v_d/\eta V_T} - 1\right] \tag{4.180}$$

$$v_d = v - Ri \tag{4.181}$$

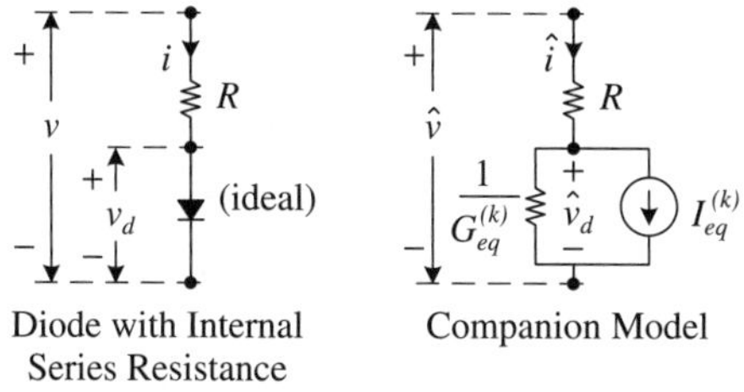

Diode with Internal
Series Resistance   Companion Model

Figure 4.25: A diode with series resistance and its companion model.

The companion model parameters are, for the ideal diode portion:

$$G_{eq}^{(k)} = \frac{I_{sat}}{\eta V_T} e^{v_d^{(k)}/\eta V_T} \tag{4.182}$$

$$I_{eq}^{(k)} = i^{(k)} - G_{eq}^{(k)} v_d^{(k)} = I_{sat}\left[e^{v_d^{(k)}/\eta V_T} - 1\right] - G_{eq}^{(k)} v_d^{(k)} \tag{4.183}$$

both of which depend on $v_d^{(k)}$, and it has terminal characteristics that are captured by the (system of) two equations:

$$\hat{i} = I_{eq}^{(k)} + G_{eq}^{(k)} \hat{v}_d \tag{4.184}$$

$$\hat{v}_d = \hat{v} - R\hat{i} \tag{4.185}$$

both of which are *linear* equations, so that we can use variable elimination in order to achieve node elimination, as follows. In order to eliminate the internal node from the circuit, we eliminate the variable $\hat{v}_d$ from the equations, by plugging (4.185) into (4.184), so that:

$$\hat{i} = I_{eq}^{(k)} + G_{eq}^{(k)}\left(\hat{v} - R\hat{i}\right) \tag{4.186}$$

which leads to:

$$\hat{i} = \frac{I_{eq}^{(k)}}{1 + G_{eq}^{(k)} R} + \frac{\hat{v}}{R + \frac{1}{G_{eq}^{(k)}}} \tag{4.187}$$

As a result, we can use the new companion model for the diode, shown in Fig. 4.26, which does *not* include an internal node, and in which:

$$G_{eq}^{(k)} = \frac{I_{sat}}{\eta V_T} e^{v_d^{(k)}/\eta V_T}$$

$$I_{eq}^{(k)} = i^{(k)} - G_{eq}^{(k)} v_d^{(k)}$$

$$i^{(k)} = I_{sat}\left[e^{v_d^{(k)}/\eta V_T} - 1\right]$$

Notice that the evaluation of these model parameters requires the internal (eliminated) variable $v_d^{(k)}$. This issue is dealt with as follows:

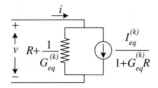

Figure 4.26: A transformed companion model for the diode, in which the internal node has been eliminated.

1. Given the companion model based on the k-th solution $(v^{(k)}, v_d^{(k)})$, solve the MNA system and find the $(k + 1)$-st solution $v^{(k+1)}$.

2. Recover the value of the eliminated variable $\hat{v}_d$, at the $(k + 1)$-st solution point, by combining (4.185) and (4.187), using $\hat{v} = v^{(k+1)}$, so that:

$$v_d^{(k+1)} = v^{(k+1)} - R \left[\frac{I_{eq}^{(k)}}{1 + G_{eq}^{(k)} R} + \frac{v^{(k+1)}}{R + \frac{1}{G_{eq}^{(k)}}} \right] \tag{4.188}$$

3. Use this to build the companion model at the $(k + 1)$-st solution point, and this process is repeated until convergence.

How do we start this process? In other words, how do we set the value of the very first $v_d^{(k)} = v_d^{(0)}$? It is typical to use the initialization $v_d^{(0)} = v^{(0)}$.

If internal node elimination is applied to every diode, the size of the MNA system is reduced by one variable for every diode, which is obviously desirable. For transistors, the gains are more significant; thus, this is a winning strategy and is always employed in practice.

Systematic Internal Nodes Elimination   McCalla (1988) gives the lengthy procedure by which internal node elimination can be done for the BJT Ebers-Moll model with series resistance. A sequence of Y-Δ and Thévenin-Norton transformations are applied to remove all internal nodes. For a fresh new device model, one would have to look for such smart combinations of just the right transformations to get the job done. Alternatively, as we now propose, it is possible to describe a systematic procedure for achieving this in the general case.

Recall, node elimination in the circuit graph is identical to Gaussian elimination of the corresponding variable in the system equations. To see how this works in the simple case of the diode, we apply signals to the diode according to the bias configuration we saw earlier for MTEs:

1. Drive the group $\mathcal{I}$ terminals by voltage sources and the group $\mathcal{V}$ terminals by current sources.

2. Monitor the currents in the group $\mathcal{I}$ terminals and the voltages on the group $\mathcal{V}$ terminals.

For the diode, this bias configuration leads to the circuit shown in Fig. 4.27. The MNA equations for this circuit capture the companion model terminal characteristics $i = \hat{h}(v)$ and lead to:

$$\frac{v_d}{R} + i = \frac{v}{R} \tag{4.189}$$

$$\left(G_{eq}^{(k)} + \frac{1}{R} \right) v_d = \frac{v}{R} - I_{eq}^{(k)} \tag{4.190}$$

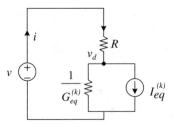

Figure 4.27: A diode with series resistance, biased in the standard bias arrangement for multiterminal elements.

We can eliminate variable v_d by multiplying the 1st equation (4.189) by the term $(1 + G_{eq}^{(k)} R)$ and subtracting from the 2nd (4.190), which gives:

$$-\left(1 + G_{eq}^{(k)} R\right) i = -G_{eq}^{(k)} v - I_{eq}^{(k)} \qquad (4.191)$$

or:

$$i = \frac{I_{eq}^{(k)}}{1 + G_{eq}^{(k)} R} + \frac{v}{R + \frac{1}{G_{eq}^{(k)}}} \qquad (4.192)$$

which is the same result we saw above. Thus, the general strategy becomes to symbolically eliminate all internal nodes from the MNA system, under the appropriate bias configuration.

Symbolic Block-GE for Node Elimination   In the general case, under the appropriate bias configuration for the companion model of a given MTE, we write the MNA equations in the form:

$$\begin{bmatrix} A_{11}^{(k)} & A_{12}^{(k)} \\ A_{21}^{(k)} & A_{22}^{(k)} \end{bmatrix} \begin{bmatrix} z \\ y \end{bmatrix} = \begin{bmatrix} x_1 \\ x_2 \end{bmatrix} + \begin{bmatrix} s_1^{(k)} \\ s_2^{(k)} \end{bmatrix} \qquad (4.193)$$

where $A_{ij}^{(k)}$ and $s_i^{(k)}$ depend on companion model parameters, z is the vector of the internal variables to be eliminated, y is the vector of the external variables to be kept, and x_1 and x_2 are vectors that contain the applied external bias. Note that z can consist of both node voltages and/or branch currents. Our purpose is to eliminate the variables in z so as to reduce the system to a an affine functional relationship $y = \hat{h}(x_1, x_2)$. This can be easily done using block-GE, which we studied earlier, and leads to:

$$y = \left(A_{22}^{(k)} - A_{21}^{(k)} A_{11}^{(k)^{-1}} A_{12}^{(k)} \right)^{-1} \left[x_2 + s_2^{(k)} - A_{21}^{(k)} A_{11}^{(k)^{-1}} \left(x_1 + s_1^{(k)} \right) \right]$$

The required matrix inversions would have to be done symbolically, up-front, possibly using determinants and Cramer's rule, as required. Note, these matrices would typically be small, probably no larger than 4×4 or similar; it may be laborious but is worth the effort. Once this is done, we have a resulting expression for the terminal responses y as a function of the applied terminal bias x, similar to (4.187) of the diode. We use this to build a new companion model that contains no internal nodes, in much the same way as we did for the diode.

The eliminated internal variables may need to be computed so as to find the parameters of the companion model, as we saw with the $v_d^{(k)}$ variable in the case of the diode, using:

$$z^{(k+1)} = A_{11}^{(k)-1} \left(x_1^{(k+1)} + s_1^{(k)} - A_{12}^{(k)} y^{(k+1)} \right) \qquad (4.194)$$

and initialization may be done using some $z^{(0)}$, either arbitrarily set, or derived in some simple way from $x^{(0)}$ or $y^{(0)}$, as appropriate for that device. The lengthy symbolic expressions that result from the above elimination become part of the *model evaluation routines* in a circuit simulator.

4.4   QUASI-NEWTON METHODS IN SIMULATION

As mentioned earlier, Newton's method is a good *local* method, and must be augmented with some *global* strategy to ensure convergence in general. Methods that implement such global convergence strategies are referred to as *quasi-Newton methods*. In circuits, convergence problems may be classified as two types:

1. Non-convergence due to having taken too large a Newton step, which can arise both during DC Analysis and during Transient Analysis.
2. Non-convergence due to having used an initial solution that is too far away from the true solution, which mainly arises during DC Analysis.

The second type of problem (being too far away to begin with) is much more problematic and requires more drastic measures in practice. Correspondingly, solutions for non-convergence in circuits fall into two classes:

1. Step-size limiting methods to ensure that dangerously large steps are avoided, sometimes referred to as *damping* in the circuits literature.
2. Continuation, homotopy, and related heuristic methods to overcome the severe problems that arise during DC Analysis.

These classes are in fact orthogonal, in the sense that one can use step-size limiting schemes as part of an implementation of, say, a homotopy approach. We will now describe these two classes of solutions.

4.4.1 Damping Methods

We first consider the problems arising from having taken too large a step in Newton's method for solving $f(x) = 0$. These problems arise both in DC Analysis and in Transient Analysis. As we saw earlier, a general stepping strategy, outside any specific application domain, is to ensure that the norm of the function f is reduced in every Newton iteration. A potential step that is found to increase the norm is rejected. Instead, a shorter step is taken, typically in the same direction. Such general strategies were tested in circuit simulation in the early 1970s. No problem-specific knowledge was used; circuit equations were simply treated as a set of general nonlinear equations. Other approaches were also studied, which used problem-specific knowledge to tailor the strategies to circuits. It was found that, although both approaches lead to improved convergence, the circuit-specific techniques work faster in practice. We will focus on the circuit-specific techniques.

Early on, these techniques were focused on the key problem of *overflow* that results from the exponential characteristics of pn-junctions. If too large a step is taken, so that the diode voltage $v_d^{(k+1)}$ is large, then the corresponding diode current can become larger than the machine limits. Of course, overflow is a generic numerical algorithms issue, irrespective of whether one is solving linear or nonlinear equations. However, in the context of Newton's method, the steep exponentials of diodes and BJTs can easily cause overflow that leads to divergence of the iterations. In the late 1970s, those techniques that had been originally developed for pn-junctions were then adapted to general circuits, as we will see below.

Overflow in pn-Junctions Overflow can arise due to the exponential diode characteristic, and appropriate circuit-based solutions have been devised for this circuit-specific problem. Large voltage steps and the corresponding extremely large currents are only temporary values that arise during the iterative Newton's method. As such, they are sometimes referred to as being "non-physical." Being "non-physical," one possible remedy is to take some "liberties" with the characteristics and apply some approximation. Generally, one introduces gentler slopes as approximations to the nonlinear exponential characteristics in steep physically-unrealistic regions.

For example, there is a difficulty with pn-junctions that arises in the 3rd quadrant, i.e., where $v_d < 0$ and $i_d < 0$. With the current nearly constant at I_{sat} in that region, the tangent is nearly horizontal and the conductance in the diode companion model becomes nearly zero, which can cause the Jacobian to be nearly singular. One way to avoid this problem is to use a secant (line through the origin) instead of a tangent, as explained in Pillage et al. (1995). Effectively, an alternative companion model is used for the diode when $v_d < 0$, consisting of only an equivalent conductance:

$$G_{eq}^{(k)} = \frac{i_d^{(k)}}{v_d^{(k)}} \qquad (4.195)$$

Such special case remedies can be devised for different situations. However, the common and more general approach is to use a step limiting algorithm, in which the stepping strategy is circuit-motivated, as we will see below. Limiting the step size is sometimes called *damping* in the circuits literature. It is a strategy that aids convergence, but at the cost of additional Newton iterations.

The essence of the problem is that a large positive voltage excursion in $v_d^{(k+1)}$ produces overflow in $i_d^{(k+1)}$. To properly address this issue, we introduce some notation, relative to the k-th Newton iteration:

- Let $v_d^{(k)}$ be the present value of the diode voltage.
- Let $\hat{v}_d$ be the candidate next value of the diode voltage, resulting from solving the linearized MNA system using the companion models.
- Let $s_N^{(k)} = \hat{v}_d - v_d^{(k)}$ be the full Newton step in v_d.
- Let $v_d^{(k+1)}$ be the actual value chosen for the updated solution.
- Let $s^{(k)} = v_d^{(k+1)} - v_d^{(k)}$ be the actual step taken.

Several damping techniques have been proposed and tested to address the overflow problem, as described in McCalla (1988). In one strategy, for diode voltages $v_d^{(k)}$ greater than $10V_T$, the excursion is limited to $v_d^{(k)} \pm 2V_T$. This was used in the CANCER program by Nagel and Rohrer (1971). Another method, implemented in the SLIC program by Idleman et al. (1971), uses the hyperbolic tangent to limit the step size:

$$v_d^{(k+1)} = v_d^{(k)} + 10V_T \tanh\left(\frac{\hat{v}_d - v_d^{(k)}}{10V_T}\right) \tag{4.196}$$

which limits the maximum excursion to $v_d^{(k)} \pm 10V_T$, and can be written more compactly as:

$$s^{(k)} = 10V_T \tanh\left(\frac{s_N^{(k)}}{10V_T}\right) \tag{4.197}$$

However, a more popular method, used in typical SPICE simulators, is motivated by the so-called *current/voltage iteration scheme*, which we now describe.

Current/Voltage Iteration Scheme   Normally, once the linearized circuit is solved for $\hat{v}_d$, we set the next voltage value as $v_d^{(k+1)} = \hat{v}_d$ and proceed with Newton's method. The next iteration uses the diode equation to find $i_d^{(k+1)}$ based on $v_d^{(k+1)}$:

$$i_d^{(k+1)} = I_{sat}\left(e^{v_d^{(k+1)}/\eta V_T} - 1\right) \tag{4.198}$$

in order to build the companion model for that iteration. Meanwhile, the total companion model current, $\hat{i}_d$, is not computed because, typically, v_d is the MNA system variable, not i_d. As described above, this standard approach can run into problems, because the current value $i_d^{(k+1)}$ often ends up being extremely large, causing overflow and subsequent non-convergence. The problem is that the full Newton step is too large. Instead, to overcome this problem, the step size is checked and a shorter step is possibly taken, based on a pre-set "critical voltage" value, $v_{crit} > 0$, such as by the following process:

1. If $\hat{v}_d \leq v_{crit}$, then we accept the full Newton step as is. Otherwise, we want to choose a step size that limits the diode current, as follows.
2. Compute the missing $\hat{i}_d = G_{eq}^{(k)} \hat{v}_d + I_{eq}^{(k)}$ from the companion model.
3. Since the last known good (not overflowed) value of current is $\hat{i}_d$, then set $i_d^{(k+1)} = \hat{i}_d$, and use the diode equation to find $v_d^{(k+1)}$, discarding $\hat{v}_d$:

$$v_d^{(k+1)} = \eta V_T \ln\left(1 + \frac{i_d^{(k+1)}}{I_{sat}}\right) \tag{4.199}$$

As a result, the new diode voltage can be shown to be:

$$v_d^{(k+1)} = v_d^{(k)} + \eta V_T \ln\left(1 + \frac{\hat{v}_d - v_d^{(k)}}{\eta V_t}\right) \tag{4.200}$$

or:

$$\frac{s^{(k)}}{\eta V_T} = \ln\left(1 + \frac{s_N^{(k)}}{\eta V_T}\right) \tag{4.201}$$

This is effectively a *logarithmic damping scheme* which, due to the logarithm, weakly limits small steps and strongly limits large steps. Effectively, we have scaled back the step size so that the new operating point has $i_d^{(k+1)} = \hat{i}_d$, instead of the usual $v_d^{(k+1)} = \hat{v}_d$. When we do this, we are said to be (in the circuits literature) using a "current iteration" instead of a "voltage iteration."

But this raises the question: if a current iteration is better than a voltage iteration, shouldn't we always use it, even for small or negative $\hat{v}_d$? The answer is "no" because, for $v_d \ll 0$, we have the reverse situation: a small current change can produce a large voltage change, due to:

$$\frac{dv_d}{di_d} = \frac{\eta V_T}{I_{sat}} e^{-v_d/\eta V_T} \tag{4.202}$$

Instead, a current iteration should be used when $v_d > v_{crit} > 0$ and a voltage iteration used when $v_d < v_{crit}$.

What is a good value of v_{crit}? Nagel (1975) reports that empirical studies have shown that a good value to use is the voltage where the diode characteristic has the smallest *radius of curvature*, so that:

$$v_{crit} = \eta V_T \ln\left(\frac{\eta V_T}{\sqrt{2} I_{sat}}\right) \approx 0.734 \text{ V} \tag{4.203}$$

where we have used values of $V_T = 26 \text{ mV}$, $\eta = 1$, and $I_{sat} = 10^{-14} \text{ A}$.

Radius of Curvature   Intuitively, *curvature* is the extent to which a geometrical object deviates from "flatness," and the *radius of curvature* is the reciprocal of curvature. For a straight line, the curvature is 0 and the radius of curvature is ∞. For a circle, the radius of curvature is equal to the circle's radius. In general, the radius of curvature is the radius of the circle that best approximates the curve in a neighborhood around a point on that curve. For a general curve in the plane, defined by the mapping $x \rightarrow y$, it can be shown that the radius of curvature at x is given by:

$$r(x) = \frac{1}{\left|\frac{d^2y}{dx^2}\right|} \left| \left[1 + \left(\frac{dy}{dx}\right)^2\right]^{3/2} \right| \tag{4.204}$$

Generalized Damping Schemes   The above current/voltage iteration scheme was derived for pn-junctions and has been used in SPICE for both diode and BJT circuits. It has been adapted to other devices as well, and has been extended to the multivariable case, as in Ho et al. (1977). Basically, viewed as a formula for general logarithmic damping, this scheme leads to a general-purpose step limiting scheme, used in SPICE, based on:

$$s^{(k)} = \frac{\gamma}{k}\mathrm{sgn}(s_N^{(k)}) \ln\left(1 + k\left|s_N^{(k)}\right|\right) \tag{4.205}$$

where sgn $(a) \in \{-1, +1\}$ gives the *sign* of $a \in \mathbb{R}$, and $s^{(k)}$ and $s_N^{(k)}$ are the steps in a component of the solution vector, so that for the i-th component:

$$s_N^{(k)} = \hat{x}_i - x_i^{(k)} \quad \text{and} \quad s^{(k)} = x_i^{(k+1)} - x_i^{(k)} \tag{4.206}$$

and where $k > 0$ replaces the constant $1/\eta V_T$, which becomes meaningless in the general case, and $1 < \gamma < 1.5$ is an empirical constant.

The parameters k and γ control the rate of step limiting. Larger k values more severely limit the step size, and good values of k have been sought in empirical studies. For BJTs, k should be in the range 12–20, while MOSFETs work best with a k in the range 2–5. BJTs, with their exponential characteristics, present more difficulties in Newton convergence than MOSFETs, which follow a square law. A value of $\gamma = 1.3$ seems to work well in practice, at least for BJTs. It is also possible to vary k during the Newton iterative process, typically decreasing it linearly with the iteration count. Stricter limits on step size are applied early in the process, when we would probably be still far from the true solution. The settings and variations of these parameters in commercial simulators are heuristic measures, based on practical experience.

In extending this scheme to multiple dimensions, one has two options:

1. Keep the step (the vector) in the same (Newton) direction. Thus, apply the same damping factor to the whole solution vector, so that $s^{(k)}/s_N^{(k)}$ is the same for all components of the vector. In this case, the damping factor $\lambda \triangleq s^{(k)}/s_N^{(k)}$ is based on the most problematic device in the circuit.

2. Apply a different damping factor to each component of the solution vector, thus using a direction possibly different from the Newton direction.

Strictly speaking, keeping to the Newton direction is a good strategy because it is known to be a descent direction for the norm of the cost function. However, typical SPICE implementations actually adopt the second option, where each component of the vector is scaled individually. Indeed, typical SPICE implementations often build the step limiting code into the device model evaluation routines, so it becomes device-specific.

4.4.2 Overview of More General Methods

The second class of convergence problems, due to having started with an initial solution which is too far from the true solution, arise mainly during DC Analysis. Commercial simulators use a battery of techniques to combat this class of problems, which generally fall under the heading of *continuation methods* or *homotopy methods*. We will study three popular methods in this class: *source stepping*, *Gmin stepping*, and *pseudo-transient*.

These methods include a Newton loop at their core, and thus they can all make use of the step-limiting strategies covered above. In addition, they employ various techniques that "wrap around" the Newton loop and make use of it, as we will see in the following. Note that any user-provided initial voltage specifications for certain nodes can be incorporated in the network as additional independent voltage sources.

Continuation and Homotopy Methods   In general, a *continuation method* is an algorithm for solving the *parameterized* nonlinear system $h(x, \lambda) = 0$, where x is a vector and λ is a scalar. A specific type of continuation method is concerned with the case where the parameterization represents a *homotopy*: two continuous functions are called *homotopic* if one of them can be "continuously deformed" into the other, and such a deformation is called a *homotopy* between the two functions. Formally, one uses a scalar parameter $\lambda \in [0, 1]$ that serves to deform a function $f(x)$ into $g(x)$ and we define a *homotopy* between f and g as the continuous function $h(x, \lambda)$, such that:

$$h(x, 0) = f(x), \forall x \qquad \text{and} \qquad h(x, 1) = g(x), \forall x$$

For the solution of the nonlinear algebraic system $f(x) = 0$, with a solution x^*, homotopy is applied by constructing, for some *known/given* x_0, a function:

$$h(x, \lambda) = f(x) - (1 - \lambda)f(x_0) \tag{4.207}$$

so that, for $\lambda = 0$, the equation $h(x, \lambda) = 0$ has the known solution x_0 and, for $\lambda = 1$, the equation $h(x, \lambda) = 0$ has the sought solution x^*. In other words:

$$h(x_0, 0) = 0 \qquad \text{and} \qquad h(x^*, 1) = 0 \tag{4.208}$$

The solution approach becomes to start with $\lambda = 0$ and the known solution x_0, then to gradually increase λ, while *tracking* the solution from x_0 to x^*. Even if x_0 is far from x^*, one hopes that the gradual steps in λ would take us reliably from x_0 to x^*. The approach is not without its pitfalls and complexities, as there are issues of step size selection, bifurcations, and multiple solutions.

4.4.3   Source Stepping

While not the most efficient, the following scheme is simple to implement and quite popular; it is the fall-back approach used in SPICE when direct DC Analysis or DC sweep, using a plain Newton's method, fails to converge. Recall, the non-linear MNA equations are:

$$Gx + Hg(x) = s \qquad (4.209)$$

where the RHS vector s contains all (and only) the contributions from the independent sources; it is called the "source vector." Most circuits have the useful property that if all independent sources are turned "off," then all voltages and currents in the circuit become zero. Thus, when $s = 0$, $x_0 = 0$ is a solution of the system. Note, "turning off" a voltage source means setting its *voltage* to zero, while for a current source it means setting its *current* to zero.

For circuits with this property, *source stepping* consists of first turning off all independent sources, so that $x = 0$ becomes a solution, then ramping them up to their full value while solving the DC system at every point. Solving the DC circuit may be done using a plain Newton's method, with the solution at the previous source setting as an initial solution.

By using small enough increments (steps) in the source values, it is hoped that Newton convergence will be achieved at every point. If it fails to converge, one can retry with a smaller step size. Alternatively, one can try "bringing up" the different sources separately, or at different rates, etc. There are many such variations in the literature. In the following, we will see that a) in fact *any* circuit can be transformed so as to have the above property, and b) that source stepping is a standard homotopy type method.

Offsets   Considering the nonlinear DC MNA system (4.209), let $f(x) \triangleq Gx + Hg(x) - s$, as usual, so that the system to be solved is in the standard form:

$$f(x) = 0 \qquad (4.210)$$

When $x = 0$, this system evaluates to:

$$f(0) = Hg(0) - s \qquad (4.211)$$

Thus, if $s = 0$ <u>and</u> if $g(0) = 0$, then $x_0 = 0$ is a valid solution of the system. Therefore, in order for $x_0 = 0$ to be a valid solution when $s = 0$, we require that every nonlinear CVS or CCS must evaluate to zero when all its controlling variables (MNA voltages and/or currents) are set to zero ($g_i(0) = 0$). For the i-th nonlinear element, we will refer to $g_i(0)$ as it's *offset*, and we will write:

$$g(x) = g(0) + \overline{g}(x) \tag{4.212}$$

where, clearly, $\overline{g}(0) = 0$. It is a simple matter to pre-process the circuit so that the offset of every nonlinear source is made explicit, as an independent source. With this, effectively, the MNA system becomes:

$$Gx + H\overline{g}(x) = s - Hg(0) \tag{4.213}$$

where the source vector s has been augmented with all the new independent sources (the offsets) $g(0)$, and all nonlinear elements now have zero offsets. Going further, we denote $\overline{s} \triangleq s - Hg(0)$, and write:

$$Gx + H\overline{g}(x) = \overline{s} \tag{4.214}$$

as the new system to be solved. This system is such that, if $\overline{s} = 0$ then $x = 0$ is a valid solution. Thus, for *any* circuit, if the independent sources <u>and</u> the offsets are turned off, then $x = 0$ becomes a valid solution of the MNA system.

Source Stepping as Homotopy   With $f(x) = Gx + H\overline{g}(x) - \overline{s}$, we want a solution x^* for $f(x) = 0$, and with $x_0 = 0$ we know that $f(x_0) = -\overline{s}$. To solve this system, we construct the standard homotopy:

$$h(x, \lambda) = f(x) - (1 - \lambda)f(x_0) \tag{4.215}$$

$$= Gx + H\overline{g}(x) - \lambda\overline{s} \tag{4.216}$$

so that, by solving $h(x, \lambda) = 0$ while stepping λ from 0 to 1, we are effectively solving $f(x) = 0$ while stepping the sources in $\overline{s}$ from 0 to their full values. In practice, most nonlinear elements have zero offsets but, as we have seen, even if the offsets are non-zero, source stepping can be easily applied. Effectively, source stepping is a DC Sweep that simultaneously sweeps *all* the independent sources (and offsets) in the network. It takes more iterations than solving the problem directly using a plain Newton's method, but has more reliable convergence.

Source stepping is not guaranteed to converge, and much theoretical work has been done on this and other aspects of homotopy methods. In practice, there is a need for many heuristics to guide the solution, and complex theoretical techniques may not always be the best choice. The key reason for this is the severe nonlinearity that one sees at the switching threshold in logic circuits. This threshold is the voltage value at the input of a logic gate at which the gate output makes a transition from one logic value to another. At that point, the high gain of fast logic circuits translates to a very steep response surface in $h(x, \lambda)$, which causes convergence problems. While there are much "fancier" methods, such simple homotopy methods, as source stepping, are often employed in simulation, typically with many heuristics.

If source stepping fails to converge, there are other options, other homotopies, such as Gmin stepping and pseudo-transient, which we explore next.

4.4.4 Gmin Stepping

Another homotopy, called *Gmin stepping* is as follows:

1. A large conductance (called Gmin), such that, say $100\,\Omega = 1/G_{\min}$, is connected from every node to ground. These high conductances "swamp" any large resistance in the elements, so that the circuit solution x_0 has every node voltage at very close to 0. This solution is easily found using a plain Newton's method, starting with an initial solution of $x = 0$.
2. Then, the Gmin value is *stepped down* in small increments to some very small value, corresponding to a large resistance of, say $10^{12}\,\Omega = 1/G_{\min}$. At every step, the circuit is solved using a plain Newton's method, using the solution at the previous step as a starting value.
3. The final solution is the DC solution of the original circuit.

Having a small $G_{\min}$ attached to every node gives a probably more realistic circuit model, and is not really an "approximation." Obviously, one can do one more (final) step and solve the circuit with a Gmin of 0, but this is often pointless. It is easy to see that this method is a homotopy, with:

$$h(x, \lambda) = (\lambda G_{\min} I + G) x + Hg(x) - s \qquad (4.217)$$

so that, with $\lambda = 1$, we can easily find a solution x_0, which is close to zero and, with $\lambda = 0$, we get the desired solution x^* of the original system.

There are other "flavors" of Gmin stepping, in which the conductances are connected across pn-junctions, or similar. It is a fairly simple technique to implement, similar to source stepping, and is available in most commercial simulators. As with source stepping, if it fails to converge, one can retry with a smaller step size. If it does not converge after several trials, one may resort to more "heavy duty" methods, such as pseudo-transient, which we now explore.

4.4.5 Pseudo-Transient

Both source stepping and Gmin stepping require disabling the dynamic elements (optionally adding resistance from every node to ground). Source stepping requires modification of the independent sources, while Gmin stepping leaves the sources as they are but modifies the network. Both methods use stepping of a certain λ, as the homotopy parameter, and solve the DC circuit at every point using a plain Newton's method. Pseudo-transient is another homotopy, where the homotopy parameter is *time*, and which has the salient features that a) the independent sources and the dynamic elements are left intact and b) additional dynamic elements are added to the network.

The idea is to first modify the network by adding dynamic elements, in such a way that a prespecified state x_0 becomes a valid initial state, and then to carry out a *transient simulation* of the network, starting from x_0, until a DC steady state is (hopefully) reached. During this transient simulation, the values of the independent sources are kept *fixed*, at their prespecified values for the intended DC Analysis run. Hence the name, *pseudo*-transient. If we successfully reach a DC steady state, then that steady state *is* the desired DC solution.

In pseudo-transient, the circuit dynamic elements help damp out the oscillations frequently encountered during the early phase of DC Analysis. It can be slow, reportedly 2–10 times slower than plain Newton's method. But, it has proven to be quite successful in practice. Nevertheless, the scheme can fail, for various reasons:

1. At some time point, the Newton loop may not converge, even for the smallest allowable time-step.
2. We may not reach a DC steady state but, instead, an oscillatory response may be generated by the transient simulation.
3. The simulation may not reach a DC steady state during the time budget allotted to this pseudo-transient run.

There are variations on the basic pseudo-transient method. One simpler possibility is to not modify the network in any way, but to ramp up the values of the independent sources over time. This would be a combination of source stepping and pseudo-transient. Another variant includes adding a nonlinear capacitor from every node to ground, whose value is decreased over time.

During the transient simulation part of pseudo-transient, and because we only care about the final steady state solution, then:

1. We can take large time-steps, with little regard to numerical errors, *provided* we eventually converge to a DC steady state. This gives significant speed-up relative to regular Transient Analysis.
2. We can modify the dynamic elements, in an arbitrary fashion, *provided* we eventually converge to a DC steady state. This, crucially, is the key to establishing the existence of a valid initial state.

The second point leads to the key construction behind the approach, and will be explained in the following.

Valid Initial State   A key question in connection with pseudo-transient is: what does it mean for a certain state x_0 to be a *valid* initial state, at $t = 0$, of the dynamic MNA system:

$$Gx(t) + Hg(x(t)) + D(x)x'(t) = s \qquad (4.218)$$

The obvious answer is that x_0 must be such that the following equation is true:

$$Gx_0 + Hg(x_0) + D(x_0)x'(0) = s \qquad (4.219)$$

i.e., there must exist an assignment of $x'(0)$ that balances this equation. We do not actually need to *know* the slopes $x'(0)$ in order to use pseudo-transient; we only need to know that a valid $x'(0)$ exists, for a given x_0. The subsequent transient simulation of the network, starting from x_0, implicitly discovers these slopes "on the fly," as part of numerical integration.

It is possible for a certain assignment of $x'(0)$ to be *inconsistent*, in the following sense. Note that KVL and KCL can be expressed in *differential* forms, both locally and globally. Locally, if $u_1(t)$, $u_2(t)$, ..., $u_m(t)$, are potential differences across branches that form a loop, then KVL implies that $\sum_j u'_j(t) = 0$, $\forall t$. Likewise, if $i_1(t)$, $i_2(t)$, ..., $i_m(t)$, are all the branch currents that are incident on a node, then KCL implies that $\sum_j i'_j(t) = 0$, $\forall t$. Globally, the differential forms are $Ai'(t) = 0$ and $A^T u'(t) = v'(t)$. Thus, one may wonder if the implicit assignment of $x'(0)$ satisfies these differential forms of the network laws. One can go further and differentiate the MNA system equation itself, and wonder if this $x'(0)$ assignment satisfies that new higher-order system. This would lead to higher order derivatives, and to further questions about consistency, and is a fruitless path to follow. For one thing, we are *not* concerned with satisfying all the higher differential orders of the MNA system, only the basic first-order system. Secondly, questions of existence of solutions to dynamical systems under arbitrary stimulus remain an open research area. Instead, pseudo-transient applies a transient simulation of the system, in spite of the possible inconsistencies at $t = 0$. As the dynamics of the system die down over time, any inconsistencies will typically vanish and the final DC steady state is a valid DC solution.

Pseudo-Transient Construction The construction required for pseudo-transient consists of two steps:

1. Add an inductor in series with every independent voltage source and with every controlled voltage source (CVS) that has a non-zero offset.
2. Add a capacitor in parallel with every independent current source and with every controlled current source (CCS) that has a non-zero offset.

Each source is *replaced* by a combination of itself and the added L_s or C_s, as shown in Fig. 4.28. Note that, with every new inductor, we get a *new node* in the network, at the connection between the inductor and its corresponding voltage source, and this introduces a new MNA variable as the voltage on that node. As well, the currents in the new inductors become additional new MNA variables.

Notice that, if a current i_k is an MNA variable (i.e., it is the control variable to some controlled source) which, in the original network, goes through a (independent or controlled) current source, then, in the new network, this current is the total current going through the *parallel combination* of that current source

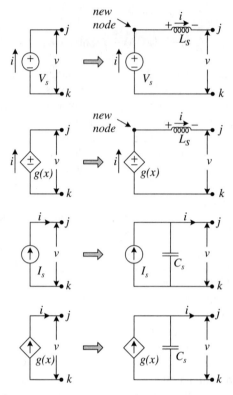

Figure 4.28: Source transformations for pseudo-transient.

and its new capacitor. One way to "access" this current is to insert a 0 V voltage source in series with the parallel combination and to label its current as i_k. Likewise, if a controlling variable $u_k = v_i - v_j$ is the voltage across a (independent or controlled) voltage source in the original network, then, in the new network, u_k becomes the voltage across the series combination of that source and its new inductor.

With these source transformations, the new MNA system is then built and the modified network is *initialized* as per the following "recipe":

1. All *pre-existing* nodes are initialized to 0 Volts and, so, *all* capacitor branch voltages are initialized to 0 Volts.
2. Every *new* node is initialized to the voltage value of its independent voltage source or the offset value of its controlled voltage source.
3. All MNA current variables are initialized to 0 Amps and *all* inductor branch currents are initialized to 0 Amps.

The transient simulation run is then applied on this network, starting from this initial condition.

In the following, we will explain why it is that the above initialization gives a valid initial state for the network. In other words, we will demonstrate the existence of an $x'(0)$ that balances the MNA equation for the case when the initial state is determined by the above settings. Intuitively, we will see that this is the case because the added dynamic elements effectively *nullify* the influence of all independent sources at time $t = 0$.

4.4.6 Justification for Pseudo-Transient

Without loss of generality, we will work with an MNA system for a *transformed* network, as follows:

1. All nonlinear elements are transformed so as to have zero offsets, by making their offsets into explicit independent sources, as we saw earlier.
2. The network is transformed so that no capacitors are connected directly to a terminal of an independent current source. This can be done without impacting the network response in any way by either a) introducing a short circuit (0 V voltage source or 0 Ω resistor) in series with the current source, or in series with the capacitor, or b) introducing a resistor in series with the current source.
3. The network is transformed so that it contains no current source cycles. This can be done without impacting the network response, by introducing either a short circuit or a resistor in series with a current source.

These transformations have no impact on the network response and are only a mathematical convenience; they allow an easier proof.

With these transformations, let the following be the MNA system of the resulting network (before addition of the new capacitors and inductors):

$$Gx(t) + Hg(x(t)) + D(x)x'(t) = s \qquad (4.220)$$

As a result of the 1st transformation, we have $g(0) = 0$. As a result of the 2nd transformation, we have the following key property: *if $x_k(t)$ is the terminal voltage of an independent current source, then $D_{ik}(x) = 0, \forall i$, in* (4.220). Thus, the columns of $D(x)$ corresponding to terminals of independent current sources are all zero. We are now interested in establishing $x_0 = 0$ as a valid initial state of the MNA system in (4.220).

Given that $g(0) = 0$, then we must find an assignment $x'(0)$ such that:

$$D(0)x'(0) = s \qquad (4.221)$$

but it is not obvious how this can be achieved. The idea behind pseudo-transient is to add new dynamic elements to the network, so that the MNA system equation can be expressed as:

$$Gx(t) + Hg(x(t)) + D(x)x'(t) + K(x)x'(t) = s \qquad (4.222)$$

where G, H, $g(\cdot)$, s, and x remain exactly the same as before, and the only change is the new matrix $K(x)$, in such a way that, at $t = 0$, we have:

$$D(0)x'(0) = 0 \quad \text{and} \quad K(0)x'(0) = s \qquad (4.223)$$

so that $x_0 = 0$ *becomes* a valid initial state of this *modified* network. Note, even though adding new dynamic elements can potentially add new nodes/currents to the original circuit and its MNA vector, but because of *how* we have transformed the network, we will see below that (4.222) can indeed be written using the same $x(t)$ as before. Under steady state, when $x'(t) = 0$, the system reduces to the DC equation for the *original* system:

$$Gx + Hg(x) = s \qquad (4.224)$$

so that the DC steady state is indeed a DC solution of the original system.

Voltage Source   Consider the MNA system (4.220) and the voltage source transformation shown in Fig. 4.29. The element equation for the original voltage source is:

$$v_j - v_k = V_s \qquad (4.225)$$

where $(v_j - v_k)$ contributes to the G matrix and V_s contributes to s. In the modified circuit, this becomes:

$$v_j - v_k + L_s i'(t) = V_s \qquad (4.226)$$

The new term $L_s i'(t)$ contributes to the new matrix $K(x)$, only, so that the modified network has an MNA equation of the form:

$$Gx(t) + Hg(x(t)) + D(x)x'(t) + K(x)x'(t) = s \qquad (4.227)$$

where G, H, $g(\cdot)$, s, and x remain exactly the same as before.[1] Note that $i(t)$, as the original voltage source current, is already a member of $x(t)$, and $D(x)$

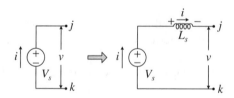

Figure 4.29: Voltage source transformation for pseudo-transient.

---

[1]Notice that, because the current in the new inductor is equal to the current in the voltage source, which is already an MNA variable, we have been able, for purpose of this proof, to accommodate the new inductors without introducing new MNA nodes and currents, thereby keeping the same variable vector $x(t)$. A practical application of pseudo-transient may introduce new MNA nodes and currents, which entails a change in $x(t)$.

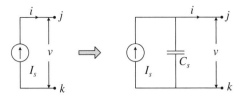

Figure 4.30: Current source transformation for pseudo-transient.

always has a 0 coefficient that multiplies $i'(t)$. Thus, at time 0, the initial condition $i'(0) = V_s/L_s$ achieves $K(x)x'(0) = s$ (in the row corresponding to $i(t)^2$), and it contributes only *zero* terms to the product $D(x)x'(0)$.

Current Source   Consider the MNA system (4.220) and the current source transformation shown in Fig. 4.30. The element equation for the original current source is:

$$i(t) = I_s \tag{4.228}$$

which contributes to s and, if this source happens to be in group 2, to G. In the modified circuit, we can view the capacitor and current source as forming a *composite* element, with the element equation:

$$i(t) = I_s - C_s \frac{d}{dt} \left(v_j(t) - v_k(t) \right) \tag{4.229}$$

or

$$i(t) + C_s \left(v'_j(t) - v'_k(t) \right) = I_s \tag{4.230}$$

The new term $C_s \left(v'_j(t) - v'_k(t) \right)$ contributes to the new matrix $K(x)$, only, so that the modified network has an MNA equation of the form:

$$Gx(t) + Hg(x(t)) + D(x)x'(t) + K(x)x'(t) = s \tag{4.231}$$

where G, H, $g(\cdot)$, s, and x remain exactly the same as before. Note that $D(x)$ has 0 coefficients that multiply $v'_j(t)$ and $v'_k(t)$, due to the 2nd transformation that we saw earlier, on page 193. Therefore, at time 0, the initial condition $v'_j(0) - v'_k(0) = I_s/C_s$ achieves $K(x)x'(0) = s$, in the rows of the MNA system corresponding to nodes j and k, but it contributes only *zero* terms to the product $D(x)x'(0)$. And, due to the 3rd transformation on page 193, we have no current source cycles, and we should be able to assign values of $v'_j(0)$ and $v'_k(0)$ around the network in order to achieve $v'_j(0) - v'_k(0) = I_s/C_s$ for every current source, without running into a conflict.

---

[2]The row of $K(x)x'(t) = s$ corresponding to $i(t)$ is $L_s i'(t) = V_s$.

Valid Initial State   As a result, we have identified the required assignments $x_i'(0)$ for all terminals of the new capacitors and currents of the new inductors, such that a) these assignments lead to $K(0)x'(0) = s$, and b) they contribute zero terms to the product $D(x)x'(0)$. For all other $x_j'(0)$, which do not correspond to voltage terminals of the new capacitors or currents of the new inductors, we simply set $x_j'(0) = 0$. In this way, we achieve $D(0)x'(0) = 0$, and the state $x = 0$ becomes a valid initial state of the system:

$$Gx(t) + Hg(x(t)) + D(x)x'(t) + K(x)x'(t) = s \qquad (4.232)$$

In a practical application of pseudo-transient, as we saw above, new nodes and new currents may be introduced, as additional MNA variables, and these must be initialized as well. For the new nodes, since the voltage at the other side of the voltage source (or CVS) is initialized to zero, the new node is initialized to V_s (or the offset value of the CVS). For the new currents, they are equal to the voltage source or CVS currents, which have been initialized to zero, so they are initialized to 0 as well. Thus, the initialization "recipe" given earlier indeed leads to a valid initial state for the modified network. We close this section with a couple of remarks:

1. The values of the added dynamic elements are somewhat arbitrary. They may be set to arbitrary fixed values, as in SPICE, where they are set at 1 H and 1 F. They may also be chosen as nonlinear elements, and their values may be changed over time, as in some modern research proposals.
2. As for the original circuit dynamic elements, they may be kept, removed, or their values modified, as seen fit to aid convergence. This is because these original elements contribute to $D(x)$ only and, since $D(x)x'(0) = 0$, they do not affect the validity of the initial state. It is often advantageous to keep them in the network because, in general, they contribute to a better conditioned Jacobian and better convergence.

This continues to be a lively research area.

Notes   Even the best known methods for solving nonlinear circuits are not *guaranteed* to always converge, and problems can arise in practice, in both DC and Transient Analysis. In Transient Analysis, one way to recover from a non-convergence problem is to reduce the time-step and try again, as done in SPICE. With a shorter time-step, the initial Newton solution is closer to the final true solution, and convergence becomes easier. However, simulators typically impose an internal lower limit on the size of the time-step. If this limit is reached, and convergence not yet achieved, SPICE will abort with an error message "Time step too small." For DC Analysis, convergence can be aided by the use of initial node voltage specification by the user, for certain key nodes in the circuit.

Additional reading is available in Pillage et al. (1995), chapter 10, in Vlach and Singhal (1994), chapter 12, and in McCalla (1988), chapter 4. For additional coverage of solution methods for nonlinear equations in general, consult the texts by Kelley (1995), Burden and Faires (2005), Bartle (1976), Dennis and Schnabel (1996), Chua and Lin (1975), and Press et al. (2007).

Problems

4.1. Consider the function $f(x) = x + 2x^{1/3} - 4$, with $x \in \mathbb{R}$.

(a) Starting with $x_0 = 27$, use the fixed point method to find a fixed point of $g(x)$.

(b) Starting with the same $x_0 = 27$, apply Newton's method to find a solution of $f(x) = 0$.

(c) Starting with $x_0 = 0.001$, apply Newton's method again to find a solution of $f(x) = 0$.

4.2. Consider the function:

$$f(x) = \begin{cases} \sqrt{x} & x \geq 0 \\ -\sqrt{|x|} & x \leq 0 \end{cases}$$

(a) Show that Newton's method, applied to this function, gives an oscillating sequence $x_0, -x_0, x_0, -x_0, \ldots$, for any initial candidate solution $x_0 \neq 0$, ignoring roundoff.

(b) Does this function satisfy the standard assumptions for convergence of Newton's method?

4.3. Let $f : \mathbb{R}^n \to \mathbb{R}^n$ have a nonsingular Jacobian $J_f(x)$, for all $x \in \mathbb{R}^n$, and let $f(x^*) = 0$, for some $x^* \in \mathbb{R}^n$.

(a) If f is affine and invertible, show that x^* is unique and that Newton's method would find it in a single iteration.

(b) If f_i is affine for only some i, show that, for any initial candidate solution $x^{(0)}$, Newton's method would give $f_i(x^{(k)}) = 0$ for all $k \geq 1$.

4.4. Let $f : \mathbb{R}^n \to \mathbb{R}^n$, let $J_f(x)$ be its Jacobian, let x^* be a solution of $f(x) = 0$, and let $\Phi(x)$ be an $n \times n$ matrix. Let $g(x) = x - \Phi(x)f(x)$ and $J_g(x)$ be its Jacobian. Prove that $J_g(x^*) = 0$ if and only if $\Phi(x^*)J_f(x^*) = I$, where I is the identity matrix.

4.5. Give the companion model for a nonlinear resistor with the following element equation:

$$v = \begin{cases} i^2 & i \geq 0 \\ -i^2 & i \leq 0 \end{cases}$$

4.6. If $v \in \mathbb{R}^p$ is a vector, we denote by $|v|$ the vector formed by replacing each entry of v by its absolute value. If $v, w \in \mathbb{R}^p$ and if $|v_i| \leq |w_i|$ for all

$i \in \{1, 2, \ldots, p\}$, then we write that $|v| \leq |w|$. By definition of p-norms, it is clear that $\|v\| = \||v|\|$, for any $v \in \mathbb{R}^p$. This is expressed by saying that p-norms are *absolute*. It follows that p-norms are also *monotone*, which is to say that $|v| \leq |w|$ implies that $\|v\| \leq \|w\|$, for any $v, w \in \mathbb{R}^p$.

(a) For any vector p-norm $\|\cdot\|_p$, prove that:

$$\forall v \in \mathbb{R}^n : (1/n)\|v\|_1 \leq \|v\|_p \leq n^{1/p}\|v\|_\infty$$

(b) For any matrix $A \in \mathbb{R}^{m \times n}$, let $v(A)$ be a vector in $\mathbb{R}^{mn}$ consisting of all the entries of A. Then, for any vector p-norm and the corresponding induced matrix p-norm, prove that:

$$\|A\| \leq n\|v(A)\|$$

(c) Let $f : D \subset \mathbb{R}^n \to \mathbb{R}^m$, where D is open and convex. If all the 2nd derivatives $\partial^2 f_k / \partial x_i \partial x_j$ are continuous and bounded by M in D, prove that the Jacobian $J(x)$ is Lipschitz continuous in D, with a Lipschitz constant $\gamma = n^3 M$.

4.7. Show that the diode i-v characteristic has the smallest radius of curvature at:

$$v_d = \eta V_T \ln\left(\frac{\eta V_T}{\sqrt{2}I_{sat}}\right)$$

4.8. (Computer Project) Based on the linear solver that was developed previously in problem 3.16, write a C or C++ implementation of a DC solver for nonlinear resistive circuits, using Newton's method based on the use of companion models and element stamps. Your implementation should be general, in the sense that it should accept any linear or nonlinear circuit description consisting of any combination of linear resistors, independent voltage and current sources, diodes, BJTs, and MOSFETs. You will model diodes, BJTs, and MOSFETs using the simple models given in the text, with no series resistance. Thus, the diode has the standard exponential diode model (4.74), the BJT has the Ebers-Moll model we saw earlier in Fig. 4.20, and MOSFETs have the simplest (quadratic) model we saw above in (4.172), which is suitable for long-channel devices.
As a first step in your solution, you should scan the element list and create the matrix G that is the contribution to the Jacobian by the *linear* elements, the vector s that is the contribution to the RHS vector by the *linear* elements, and the vector $g(x)$ of nonlinear functions that correspond to each nonlinear CVS or CCS in the network. When using Newton's method, you should build the Jacobian $J_f(x^{(k)}) \triangleq G + H J_g(x^{(k)})$ by using element stamps, instead of by using the defining expression $G + H J_g(x^{(k)})$ directly. In each Newton iteration, you should build the

system $J_f(x^{(k)})x^{(k+1)} = s^{(k)}$ by creating fresh copies of G and s and then adding the stamps due to the nonlinear elements.

In order to determine when to stop the Newton iterations, you should check both the size (norm) of the steps in x and the value (norm) of the function $f(x) = Gx + Hg(x) - s$, using a relative tolerance of 0.1% and an absolute tolerance of $1\,\text{mV}$ (for voltages) and $1\,\mu\text{A}$ (for currents). To

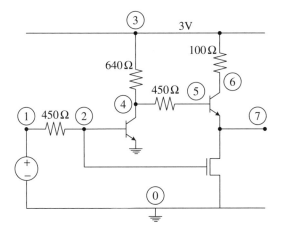

Figure 4.31: A nonlinear test circuit.

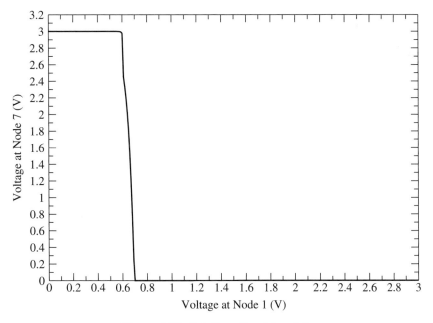

Figure 4.32: Solution of problem 4.8.

evaluate the function $f(x)$, you need to know the matrix H, which you can infer from the incidence relations in the circuit, and you need to evaluate $g(x)$. In order to improve the chances of convergence in Newton's method, you should use generalized damping with $\gamma = 1.3$ and $k = 16$.

Use your code to perform a DC Sweep of the circuit shown in Fig. 4.31, based on the following parameters. For the MOSFET, $V_t = 0.6\,\text{V}$, $\lambda = 0.01/\text{V}$, and $\beta = 0.5\,\text{mA/V}^2$. For the BJT, $\alpha_F = 0.99$, $\alpha_R = 0.02$, $I_{es} = 2 \times 10^{-14}\,\text{A}$, $I_{cs} = 99 \times 10^{-14}\,\text{A}$, and $V_{Tc} = V_{Te} = 26\,\text{mV}$. Generate a plot of the output DC voltage (at node 7) versus the input DC voltage (at node 1), as it is swept from 0 V to 3 V, in steps of 0.01 V. The correct solution is shown in Fig. 4.32.

CHAPTER 5

Solution of Differential Circuit Equations

In the presence of dynamic (L and C) elements, the network equations can be formulated as a system of differential equations. Solving such systems is not easy, and gets harder when the dynamic elements are nonlinear. As we will see, the practical approach for solving differential equations is to repeatedly *discretize* them and solve the resulting *algebraic* equations. The need to solve a dynamic network arises as part of the `Transient Analysis` mode of standard circuit simulators. We will study dynamic elements and the resulting dynamic MNA equations, general solution methods, and their application to circuit simulation.

5.1 DIFFERENTIAL NETWORK EQUATIONS

Considering the formulation of the network equations, it should be clear that KCL and KVL remain as linear *algebraic* relationships. The network equations become differential due only to the *dynamic element equations*.

5.1.1 Dynamic Elements

The most basic dynamic elements are the familiar two-terminal capacitors and inductors, be they linear or nonlinear, which we will now review. As well, and in order to incorporate internal dynamic elements of multiterminal elements (MTE), we will study some generalizations of the basic L and C elements.

Capacitors If $q(t)$ is the charge on a capacitor, then recall that capacitor current is the rate of change of charge, so that:

$$i(t) = \frac{d}{dt}q(t) = q'(t) \tag{5.1}$$

Circuit Simulation, by Farid N. Najm
Copyright © 2010 John Wiley & Sons, Inc.

As well, recall that the charge q is a function of the voltage v across the capacitor, so that:

$$i(t) = \frac{dq}{dv}\frac{dv}{dt} = C(v)\frac{dv}{dt} = C(v)v'(t) \tag{5.2}$$

where $C(v) \triangleq dq/dv$ is the *capacitance*. Note that the dependence of v on t is implicit when we write $C(v)$, but is omitted to simplify the notation. If dq/dv is a constant, independent of v, then the capacitor is said to be *linear* and its capacitance is fixed, otherwise it is *nonlinear* and its capacitance is a function of voltage.

More generally, the charge can be a function of other voltages or currents in the network, so that, if $x(t)$ is the MNA vector, then:

$$i(t) = \frac{d}{dt}q\,(x(t)) = \sum_j \frac{\partial q}{\partial x_j}\frac{dx_j}{dt} \tag{5.3}$$

In practice, where this arises, such as in MOSFETs, the charge is typically a function of the MOSFET terminal voltages. If we let $C_j(x) \triangleq \partial q/\partial x_j$, then:

$$i(t) = \sum_j C_j(x)x'_j(t) \tag{5.4}$$

When x_j is not a terminal voltage of this capacitor, C_j is referred to as a *mutual capacitance*, otherwise it is the *self-capacitance*. Equivalently, one can view the expression (5.4) as representing the current through a parallel collection of several *dynamic controlled current sources* (DCCS), to ground:

$$i(t) = \sum_j p_j(x, x'_j) \tag{5.5}$$

However, the best way to handle mutual capacitance is to consider it as part of the device model of a multiterminal element. This ensures that charge conservation is maintained, as we will see later. Therefore, we will mainly be concerned with self, not mutual, capacitance, except for brief coverage later on under multiterminal elements.

Inductors   If $\phi(t)$ is the flux in an inductor, then recall that voltage across the inductor is the rate of change of flux, so that:

$$v(t) = \frac{d}{dt}\phi(t) = \phi'(t) \tag{5.6}$$

As well, recall that the flux ϕ is a function of the current i in the inductor, so that:

$$v(t) = \frac{d\phi}{di}\frac{di}{dt} = L(i)\frac{di}{dt} = L(i)i'(t) \tag{5.7}$$

where $L(i) \triangleq d\phi/di$ is the *inductance*. Note that the dependence of i on t is implicit when we write $L(i)$, but is omitted to simplify the notation. If $d\phi/di$ is a constant, independent of i, then the inductor is said to be *linear* and its inductance is fixed, otherwise it is *nonlinear* and its inductance is a function of current.

More generally, the flux can be a function of other voltages or currents in the network, so that, if $x(t)$ is the MNA vector, then:

$$v(t) = \frac{d}{dt}\phi(x(t)) = \sum_j \frac{\partial \phi}{\partial x_j}\frac{dx_j}{dt} \tag{5.8}$$

If we let $L_j(x) \triangleq \partial\phi/\partial x_j$, then:

$$v(t) = \sum_j L_j(x)x'_j(t) \tag{5.9}$$

When x_j is not the current in this inductor, L_j is referred to as a *mutual inductance*, otherwise it is the *self-inductance*. Equivalently, one can view the expression (5.9) as representing the voltage across a series collection of several *dynamic controlled voltage sources* (DCVS):

$$v(t) = \sum_j p_j(x, x'_j) \tag{5.10}$$

However, as with capacitance, we will mostly be concerned with self, not mutual, inductance.

5.1.2 Dynamic MNA Equations

We saw earlier that the dynamic MNA equations in the *linear* case can be written as:

$$Gx(t) + Dx'(t) = s(t) \tag{5.11}$$

where G is the MNA system matrix and D is a constant matrix arising from the contributions of all the dynamic L and C elements. We also saw, in the case of nonlinear DC circuits, that the MNA equations can be written as:

$$Gx + Hg(x) = s \tag{5.12}$$

where H is a matrix whose entries are either 0 or ± 1 and $g(x)$ is a vector of all the element functions $g_j(\cdot)$ of the nonlinear controlled current source (CCS) and controlled voltage source (CVS) elements. Combining the above, if we have general nonlinear resistive elements and only *linear* dynamic elements, then the MNA system becomes:

$$Gx(t) + Hg(x) + Dx'(t) = s(t) \tag{5.13}$$

where the dependence of x on t is implicit when we write $g(x)$, but is omitted in order to simplify the notation. Finally, in the presence of nonlinear two-terminal dynamic elements, this becomes:

$$Gx(t) + Hg(x) + D(x)x'(t) = s(t) \qquad (5.14)$$

where $D(x)$ is a matrix that includes contributions from all the linear and non-linear (self and mutual) capacitance and inductance terms.

More generally, it is easy to see that, as was done in the case of nonlinear resistive circuits, we can write the dynamic MNA system as:

$$Gx(t) + Hg(x) + Ep(x, x') = s(t) \qquad (5.15)$$

where E is a matrix whose entries are either 0 or ± 1 and $p(x, x')$ is a vector of the functions $p_j(\cdot)$ of all the DCCS and DCVS elements. This form (5.15) subsumes the previous two special cases (5.13) and (5.14), but is only of theoretical interest; we will see that (5.14) is the form that will be most useful.

5.1.3   DAEs and ODEs

An equation $x'(t) = f(x, t)$ where $x : \mathbb{R} \to \mathbb{R}$ and $f : \mathbb{R} \times \mathbb{R} \to \mathbb{R}$ is called a *first-order ordinary differential equation* (ODE). The equation is *ordinary* because it has an ordinary derivative, as opposed to a partial derivative. It is of *first order* because it includes only the first derivative. For our purposes, $t \in \mathbb{R}$ represents time, but, in general, it can be otherwise. When x and f are vectors, and t a scalar, then

$$x'(t) = f(x, t) \qquad (5.16)$$

is a *system* of ordinary differential equations, or simply an *ODE system*. As we will see, ODEs have a rich history, with the earliest techniques for their numerical solution dating back to the late 1800s.

Unfortunately, in general, circuit equations are *not* ODEs. Only under certain conditions can the dynamic MNA system be transformed to an ODE, a form which is called the *state variable representation* in the circuits literature. For example, if the system is of the form (5.13) or (5.14), and if D or $D(x)$ is nonsingular, then we can easily transform the MNA system to an ODE system, but this is not guaranteed in general. There are methods, based on the construction of a *hybrid port matrix*, as described in Chua and Lin (1975), to convert any circuit to a state variable representation, but these methods are computationally expensive and generally not suitable for use in circuit simulation.

In general, then, the dynamic MNA system is of the form:

$$\mathfrak{F}(x, x', t) = 0 \qquad (5.17)$$

where $\mathfrak{F} : (\mathbb{R}^m \times \mathbb{R}^m \times \mathbb{R}) \to \mathbb{R}^m$, which is called a system of *differential-algebraic equation* (DAE), or a DAE system. DAEs are more difficult to solve, and techniques for their solution are not yet fully understood and developed; they remain an active area of research. Indeed, in contrast to ODEs, as we will see, there is no general theorem that guarantees existence and uniqueness of solutions for DAEs.

Loosely speaking, integration is a *smoothing* process, while differentiation is the opposite; one may call it an *antismoothing* process, as in Burden and Faires (2005). An ODE, like $x'(t) = Ax(t) + b(t)$ involves integration of the right hand side, so that the solution $x(t)$ is, again loosely speaking, smoother than $b(t)$. A DAE, on the other hand, involves both differentiation and integration. By taking derivatives of a DAE, one can often transform it into an ODE. The minimum number of differentiations required to do this is called the *index* of the DAE. As a simple example, given in Ascher and Petzold (1998), consider the following DAE system, where the dependence of x_i on t is suppressed to simplify the notation:

$$\begin{aligned} x_1 &= a(t) \\ x_2 &= x_1' \end{aligned} \tag{5.18}$$

We can differentiate the first equation to get $x_1' = a'(t)$. We can also differentiate the second equation, $x_2' = x_1''$, so that $x_2' = a''(t)$, leading to the ODE system:

$$\begin{aligned} x_1' &= a'(t) \\ x_2' &= a''(t) \end{aligned} \tag{5.19}$$

Because two differentiations of $a(t)$ were needed, this is an index 2 DAE. The notion of index is useful to understand the mathematical structure, and difficulty, of DAE systems, but it is not helpful as far as solving them. For the general MNA system, a DAE, the index depends on the type of circuit considered; it is often 0 or 1, but can be higher. In general, lower-index DAEs are easier to solve and, for DAEs of index greater than 2, one often attempts to reduce their index before solving them. But there is no general systematic technique for doing so.

Even though there are no fully developed solution methods for DAEs, it is possible to use some ODE solution methods to solve DAEs, and this often works well in practice, but there are no guarantees. However, many of the attractive properties (accuracy, stability) that these methods have when used on ODEs are lost when they are applied to DAEs. There are in fact DAEs for which even the best known ODE methods lead to unstable behavior. One such strategy (of adapting ODE solution methods to DAEs) is called *direct discretization*, and is the method of choice in circuit simulation, and many other fields where DAEs arise.

Therefore, our plan of work in the remainder of this chapter will be to study ODE solution methods in some detail, with emphasis on those methods that *can* be applied to DAEs by means of direct discretization, and then to illustrate how these methods can be applied to DAEs in general, and to circuit simulation in particular.

5.2 ODE SOLUTION TECHNIQUES

In this section, and the next two, we will describe numerical methods for solving
ODEs. The treatment is general, and not specific to circuit simulation. However,
much care has been taken to limit the discussion to those techniques and concepts that are essential to circuit simulation. This material is based on a number
of sources, including primarily Lambert (1991), as well as Ascher and Petzold
(1998), Burden and Faires (2005), and Ralston and Rabinowitz (2001).

For the following material, it will be useful to introduce the $0(h)$-notation. We
are already familiar with the $\mathcal{O}(h)$-notation, which is useful to capture behavior
at large h, and we now define notation that is useful for small h.

Definition 5.1. If $r \in \mathbb{R}$ and $F(r)$ is either a scalar or a vector, and if there is
a constant $K > 0$ such that $\|F(r)\| \le Kh^p$ for sufficiently small $|h|$, then we
write:

$$F(h) = 0(h^p) \tag{5.20}$$

5.2.1 ODE Systems and Basic Theorems

Recall, a first-order ODE system is the m-dimensional system:

$$x'(t) = f(x, t) \tag{5.21}$$

where $t \in \mathbb{R}$, $x : \mathbb{R} \to \mathbb{R}^m$, and $f : \mathbb{R}^m \times \mathbb{R} \to \mathbb{R}^m$. The *general solution* of such
a system contains, in general, m arbitrary constants. For example, after Lambert
(1991), the system:

$$x_1' = \frac{x_1}{t} + tx_2 \tag{5.22}$$

$$x_2' = \left(x_2^2 - 1\right) \frac{t}{x_1} \tag{5.23}$$

has the general solution:

$$x_1(t) = \frac{t}{c_1} \cos\left(c_1 t + c_2\right) \tag{5.24}$$

$$x_2(t) = -\sin\left(c_1 t + c_2\right) \tag{5.25}$$

for arbitrary constants c_1 and c_2, provided $c_1 \ne 0$. If an additional set of m
side conditions are imposed on the system, they can serve to select a specific solution with specific c_1 and c_2 values. If these m conditions are that
$x_1(t), x_2(t), \dots, x_m(t)$ take given values at the *same* initial time-point t_0, then
the resulting problem is called an *initial value problem* (IVP):

$$\begin{aligned} x'(t) &= f(x, t) \\ x(t_0) &= x_0 \end{aligned} \tag{5.26}$$

and we are typically interested in the solution for $t \geq t_0$, up to some $t \leq t_f$. For the time-domain solution of circuit equations, we are primarily interested in solution methods for IVPs.

Notice that $x(t)$ solves the IVP (5.26) if and only if:

$$x(t) = x(t_0) + \int_{t_0}^{t} f(x(\tau), \tau) \, d\tau \tag{5.27}$$

Thus, as we will see, there is a strong connection between numerical solution methods for ODEs and methods for *numerical integration*. Indeed, the act of solving an ODE is often referred to as numerical integration, and ODE solution methods are often called numerical integration methods.

Existence and Uniqueness   Not all IVPs possess a unique solution, nor indeed any solution at all. For example, with $m = 1$, the IVP $x' = 3x + 2$, $x(0) = 5$, has the *unique* solution $x(t) = (17/3)e^{3t} - (2/3)$. As another 1-dimensional example, the IVP:

$$x' = \begin{cases} \sqrt{x}, & \text{if } x \geq 0, \\ 0, & \text{otherwise.} \end{cases} \tag{5.28}$$

$$x(0) = 0 \tag{5.29}$$

has the solution:

$$x(t) = \begin{cases} \dfrac{(t-c)^2}{4}, & \text{for } t \geq c, \\ 0, & \text{for } 0 \leq t \leq c. \end{cases} \tag{5.30}$$

where $c > 0$ is arbitrary, so that the solution is *not unique*. The following theorem gives a local existence result:

Theorem 5.1. *(Peano Existence Theorem) If $f(x, t)$ is continuous at (x_0, t_0), then there exists a solution to the IVP $x' = f(x, t)$, $x(t_0) = x_0$, over some interval that contains t_0.*

Peano's theorem does not guarantee a unique solution, as shown by an example in Chua and Lin (1975). For a local existence and uniqueness result, we have access to the following:

Theorem 5.2. *(Picard-Lindelöf Theorem) Let $\mathcal{D} \subset \mathbb{R}^m$ and $\mathcal{I} \subset \mathbb{R}$ be closed and let $f : \mathcal{D} \times \mathcal{I} \rightarrow \mathbb{R}^m$ be bounded, continuous in t, $\forall t \in \mathcal{I}$, and Lipschitz continuous in x, $\forall x \in \mathcal{D}$. Then, for any $(x_0, t_0) \in \mathcal{D} \times \mathcal{I}$, there exists a unique solution $x(t)$ of the IVP $x' = f(x, t)$, $x(t_0) = x_0$, over some interval $\mathcal{I}' \subset \mathcal{I}$, with $t_0 \in \mathcal{I}'$. Moreover, the solution $x(t)$ is continuous with respect to (x_0, t_0).*

Loosely speaking, the final statement, of continuity with respect to (x_0, t_0), means that the system is not "overly sensitive" to its initial conditions. This is a desirable property because, otherwise, a numerical solution of such a system would be unreliable, due to inevitable roundoff error. We will have more to say about this property of ODE systems.

Well-Posedness   Given the IVP $x' = f(x, t)$, $x(t_0) = x_0$, and with $t \in [t_0, t_f]$, we define the following *perturbed* IVP:

$$\hat{x}' = f(\hat{x}, t) + \delta(t)$$
$$\hat{x}(t_0) = x_0 + \delta_0$$
$$(5.31)$$

where $\delta : \mathbb{R} \to \mathbb{R}^m$ and $\delta_0 \in \mathbb{R}$ constitute a *perturbation* $(\delta(t), \delta_0)$ of the system, and where the solution $\hat{x}(t)$ is called the *perturbed solution*. Due to inevitable roundoff errors, numerical methods are, in practice, always solving a *perturbed* version of the intended problem. Thus, we are mostly interested in small magnitude perturbations, such as would be representative of numerical or roundoff errors. The following definition, after Lambert (1991), is useful to capture a type of stability under perturbations:

Definition 5.2. Let $(\delta(t), \delta_0)$ and $(\delta^*(t), \delta_0^*)$ be any two perturbations of the IVP $x' = f(x, t)$, $x(t_0) = x_0$, and let $\hat{x}(t)$ and $\hat{x}^*(t)$ be the resulting perturbed solutions. If there exists a $S > 0$ such that, for all $t \in [t_0, t_f]$, we have:

$$\|\hat{x}(t) - \hat{x}^*(t)\| \leq S\epsilon, \text{ whenever } \|\delta(t) - \delta^*(t)\| \leq \epsilon \text{ and } \|\delta_0 - \delta_0^*\| \leq \epsilon,$$

then the IVP is said to be *totally stable*, or *well-posed*.

Thus, loosely speaking, a problem is totally stable if small perturbations in the data produce correspondingly small perturbations in the solution. Note that this is a property of the *problem* and not of any numerical method for solving it; a problem with this property is also said to be *well-posed*. If a problem is *not* well-posed, i.e., *not* totally stable, then any numerical solution for it is totally unreliable. We are only interested in problems that are totally stable or well-posed.

Standard Theorem   The following **standard theorem** gives sufficient conditions for existence and uniqueness of a global solution of an IVP, as well as for total stability:

Theorem 5.3. *(Standard Theorem) Let* $\mathcal{D} = \{(x, t) | t_0 \leq t \leq t_f \text{ and } x \in \mathbb{R}^m\}$ *where* t_0 *and* t_f *are finite, and let* $f(x, t)$ *be continuous in* t *on* $\mathcal{D}$ *and Lipschitz continuous in* x *on* $\mathcal{D}$. *Then, for any* $x_0 \in \mathbb{R}^m$, *there exists a unique solution* $x(t)$ *of the IVP* $x' = f(x, t)$, $x(t_0) = x_0$, *where* $x(t)$ *is continuous and differentiable in* $\mathcal{D}$. *Moreover, the IVP is well-posed (totally stable).*

In all the subsequent work, we will always assume that the hypotheses of this theorem apply to the systems under study.

Linear Systems with Constant Coefficients   The ODE system $x' = f(x, t)$ is said to be *linear* if $f(x, t)$ takes the form:

$$f(x, t) = A(t)x + b(t) \tag{5.32}$$

where $A(t)$ is an $m \times m$ matrix (of time-functions) and $b : \mathbb{R} \to \mathbb{R}^m$. Furthermore, if $A(t) = A$ is independent of time, then the system is said to be *linear with constant coefficients*:

$$x' = Ax + b(t) \tag{5.33}$$

The study of the (special case) of linear constant-coefficient systems will turn out to be useful in order to study the stability of general numerical methods. Associated with a system $x' = Ax + b(t)$ is the so-called *homogeneous* system:

$$x' = Ax \tag{5.34}$$

If $\tilde{x}(t)$ is the *general* solution of (5.34) (which allows for m arbitrary constants) and $\psi(t)$ is a *particular* solution of (5.33), then one can show that:

$$x(t) = \psi(t) + \tilde{x}(t) \tag{5.35}$$

is the *general* solution of (5.33).

Suppose that A has m distinct eigenvalues, $\lambda_1, \lambda_2, \ldots, \lambda_m$, with the corresponding eigenvectors $q_1, q_2, \ldots, q_m$. It is easy to check by substitution that $e^{\lambda_i t} q_i$ is a particular solution of (5.34), and one can prove that:

$$\tilde{x}(t) = \sum_{i=1}^{m} c_i e^{\lambda_i t} q_i \tag{5.36}$$

is the general solution of (5.34), where the c_i are arbitrary constants. As a result, the general solution of the linear ODE system with constant coefficients (5.33) is given by:

$$x(t) = \psi(t) + \sum_{i=1}^{m} c_i e^{\lambda_i t} q_i \tag{5.37}$$

5.2.2   Overview of Solution Methods

Exact solutions to ODE systems can be found only in certain limited cases. In the general case, we must resort to *numerical* solution methods. Numerical methods can only provide *particular* solutions, i.e., a solution for a given initial value assignment. Numerical methods start with the known value of x at $t = t_0$ and then, as Butcher (2000) puts it, "*their basic approach is to extend the set of t values for which an approximation to $x(t)$ is known, in a step-by-step fashion,*" up to some $t = t_f$.

Numerical methods also involve the use of *discretization*. The continuous interval $[t_0, t_f]$ is replaced by the discrete point set $\{t_n\}$:

$$t_0 < t_1 < t_2 < \cdots < t_{n-1} < t_n < \cdots \tag{5.38}$$

where $h_n \triangleq t_n - t_{n-1}$ is called the *step-size* or the *time-step*. In many cases, we will work with a *fixed time-step* h, so that $t_n = t_0 + nh$. Numerical methods provide approximate values of the solution at the time-points t_n. Thus, we distinguish between the *exact solution* to the IVP over $[t_0, t_f]$ and the *approximate* or *numerical solution* over $\{t_n\}$.

In order to simplify the notation, it will be useful to denote the (numerically found) approximate solution at time t_n by x_n, so that our aim is to achieve:

$$x_n \approx x(t_n) \tag{5.39}$$

To avoid confusion, we will denote the components of a vector x by $x_{(j)}$ rather than x_j; we will only rarely have need to refer to $x_{(j)}$. Thus, we are interested in numerical methods that produce a sequence $\{x_n\}$ that approximates the exact solution evaluated at $\{t_n\}$, i.e., that approximates the sequence $\{x(t_n)\}$.

History and Classification   Numerical methods for solving ODEs, specifically IVPs, have a rich history, summarized in Butcher (2000), and the earliest solutions were proposed in the late 1800s. There are two classes of methods:

1. Linear multistep (LMS) methods, due originally to work by J. C. Adams, published in a paper by F. Bashforth and J. C. Adams in 1883, leading to the Adams-Bashforth methods, with later work by F. R. Moulton, published in 1926, leading to the Adams-Moulton methods.
2. One-step methods, called Runge-Kutta methods, due originally to the work of C. Runge, published in 1895, further developed in a work by K. Heun, dated 1900, and a work by W. Kutta, published in 1901.

Multistep methods make use of the previously found values $x_n, x_{n-1}, \ldots$ in order to produce a value of x_{n+1}. In contrast, Runge-Kutta methods make use of the previously computed x_n to produce a value of x_{n+1}. They do make use of one or more evaluations of the function at intermediate points between t_n and t_{n+1} in order to improve accuracy. However, these function evaluations are then discarded and are not reused in making future steps.

Linear multistep methods are more computationally efficient than Runge-Kutta methods, and are more suitable for solving *stiff systems*.[1] Thus, we will focus on linear multistep methods; these are the methods that are typically used in circuit simulators, SPICE included. The modern analysis schemes for linear multistep methods are due largely to the work of G. Dahlquist in the 1950s. The study of

---

[1]We will return later on to the topic of *stiffness*; circuit equations often turn out to be *stiff systems*.

stiff systems dates back to the early work of C. F. Curtiss and J. O. Hirschfelder, published in 1952, and practical methods for solving them were developed by Gear (1971). Gear gave a detailed study of the backward differentiation formulas (BDF) which are among the best modern methods for solving stiff systems.

General Form Following the notation of Lambert (1991), almost all numerical methods for IVPs can be written in the following general form:

$$\sum_{j=-1}^{k-1} \alpha_j x_{n-j} = h\phi_f(x_{n+1}, x_n, \ldots, x_{n-k+1}, t_n, h) \tag{5.40}$$

where the time-step h has been assumed fixed, $k \geq 1$ is the *step-number*, and $\phi_f(\cdot)$ is a function that depends on the system function $f(\cdot)$. The above (5.40) is called a k-step method; if $k = 1$, it is a *one-step method*, otherwise it is a *multistep method*. Multistep methods require some start-up scheme, which we discuss later on. If the specific $\phi_f(\cdot)$ being used does not actually depend on x_{n+1}, then the method is called *explicit*, otherwise it is said to be *implicit*. Implicit methods are, in general, more powerful but harder to solve, and typically require the solution of a nonlinear equation to find x_{n+1}.

Linear multistep (LMS) methods are characterized by a $\phi_f(\cdot)$ which is a *linear* function of the values of $f(\cdot)$ at the current and previous time-points, so that:

$$\sum_{j=-1}^{k-1} \alpha_j x_{n-j} = h \sum_{j=-1}^{k-1} \beta_j f(x_{n-j}, t_{n-j}) \tag{5.41}$$

where $\alpha_{-1} \triangleq 1$ and where, as a technical condition that does not reduce the generality of the formula in any way, we require that $|\alpha_{k-1}| + |\beta_{k-1}| \neq 0$. It is customary to introduce the notation $f_i \triangleq f(x_i, t_i)$, and to write the LMS method as:

$$\sum_{j=-1}^{k-1} \alpha_j x_{n-j} = h \sum_{j=-1}^{k-1} \beta_j f_{n-j} \tag{5.42}$$

where, if $\beta_{-1} = 0$, the method is *explicit*, otherwise it is *implicit*.

5.2.3 Three Basic Methods: FE, BE, and TR

We will now study three simple and basic linear multistep methods, before returning to a formal study of their general properties.

Forward Euler (FE) The *forward Euler* (FE) method, also called simply *Euler's rule*, is rarely used in practice, but is simple enough to allow a detailed analysis. In the 1-dimensional case, we can write a Taylor series expansion at t_n:

$$x(t) = x(t_n) + (t - t_n)x'(t_n) + \frac{(t - t_n)^2}{2}x''(\xi) \tag{5.43}$$

for some ξ between t_n and t. Then, at $t = t_{n+1} = t_n + h$, we have:

$$x(t_{n+1}) = x(t_n) + hx'(t_n) + \frac{h^2}{2}x''(\xi) \qquad (5.44)$$

In the m-dimensional case, we can write this as:

$$x(t_{n+1}) = x(t_n) + hx'(t_n) + 0(h^2) \qquad (5.45)$$

This is the motivation for Euler's rule, or the forward Euler (FE) method, for solving the IVP $x' = f(x, t)$, as:

$$x_{n+1} = x_n + hf(x_n, t_n) \qquad (5.46)$$

or, simply:

$$x_{n+1} = x_n + hf_n \qquad (5.47)$$

Notice that this is a linear multistep method, with $\alpha_{-1} = 1$, $\alpha_0 = -1$, $\beta_0 = 1$, and all other coefficients at 0. It is an *explicit* method, providing an explicit formula for computing x_{n+1}.

Euler's rule ignores the $0(h^2)$ remainder term, but how accurate is it? In other words, how good is the approximation $x_n \approx x(t_n)$ resulting from its use? We can expect that "fresh" error may be introduced at every step, due to truncation of the Taylor series expression. This is called the *local truncation error* (LTE) and will be defined later. We can also expect that error may accumulate as we move forward in time, leading to a *global error* which is possibly larger than the LTE. We can further expect that additional error would be introduced due to *roundoff* and finite precision arithmetic.

FE is simple enough that we have access to the following results. If $f(x, t)$ is Lipschitz continuous for all $x \in \mathbb{R}$ and all $t_0 \leq t \leq t_f$, with constant L, and if $|x''(t)| \leq M, \forall t \in [t_0, t_f]$, then, for Euler's rule, it is shown in Burden and Faires (2005) that:

$$\left| x_n - x(t_n) \right| \leq \frac{hM}{2L} \left[e^{L(t_n - t_0)} - 1 \right] \qquad (5.48)$$

Considering this result, notice that local (truncation) error is introduced at every step, and it shrinks linearly with smaller h, and that the global error accumulates over time, growing exponentially as $t_n - t_0$ increases. Because the local error is reduced for smaller h, we are motivated to use smaller time-steps. However, for very small time-steps, roundoff error becomes problematic; the above result actually does not address roundoff, but the next one does.

If the roundoff error introduced in each application of FE is no larger than $\delta > 0$, then the above result is modified, as in Burden and Faires (2005), to give:

$$\left| x_n - x(t_n) \right| \leq \frac{1}{L} \left(\frac{hM}{2} + \frac{\delta}{h} \right) \left[e^{L(t_n - t_0)} - 1 \right] + \delta e^{L(t_n - t_0)} \qquad (5.49)$$

The behavior of this error bound is more complex; it does not decrease uniformly with h. Instead it has a non-monotone behavior: it decreases as we decrease h from very large values, it decreases as we increase h from very small values, and it achieves a minimum at:

$$h = \sqrt{\frac{2\delta}{M}} \qquad (5.50)$$

In practice, however, δ is small enough relative to typical values of the LTE, that this optimal value of h is extremely small, and does not affect the practical application of FE.

Global and Roundoff Error   As we will see later on, practical methods for solving IVPs vary the size of the time-step based on the estimated local truncation error. Strictly speaking, this is not enough. One would like to be able to also monitor error accumulation over time, i.e., the global error, not just the local error. As well, one would like to be able to take roundoff error into account. However, while there are some theoretical results for taking both global and roundoff errors into account, these results are not useful in practice. They give only error bounds that are extremely loose, as described in Lambert (1991). It is provable that, for linear multistep methods, the bound on the *total* error (including global and roundoff error) behaves in a way similar to FE. It decreases as h is decreased from very high values, but eventually increases, as $1/h$, as h is decreased further to very small values. However, as mentioned, these bounds are extremely loose and cannot be used for error estimation and time-step control.

   Thus, in practice, numerical methods for solving IVPs are applied while monitoring *only* the local truncation error. This means that there is no iron-clad guarantee that these methods are as accurate as we think they are. Yet these methods are quite successful in practice, and they continue to be used in many fields of science and engineering. It may be, as is the case with forward Euler, that the range of values of h where roundoff error becomes significant, compared to LTE, is so low that we should never have to worry about them; but we really do not know this for sure.

Backward Euler (BE)   In fact, as we will see later on, the problem with forward Euler is not so much its accuracy, but its stability. As an alternative, the *backward Euler* (BE) method is preferred, and is as follows. In the 1-dimensional case, writing a Taylor series expansion at t_{n+1}, then:

$$x(t) = x(t_{n+1}) + (t - t_{n+1})x'(t_{n+1}) + \frac{(t - t_{n+1})^2}{2}x''(\xi) \qquad (5.51)$$

for some ξ between t_{n+1} and t. Then, at $t = t_n = t_{n+1} - h$, we have:

$$x(t_n) = x(t_{n+1}) - hx'(t_{n+1}) + \frac{h^2}{2}x''(\xi) \qquad (5.52)$$

In the m-dimensional case, we can write this as:

$$x(t_n) = x(t_{n+1}) - hx'(t_{n+1}) + 0(h^2) \tag{5.53}$$

This is the motivation for the backward Euler (BE) method for solving the IVP $x' = f(x, t)$, as:

$$x_{n+1} = x_n + hf(x_{n+1}, t_{n+1}) \tag{5.54}$$

or, simply:

$$x_{n+1} = x_n + hf_{n+1} \tag{5.55}$$

Notice that this is a linear multistep method, with $\alpha_{-1} = 1$, $\alpha_0 = -1$, $\beta_{-1} = 1$, and all the other coefficients at 0. It is an *implicit* method, providing only an implicit formula for x_{n+1}. In general, as the function $f(\cdot)$ is nonlinear, this would require the use of a nonlinear solution method, such as Newton's method, to find x_{n+1}. Backward Euler ignores the $0(h^2)$ error term, and has comparable LTE to forward Euler, but is more stable, as we will see later on.

Trapezoidal Rule (TR)   Both FE and BE have error terms that are $0(h^2)$; better accuracy can be obtained by using "higher order" methods, a notion that we will define below. One such method, the *trapezoidal rule* (TR), is as follows. In the 1-dimensional case, using a Taylor series expansion at t_n, we can write:

$$x(t) = x(t_n) + (t - t_n)x'(t_n) + \frac{(t - t_n)^2}{2}x''(t_n) + \frac{(t - t_n)^3}{6}x'''(\xi) \tag{5.56}$$

for some ξ between t_n and t_{n+1}, which we can differentiate to get:

$$x'(t) = x'(t_n) + (t - t_n)x''(t_n) + \frac{(t - t_n)^2}{2}x'''(\xi) \tag{5.57}$$

Writing both results, (5.56) and (5.57), at $t = t_{n+1} = t_n + h$, and multiplying the first equation by 2 and the second by $(-h)$, we get:

$$2x(t_{n+1}) = 2x(t_n) + 2hx'(t_n) + h^2x''(t_n) + \frac{h^3}{3}x'''(\xi) \tag{5.58}$$

$$-hx'(t_{n+1}) = -hx'(t_n) - h^2x''(t_n) - \frac{h^3}{2}x'''(\xi) \tag{5.59}$$

Adding the two equations, then dividing by 2, we get:

$$x(t_{n+1}) - x(t_n) = \frac{h}{2}\left[x'(t_{n+1}) + x'(t_n)\right] - \frac{h^3}{12}x'''(\xi) \tag{5.60}$$

In the m-dimensional case, we can write this as:

$$x(t_{n+1}) - x(t_n) = \frac{h}{2}\left[x'(t_{n+1}) + x'(t_n)\right] + 0(h^3) \tag{5.61}$$

This is the motivation for the trapezoidal rule (TR) as:

$$x_{n+1} = x_n + \frac{h}{2}\left[f(x_{n+1}, t_{n+1}) + f(x_n, t_n)\right] \tag{5.62}$$

or, simply:

$$x_{n+1} = x_n + \frac{h}{2}(f_{n+1} + f_n) \tag{5.63}$$

The trapezoidal rule is probably the most often used solution method in circuit simulation; it is the workhorse of most SPICE engines. It is a linear multistep method with $\alpha_{-1} = 1$, $\alpha_0 = -1$, $\beta_{-1} = \beta_0 = 1/2$, and all other coefficients at 0. TR is an *implicit* method with $0(h^3)$ error and excellent stability. It is suitable for stiff systems, and has better accuracy than both FE and BE. It does have one weakness, however, as we will see later on. It is also classified as a 1-step Adams-Moulton method and thus is among the oldest of linear multistep methods, dating back to the 19th century.

The trapezoidal rule has an alternate form, as follows. Let $y(t) = x'(t)$, then TR (5.60) gives:

$$\int_{t_n}^{t_{n+1}} y(t)dt = x(t_{n+1}) - x(t_n) = \frac{h}{2}\left[x'(t_{n+1}) + x'(t_n)\right] - \frac{h^3}{12}x'''(\xi) \tag{5.64}$$

$$= \frac{h}{2}\left[y(t_{n+1}) + y(t_n)\right] - \frac{h^3}{12}y''(\xi) \tag{5.65}$$

which gives an alternate form of TR, suitable for numerical integration, as:

$$\int_{t_n}^{t_{n+1}} y(t)dt \approx \frac{h}{2}\left[y(t_{n+1}) + y(t_n)\right] \tag{5.66}$$

which approximates the integral over $[t_n, t_{n+1}]$, i.e., the area under $y(t)$ over $[t_n, t_{n+1}]$, by the area of a trapezoid; hence the name.

There are many ways of deriving the linear multistep formulas, including FE, BE, and TR. We have derived FE, BE, and TR starting from the familiar Taylor series. Higher order multistep methods can be derived in other ways, as we'll see later. But, first, we will "return to basics," so to speak, and discuss issues of accuracy and stability of general numerical methods for IVPs.

5.2.4 Quality Metrics

What does one require of a "good" numerical method? Obviously, one would like it to be *accurate* and *computationally efficient*. As we will see, this will mean that the method must also be *stable*. Linear multistep methods are the most computationally efficient methods available today, but it remains to discuss their *accuracy* and *stability*. The study of stability will also require an understanding of the notions of *convergence* and *consistency* of numerical methods. We will first discuss these issues for general numerical methods, and then specifically address linear multistep methods.

Convergence   Using again the notation of Lambert (1991), consider a numerical method in the general form:

$$\sum_{j=-1}^{k-1} \alpha_j x_{n-j} = h\phi_f(x_{n+1}, x_n, \ldots, x_{n-k+1}, t_n, h) \tag{5.67}$$

In the limit, as $h \to 0$, the discrete set of points $\{t_n\}$ approaches the continuous interval $[t_0, t_f]$. An obvious property to require of a "good" numerical method is that, as $h \to 0$, the numerical solution $\{x_n\}$ must approach the exact solution $x(t)$. A numerical method with this property is said to be *convergent*. This can be formalized as follows.

Definition 5.3. (Convergence) A numerical method is said to be *convergent* if, for all IVPs satisfying the hypotheses of Theorem 5.3, we have that:

$$\lim_{h \to 0} \left(\max_{t_n \in [t_0, t_f]} \|x(t_n) - x_n\| \right) = 0 \tag{5.68}$$

This is an *asymptotic* notion—it describes what happens in the limit as $h \to 0$, and is not helpful otherwise. Nevertheless, it is a good starting point. We obviously would not be interested in a numerical method which, *even in the limit as $h \to 0$*, does not produce a result that approaches $x(t)$! What properties must a numerical method satisfy, in order for it to be convergent? We consider this next.

Consistency   Consider a numerical method in the general form:

$$\sum_{j=-1}^{k-1} \alpha_j x_{n-j} = h\phi_f(x_{n+1}, x_n, \ldots, x_{n-k+1}, t_n, h) \tag{5.69}$$

Being a scheme for generating future values of the sequence $\{x_n\}$ from its past values, this formula is effectively a *difference equation*. In general, it is a *nonlinear difference equation*, because the functions $f(\cdot)$ and $\phi_f(\cdot)$ may be nonlinear. Thus, we are using a difference equation as a proxy for a differential equation. In order for the method to be convergent, one would expect that one requirement would be that the difference equation must be, as Lambert (1991) puts it, "*a sufficiently accurate representation of the differential system.*" Suppose it were the case that, if $x_{n-k+1}, \ldots, x_n$ in (5.69) are replaced by the exact $x(t_{n-k+1}), \ldots, x(t_n)$, then (5.69) would yield an $x_{n+1} = x(t_{n+1})$. In such a case, the difference equation would be an "*infinitely accurate*" representation of the differential equation. The assumption that all previous values are exact is obviously artificial and hypothetical; it is referred to as a *localizing assumption*. Nevertheless, it motivates our interest in the *residual*, defined as:

$$R_{n+1} \triangleq \sum_{j=-1}^{k-1} \alpha_j x(t_{n-j}) - h\phi_f(x(t_{n+1}), x(t_n), \ldots, x(t_{n-k+1}), t_n, h) \tag{5.70}$$

Thus, the residual is the result of applying the numerical method to the true solution $x(t)$. Having a zero residual, $R_{n+1} = 0$, would mean that the method is exact under the localizing assumption considered above. Except for a vanishing higher-order term, as we will see later on, the residual is essentially equal to the local truncation error.

In order for the difference equation to be, in some sense, consistent with the differential equation, we will insist on a small residual. We must, essentially, insist that the local truncation error approach zero, and fast enough, at each step, as the time-step approaches zero. This can be formalized as follows.

Definition 5.4. (Consistency) A numerical method is said to be *consistent* if, for all IVPs satisfying the hypotheses of Theorem 5.3, we have that:

$$\lim_{h \to 0} \left(\max_{t_n \in [t_0, t_f]} \left\| \frac{1}{h} R_{n+1} \right\| \right) = 0 \tag{5.71}$$

Consistency is, again, an *asymptotic* property and, by itself, it does not guarantee convergence, but it is one key ingredient. It can be shown that convergence implies consistency, but the converse is not true. There is one other required ingredient, as we will see below, but first we will examine consistency a little more closely.

Definition 5.5. For the general numerical method (5.67), we define the *first characteristic polynomial*, in $z \in \mathbb{C}$, as:

$$\rho(z) = \sum_{j=-1}^{k-1} \alpha_j z^{k-j-1} = \alpha_{-1} z^k + \alpha_0 z^{k-1} + \cdots + \alpha_{k-2} z + \alpha_{k-1} \tag{5.72}$$

As $h \to 0$, and the right-hand side of the method (5.67) vanishes, due to the term $h\phi_f(\cdot)$, its properties are pretty much determined by the α_j terms on the left-hand side. Thus, it should be useful to study this first characteristic polynomial, $\rho(z)$. Indeed, it can be shown that a numerical method is consistent if and only if, for any $t \in [t_0, t_f]$, the following two conditions hold:

$$\rho(1) = 0 \tag{5.73}$$

$$\phi_f(x(t), x(t), \ldots, x(t), t, 0) = \rho'(1) \times f(x(t), t) \tag{5.74}$$

The term $\phi_f(x(t), x(t), \ldots, x(t), t, 0)$ is the result, in the limit, of the following process. Let $h \to 0$, and consider the index $n \to \infty$ such that $t_n \to t$, where t is a fixed preselected time value. In the limit, all the time-points, $t_{n+1}, t_n, t_{n-1}, \ldots, t_{n-k+1}$, approach t, and all the values $x_{n+1}, x_n, x_{n-1}, \ldots, x_{n-k+1}$ approach $x(t)$.

Examples   We give a couple of examples, based on test cases given in Lambert (1991), one of which turns out to be consistent, and the other not. Consider the numerical method:

$$x_{n+1} + x_n - 2x_{n-1} = \frac{h}{4}(f_{n+1} + 8f_n + 3f_{n-1}) \tag{5.75}$$

which is a 2-step method ($k = 2$) for which $\rho(z) = z^2 + z - 2$ and $\rho'(z) = 2z + 1$, so that $\rho(1) = 0$, as it should be for consistency, and $\rho'(1) = 3$. In order to check the second condition for consistency, and since:

$$\phi_f(x_{n+1}, x_n, x_{n-1}, t_n, h) = \frac{1}{4}(f_{n+1} + 8f_n + 3f_{n-1}) \tag{5.76}$$

then:

$$\phi_f(x(t), x(t), x(t), t, 0) = \frac{1}{4}\left[f(x(t), t) + 8f(x(t), t) + 3f(x(t), t)\right]$$
$$= 3f(x(t), t)$$

therefore, the method is consistent. Consider another numerical method:

$$x_{n+1} - x_n = \frac{h}{3}(3f_n - 2f_{n-1}) \tag{5.77}$$

which is a 2-step method ($k = 2$) for which $\rho(z) = z^2 - z$ and $\rho'(z) = 2z - 1$, so that $\rho(1) = 0$, as it should be for consistency, and $\rho'(1) = 1$. In order to check the second condition for consistency, and since:

$$\phi_f(x_{n+1}, x_n, x_{n-1}, t_n, h) = \frac{1}{3}(3f_n - 2f_{n-1}) \tag{5.78}$$

then:

$$\phi_f(x(t), x(t), x(t), t, 0) = \frac{1}{3}\left[3f(x(t), t) - 2f(x(t), t)\right]$$
$$= \frac{1}{3}f(x(t), t)$$

therefore, the method is *not* consistent. It is easy to show that FE, BE, and TR, are all consistent. We are now ready to examine the second ingredient that is required for convergence.

Zero-Stability   Recall that we discussed earlier a notion of *total stability*, or well-posedness, as a property of the differential system. We are now concerned with notions of stability of the numerical method, i.e., of the difference equation. Nonlinear difference equations are much harder to study than differential equations, so there are several different useful notions of stability. We will start with a notion of stability in an asymptotic sense, i.e., in relation to what happens as $h \to 0$, hence the name *zero-stability*.

Zero-stability of the difference system will be defined in a way which is analogous to the earlier definition of total stability for the differential system. This requires that we be a bit more precise regarding the starting values of the numerical method. Strictly speaking, we are interested in the difference *system*, consisting of the numerical method *and* its startup scheme:

$$\sum_{j=-1}^{k-1} \alpha_j x_{n-j} = h\phi_f(x_{n+1}, x_n, \ldots, x_{n-k+1}, t_n, h)$$

$$x_i = \eta_i(h), \quad i = 0, 1, \ldots, k-1$$

where the $\eta_i(h)$ are the starting values. We now consider the effect of *perturbations* of the function $\phi_f(\cdot)$ and of the starting values $\eta_i(h)$, and we define the *perturbed* numerical method as:

$$\sum_{j=-1}^{k-1} \alpha_j \hat{x}_{n-j} = h\left[\phi_f(\hat{x}_{n+1}, \hat{x}_n, \ldots, \hat{x}_{n-k+1}, t_n, h) + \delta_{n+1}\right]$$

$$\hat{x}_i = \eta_i(h) + \delta_i, \quad i = 0, 1, \ldots, k-1$$

where $\{\delta_n\}$ is a *perturbation* of the system, and where the solution $\{\hat{x}_n\}$ is called the *perturbed solution*. We can now give the definition of zero-stability of the difference system, after Lambert (1991).

Definition 5.6. Let $\{\delta_n\}$ and $\{\delta_n^*\}$ be any two perturbations of the difference system, and let $\{\hat{x}_n\}$ and $\{\hat{x}_n^*\}$ be the resulting perturbed solutions. If there exist constants S and h_0 such that, for all $0 < h \le h_0$, we have:

$$\|\hat{x}_n - \hat{x}_n^*\| \le S\epsilon, \forall n \quad \text{whenever} \quad \|\delta_n - \delta_n^*\| \le \epsilon, \forall n, \tag{5.79}$$

then, we say that the difference system is *zero-stable*.

Note that zero-stability is, again, an asymptotic result, and it is a property of the *numerical method*, not of the system. Recall, Lipschitz continuity of the function $f(\cdot)$ ensures the *differential system* is totally stable (well-posed) and insensitive to perturbations. Zero-stability, as a property of the *difference system* ensures that the *difference system* is likewise insensitive to perturbations. It is equivalent to saying that the difference system is well-posed. If the difference system is not zero-stable, then essentially, the solution is not *computable* using finite precision computers.

Thus, zero-stability would seem to be another excellent property that we must insist upon, but how do we achieve it? How do we guarantee that a numerical method is zero-stable? As we will now see, the answer has to do with the roots of the first characteristic polynomial.

The Root Condition   It turns out that the *roots* of the first characteristic polynomial, $\rho(z)$, are quite important for studying zero-stability.

Definition 5.7. A numerical method is said to satisfy the *root condition* if every root of the first characteristic polynomial $\rho(z)$ is either *inside* the unit circle (in the complex plane), or *on* the unit circle but with multiplicity 1 (i.e., it is a *simple* root).

Thus, the root condition requires that the magnitudes of all the roots must be less than or equal to 1, and those whose magnitudes are equal to 1 must be simple roots. We can now state the following key result:

Theorem 5.4. *A difference system is zero-stable if and only if it satisfies the root condition.*

Thus, the root condition is a key *quality metric* that we must insist upon, in order to guarantee a "good" numerical method. As an example, the numerical method in (5.77) is a 2-step method ($k = 2$), for which $\rho(z) = z^2 - z$, whose roots are 0 and 1, so that this method is zero-stable. In contrast, the method in (5.75) is a 2-step method ($k = 2$), for which $\rho(z) = z^2 + z - 2$, whose roots are 1 and -2, so that this method is *not* zero-stable. It is easy to show that FE, BE, and TR are all zero-stable.

Fundamental Theorem   We can now state the *fundamental theorem* of the study of IVPs, as:

Theorem 5.5. *A difference system is convergent if and only if it is both consistent and zero-stable.*

Thus, consistency and zero-stability are the two key ingredients that are required to ensure that a numerical method is convergent. Since they are consistent and zero-stable, FE, BE, and TR are convergent. We now restrict our attention to linear multistep (LMS) methods.

5.2.5   Linear Multistep Methods

Recall, a k-step LMS method for solving $x' = f(x, t)$ has the general form:

$$\sum_{j=-1}^{k-1} \alpha_j x_{n-j} = h \sum_{j=-1}^{k-1} \beta_j f(x_{n-j}, t_{n-j}) \tag{5.80}$$

or, using the more compact notation:

$$\sum_{j=-1}^{k-1} \alpha_j x_{n-j} = h \sum_{j=-1}^{k-1} \beta_j f_{n-j} \tag{5.81}$$

where $\alpha_{-1} \triangleq 1$ and $|\alpha_{k-1}| + |\beta_{k-1}| \neq 0$. An LMS method is *explicit* if $\beta_{-1} = 0$; otherwise, it is *implicit*. Note, the value of k is the difference between the largest and smallest index. We saw earlier that FE, BE, and TR, are examples of 1-step LMS methods. FE is explicit, while BE and TR are implicit. Here are some other *explicit* examples, a few of the *Adams-Bashforth methods*:

$$x_{n+1} = x_n + hf_n \qquad\qquad (k = 1)$$

$$x_{n+1} = x_n + \frac{h}{2}(3f_n - f_{n-1}) \qquad\qquad (k = 2)$$

$$x_{n+1} = x_n + \frac{h}{12}(23f_n - 16f_{n-1} + 5f_{n-2}) \qquad\qquad (k = 3)$$

$$x_{n+1} = x_n + \frac{h}{24}(55f_n - 59f_{n-1} + 37f_{n-2} - 9f_{n-3}) \qquad (k = 4)$$

the first of which is FE, and here are some *implicit* examples, a few of the *Adams-Moulton methods*:

$$x_{n+1} = x_n + \frac{h}{2}(f_{n+1} + f_n) \qquad\qquad (k = 1)$$

$$x_{n+1} = x_n + \frac{h}{12}(5f_{n+1} + 8f_n - f_{n-1}) \qquad\qquad (k = 2)$$

$$x_{n+1} = x_n + \frac{h}{24}(9f_{n+1} + 19f_n - 5f_{n-1} + f_{n-2}) \qquad\qquad (k = 3)$$

$$x_{n+1} = x_n + \frac{h}{720}(251f_{n+1} + 646f_n - 264f_{n-1} + 106f_{n-2} - 19f_{n-3}) \quad (k = 4)$$

the first of which is the trapezoidal rule. Collectively, the above two classes of methods are referred to simply as the *Adams methods*.

5.3 ACCURACY OF LMS METHODS

In this and the next section, we will study various issues related to *accuracy* and *stability* of LMS methods, and we will see how one can derive the LMS formulas.

5.3.1 Order

To study accuracy, we start with the notion of *order*. Inspired by the earlier definition of the residual (5.70), which for an LMS method becomes:

$$R_{n+1} = \sum_{j=-1}^{k-1} \alpha_j x(t_{n-j}) - h \sum_{j=-1}^{k-1} \beta_j f(x(t_{n-j}), t_{n-j}) \qquad (5.82)$$

$$= \sum_{j=-1}^{k-1} \alpha_j x(t_n - jh) - h \sum_{j=-1}^{k-1} \beta_j x'(t_n - jh) \qquad (5.83)$$

we now define the *linear difference operator*, as follows.

Definition 5.8. The *linear difference operator* of an LMS method, denoted $\mathcal{D}$, is an operator which, applied to an arbitrarily differentiable time function $s(t)$, produces another time function:

$$\mathcal{D}[s(t), h] \triangleq \sum_{j=-1}^{k-1} \alpha_j s(t - jh) - h \sum_{j=-1}^{k-1} \beta_j s'(t - jh) \tag{5.84}$$

The term *arbitrarily differentiable* means that $s(t)$ differentiable as often as desired. Notice that, if $\mathcal{D}$ is applied to the true solution of the system, i.e., if $s(t)$ is replaced by $x(t)$, and the resulting function is evaluated at t_n, we would get exactly the residual; we express this by the notation:

$$R_{n+1} = \mathcal{D}[x(t), h]_{t_n} \tag{5.85}$$

The linear difference operator allows us to define the notion of *order* of an LMS method, as follows. Using a Taylor series expansion of $s(\tau)$ around t, we write:

$$s(\tau) = s(t) + \sum_{q=1}^{\infty} \frac{1}{q!} s^{(q)}(t)(\tau - t)^q \tag{5.86}$$

where $s^{(q)}(t)$ is the q-th derivative of $s(\cdot)$ evaluated at t; then, taking the derivative of (5.86) with respect to τ, we get:

$$s'(\tau) = \sum_{q=1}^{\infty} \frac{1}{(q-1)!} s^{(q)}(t)(\tau - t)^{q-1} \tag{5.87}$$

where $0! \triangleq 1$. Evaluating the above two results at $\tau = t - jh$, we get:

$$s(t - jh) = s(t) + \sum_{q=1}^{\infty} \frac{1}{q!} s^{(q)}(t)(-jh)^q \tag{5.88}$$

and:

$$s'(t - jh) = \sum_{q=1}^{\infty} \frac{1}{(q-1)!} s^{(q)}(t)(-jh)^{q-1} \tag{5.89}$$

Plugging these results into the expression for $\mathcal{D}[s(t), h]$ and collecting similar terms, we get the series:

$$\mathcal{D}[s(t), h] = C_0 s(t) + C_1 h s^{(1)}(t) + \cdots + C_q h^q s^{(q)}(t) + \cdots \tag{5.90}$$

where:

$$C_0 = \sum_{j=-1}^{k-1} \alpha_j \tag{5.91}$$

$$C_1 = -\sum_{j=-1}^{k-1} j\alpha_j - \sum_{j=-1}^{k-1} \beta_j \tag{5.92}$$

$$\vdots$$

$$C_q = \frac{(-1)^q}{q!}\sum_{j=-1}^{k-1} j^q\alpha_j - \frac{(-1)^{q-1}}{(q-1)!}\sum_{j=-1}^{k-1} j^{q-1}\beta_j \tag{5.93}$$

$$\vdots$$

We now define the key notion of *order* of an LMS method:

Definition 5.9. An LMS method is said to be of *order* p if $C_0 = \cdots = C_p = 0$, but $C_{p+1} \neq 0$, and C_{p+1} is called the *error constant* of the LMS method.

Note that the above discussion employed a Taylor series expansion of $s(\tau)$ around the point t. However, this does *not* mean that the order is dependent on the point around which the Taylor series is taken. Indeed, it can be shown, as in Lambert (1991), that the order of an LMS method is *independent* of the point $t \in \mathbb{R}$ around which the Taylor series expansion is taken: for a method of order p, it turns out, $C_0 = C_1 = \cdots = C_p = 0$ irrespective of t, C_{p+1} is independent of t, while $C_{p+2}, C_{p+3}, \ldots$ are dependent on t. Thus, the order and the value of the error constant are well-defined intrinsic properties of an LMS formula.

A higher-order method would typically (but not always) have better accuracy, keeping in mind the following result, called *the first Dahlquist barrier*:

Theorem 5.6. *(The first Dahlquist Barrier) No zero-stable k-step LMS method can have order exceeding $k + 1$ when k is odd and $k + 2$ when k is even.*

Thus, while selecting a desired order for an LMS method, one must select it in relation to the chosen number of steps k.

Finally, if the true solution, $x(t)$, has derivatives up to at least $(p + 1)$, then it is clear, due to (5.85) and (5.90), that the residual for an LMS method of order p is given by:

$$R_{n+1} = C_{p+1}h^{p+1}x^{(p+1)}(t_n) + 0(h^{p+2}) \tag{5.94}$$

This expression will be useful in order to capture the local truncation error, as we will see below.

5.3.2 Consistency

Recall that we had earlier defined the first characteristic polynomial $\rho(z)$ of a general numerical method for IVPs. We now define the *second characteristic polynomial* of an LMS method, as follows.

Definition 5.10. For a general LMS method (5.80), we define the *second characteristic polynomial*, in $z \in \mathbb{C}$, as:

$$\sigma(z) \triangleq \sum_{j=-1}^{k-1} \beta_j z^{k-j-1} = \beta_{-1} z^k + \beta_0 z^{k-1} + \cdots + \beta_{k-2} z + \beta_{k-1} \qquad (5.95)$$

It is easy to verify that:

$$C_0 = \rho(1) \qquad \text{and} \qquad C_1 = \rho'(1) - \sigma(1) - (k-1)\rho(1) \qquad (5.96)$$

Due to (5.85), then an LMS method is consistent if and only if:

$$\lim_{h \to 0} \frac{1}{h} \left(C_0 x(t_n) + C_1 h x^{(1)}(t_n) + C_2 h^2 x^{(2)}(t_n) + \cdots \right) = 0 \qquad (5.97)$$

Thus, an LMS method is consistent if $C_0 = C_1 = 0$, i.e., if it has an order $p \geq 1$, so that *every LMS method with order $p \geq 1$ is consistent*. In general, it is easy to show that an LMS method is consistent if and only if $\rho(1) = 0$ and $\rho'(1) = \sigma(1)$, since $\phi_f(x(t), \cdots, x(t), t, 0) = \sigma(1) f(x(t), t)$. Furthermore, the condition $C_0 = C_1 = 0$ means that:

$$\rho(1) = 0 \qquad \text{and} \qquad \rho'(1) = \sigma(1) \qquad (5.98)$$

Notice that if $\sigma(1) = 0$, then a consistent LMS method would have $\rho(1) = \rho'(1) = 0$, so that $+1$ is a double root of $\rho(z)$ and the LMS method would not be zero-stable. Thus, all consistent zero-stable LMS methods, must have $\sigma(1) \neq 0$.

5.3.3   The Backward Differentiation Formulas

For circuit simulation, the most useful formulas are TR and the so-called backward differentiation formulas (BDF), which are all implicit. The BDFs for $k = 1, 2, \ldots, 6$, with, respectively, order $p = 1, 2, \ldots, 6$, are as follows:

$$x_{n+1} - x_n = h f_{n+1}$$

$$x_{n+1} - \frac{4}{3} x_n + \frac{1}{3} x_{n-1} = \frac{2}{3} h f_{n+1}$$

$$x_{n+1} - \frac{18}{11} x_n + \frac{9}{11} x_{n-1} - \frac{2}{11} x_{n-2} = \frac{6}{11} h f_{n+1}$$

$$x_{n+1} - \frac{48}{25} x_n + \frac{36}{25} x_{n-1} - \frac{16}{25} x_{n-2} + \frac{3}{25} x_{n-3} = \frac{12}{25} h f_{n+1}$$

$$x_{n+1} - \frac{300}{137}x_n + \frac{300}{137}x_{n-1} - \frac{200}{137}x_{n-2} + \frac{75}{137}x_{n-3} - \frac{12}{137}x_{n-4} = \frac{60}{137}hf_{n+1}$$

$$x_{n+1} - \frac{360}{147}x_n + \frac{450}{147}x_{n-1} - \frac{400}{147}x_{n-2} + \frac{225}{147}x_{n-3} - \frac{72}{147}x_{n-4} + \frac{10}{147}x_{n-5} = \frac{60}{147}hf_{n+1}$$

Notice that the first BDF (for $k = 1$ and $p = 1$) is simply BE and, it can be shown that, any BDF with order higher than 6 is zero-unstable. We will return, below, to a formal definition and derivation of the BDFs. The most commonly used BDF in circuit simulation is the second order two-step BDF, which we will denote as BDF2. This is also referred to in the circuit simulation literature as the Gear-Shichman formula or the second order Gear formula. Higher order formulas are useful because they can be more accurate, but they also bring certain complications with them, e.g., it becomes more complicated to vary the time-step.

Many numerical methods, such as the BDFs but also the Adams methods we saw earlier, come in *families*, consisting of several members of different orders. To start a higher-order LMS method, it is typical to first use the lower order members of the family to establish starting values, and then use the higher order ones. For example, to use BDF2, you start with only x_0 and use BE to get x_1, and, only then, can you start up BDF2 using x_0 and x_1.

5.3.4 Local Truncation Error

We are now ready to define the *local truncation error* (LTE). Let $\tilde{x}_{n+1}$ be the value returned by the LMS method when we artificially set $x_{n-j} = x(t_{n-j})$, for $j = 0, 1, \ldots, k - 1$ (the localizing assumption). Then, the *local truncation error* (LTE) is defined as:

$$\tau_{n+1}(h) \triangleq x(t_{n+1}) - \tilde{x}_{n+1} \tag{5.99}$$

Thus, $\tau_{n+1}(h)$ is the "fresh" error incurred when stepping from t_n to t_{n+1}, due only to the local truncation of the Taylor series of $x(t)$. But it is this error *only* under the localizing assumption!

In order to find an expression for the LTE, we now explore the relationship between the LTE and the residual, as in Lambert (1991). Firstly, based on the definition of $\tilde{x}_{n+1}$, and since $\alpha_{-1} = 1$, we can write:

$$\tilde{x}_{n+1} + \sum_{j=0}^{k-1} \alpha_j x(t_{n-j}) = h\beta_{-1} f(\tilde{x}_{n+1}, t_{n+1}) + h\sum_{j=0}^{k-1} \beta_j x'(t_{n-j}) \tag{5.100}$$

Secondly, based on the definition of the residual, we can write:

$$x(t_{n+1}) + \sum_{j=0}^{k-1} \alpha_j x(t_{n-j}) = h\beta_{-1} f(x(t_{n+1}), t_{n+1}) + h\sum_{j=0}^{k-1} \beta_j x'(t_{n-j}) + R_{n+1} \tag{5.101}$$

Subtracting the first equation from the second, we have:

$$x(t_{n+1}) - \tilde{x}_{n+1} = h\beta_{-1}\left[f(x(t_{n+1}), t_{n+1}) - f(\tilde{x}_{n+1}, t_{n+1})\right] + R_{n+1} \qquad (5.102)$$

Using the mean value theorem, we can write:

$$f(x(t_{n+1}), t_{n+1}) - f(\tilde{x}_{n+1}, t_{n+1}) = \overline{J}\left(x(t_{n+1}) - \tilde{x}_{n+1}\right) \qquad (5.103)$$

where $\overline{J}$ is a "special" evaluation of the Jacobian of $f(\cdot)$, obtained by evaluating each row i at a different point ξ_i along the line segment from $x(t_{n+1})$ to $\tilde{x}_{n+1}$ in $\mathbb{R}^m$. Therefore, the key relationship between the LTE and the residual is:

$$\left(I - h\beta_{-1}\overline{J}\right)\tau_{n+1}(h) = R_{n+1} \qquad (5.104)$$

As a result, for an explicit LMS method, with $\beta_{-1} = 0$, the LTE is equal to the residual, while for an implicit LMS method, the same is true with some approximation. In general, for an LMS method of order p, and if the solution is differentiable up to at least $(p + 1)$, then it follows from (5.104) and (5.94) that:

$$\text{LTE} = \tau_{n+1}(h) = C_{p+1}h^{p+1}x^{(p+1)}(t_n) + 0(h^{p+2}) \qquad (5.105)$$

so that, in general:

$$\tau_{n+1}(h) - R_{n+1} = 0(h^{p+2}) \qquad (5.106)$$

and we define the *principal local truncation error* (PLTE) as:

$$\text{PLTE} = C_{p+1}h^{p+1}x^{(p+1)}(t_n) \qquad (5.107)$$

In most cases, we only care to find (or we can only estimate) the PLTE, so that the two names are often used interchangeably in the literature.

Some remarks are in order. Because the LTE is essentially the same as the residual, except for a vanishing higher-order term, it is common to use:

$$\text{LTE} = \tau_{n+1}(h) \approx R_{n+1} \approx \text{PLTE} \qquad (5.108)$$

As far as an overall bound for the LTE, the following bound is available:

$$\|\tau_{n+1}(h)\| \approx \|R_{n+1}\| \leq Gh^{p+1} \max_{\xi \in [t_0, t_f]} \|x^{(p+1)}(\xi)\| \qquad (5.109)$$

where G is a constant that depends on the coefficients α_i and β_i, although this does not seem to be of much practical use. Finally, note that some authors define the LTE to be (exactly) the residual, while some others define it as the residual divided by h.

Examples   We study the order and accuracy of the basic methods: FE, BE, and TR.

Forward Euler: With $x_{n+1} = x_n + hf(x_n, t_n)$, we have that:

$$\tilde{x}_{n+1} = x(t_n) + hx'(t_n) \tag{5.110}$$

while from the Taylor series expansion, we have:

$$x(t_{n+1}) = x(t_n) + hx'(t_n) + \frac{1}{2}h^2 x''(\xi) \tag{5.111}$$

where ξ is between t_n and t_{n+1}, so that:

$$\tau_{n+1}(h) = \frac{1}{2}h^2 x''(\xi) \tag{5.112}$$

or, equivalently, for a slightly longer Taylor series and for some other ξ between t_n and t_{n+1}:

$$\tau_{n+1}(h) = \frac{1}{2}h^2 x''(t_n) + \frac{1}{6}h^3 x'''(\xi) \tag{5.113}$$

so that FE has order 1, with $C_2 = 1/2$, and its LTE is said to be of order 2.

Backward Euler: With $x_{n+1} = x_n + hf(x_{n+1}, t_{n+1})$, we have that $\alpha_{-1} = 1$, $\alpha_0 = -1$, and $\beta_{-1} = 1$, so that:

$$C_0 = 0, \quad C_1 = 0, \quad \text{and} \quad C_2 = -\frac{1}{2} \tag{5.114}$$

so that BE has order 1, with $C_2 = -1/2$, and its LTE is of order 2, given by:

$$\tau_{n+1}(h) = -\frac{1}{2}h^2 x''(t_n) + 0(h^3) \tag{5.115}$$

Trapezoidal Rule: With $x_{n+1} = x_n + (h/2)(f_{n+1} + f_n)$, we have that $\alpha_{-1} = 1$, $\alpha_0 = -1$, $\beta_{-1} = 1/2$, and $\beta_0 = 1/2$, so that:

$$C_0 = 0, \quad C_1 = 0, \quad C_2 = 0, \quad \text{and} \quad C_3 = -\frac{1}{12} \tag{5.116}$$

so that TR has order 2, with $C_3 = -1/12$, and its LTE is of order 3, given by:

$$\tau_{n+1}(h) = -\frac{1}{12}h^3 x'''(t_n) + 0(h^4) \tag{5.117}$$

5.3.5   Deriving the LMS Methods

There is an infinite variety of possible LMS methods but, obviously, one is interested only in the good ones. So far, we have seen that quality metrics include *convergence* and *small error* (high p and small C_{p+1}). Soon, we will add *stability* as the third key quality metric. But there is one question yet to be addressed: how does one *derive* an LMS formula? How do we find the coefficients?

A k-step LMS formula can have up to $2k+1$ coefficients that need to be determined (recall, $\alpha_{-1} = 1$ is assumed). Some of these may be specified up-front, due to various considerations; for example, explicit methods have $\beta_{-1} = 0$. Otherwise, the zero/non-zero pattern of the coefficients may be captured by specifying the two characteristic polynomials $\rho(z)$ and $\sigma(z)$. For example, the Adams methods are characterized by having the simplest possible first characteristic polynomial (subject to consistency):

$$\rho(z) = z^k - z^{k-1} \tag{5.118}$$

while the BDFs are characterized by having the simplest possible second characteristic polynomial (subject to implicitness):

$$\sigma(z) = \beta_{-1} z^k \tag{5.119}$$

Once the step number k along with $\rho(z)$ and $\sigma(z)$ are specified, we know exactly how many coefficients there are to be determined. Then, we must specify the desired order, p, keeping in mind the first Dahlquist barrier. As a result, the equations $C_0 = C_1 = \cdots = C_p = 0$ provide a set of $(p+1)$ simultaneous linear equations in the unknown coefficients. This system of equations is then solved and, in general, it may have no solution, a unique solution, or an infinity of solutions. Where multiple solutions are possible, other criteria may be used to help make a selection.

A simpler alternative approach to finding the coefficients is as follows. Consider the set of polynomials $s(t) \in \{1, t, t^2, \ldots, t^p\}$. Note that they are arbitrarily differentiable, and that $s^{(q)}(t) = 0$, $\forall q > p$, and consider the linear difference operator applied to $s(t)$:

$$\mathcal{D}[s(t), h] = C_0 s(t) + C_1 h s^{(1)}(t) + C_2 h^2 s^{(2)}(t) + \cdots \tag{5.120}$$

Then, for a method of order p, $\mathcal{D}[s(t), h] = 0$, for every $s(t)$, because the first $(p+1)$ terms are zero due to $C_0 = C_1 = \cdots = C_p = 0$, while the rest are zero because they contain $s^{(q)}$ with $q > p$. Thus, by plugging the above $(p+1)$ polynomials into the LMS formula, we get a system of $(p+1)$ simultaneous linear equations in the unknown coefficients. The resulting equations apply for any t_n, so we set t_n arbitrarily. This system is solved, if possible, and the LMS method is thereby determined. We will now see how this applies to deriving the BDFs.

Deriving the BDFs   A k-step BDF is *defined* by:

$$\rho(z) = \alpha_{-1}z^k + \alpha_0 z^{k-1} + \cdots + \alpha_{k-2}z + \alpha_{k-1} \qquad (5.121)$$

$$\sigma(z) = \beta_{-1}z^k \qquad (5.122)$$

so that there are only $(k+1)$ coefficients to be determined, because $\alpha_{-1} = 1$, to build the BDF:

$$\sum_{j=-1}^{k-1} \alpha_j x_{n-j} = h\beta_{-1} f(x_{n+1}, t_{n+1}) \qquad (5.123)$$

It can be shown that a k-step BDF has order $p = k$ and an error constant:

$$C_{p+1} = \frac{-1}{(p+1)\sum_{i=1}^{p}(1/i)} \qquad (5.124)$$

Plugging the polynomials $x(t) = t^q$, for $q = 0, 1, \ldots, p$, into the LMS formula gives the following $(p+1)$ equations in $(p+1)$ unknowns:

$$\begin{cases} \displaystyle\sum_{j=-1}^{k-1} \alpha_j = 0 \\ \displaystyle\sum_{j=-1}^{k-1} \alpha_j t_{n-j}^q = h\beta_{-1}q t_{n+1}^{q-1}, \quad q = 1, 2, \ldots, p \end{cases} \qquad (5.125)$$

and, arbitrarily setting $t_n = 0$, the system of equations becomes:

$$\begin{cases} \displaystyle\sum_{j=-1}^{k-1} \alpha_j = 0 \\ 1 + \displaystyle\sum_{j=1}^{k-1} \alpha_j(-j)^q = \beta_{-1}q, \quad q = 1, 2, \ldots, p \end{cases} \qquad (5.126)$$

This system is then solved for all the coefficients.

5.3.6   Solving Implicit Methods

For an implicit LMS method, finding x_{n+1} requires solving the possibly nonlinear system:

$$x_{n+1} + \sum_{j=0}^{k-1} \alpha_j x_{n-j} = h\beta_{-1} f(x_{n+1}, t_{n+1}) + \sum_{j=0}^{k-1} \beta_j f(x_{n-j}, t_{n-j}) \qquad (5.127)$$

or:

$$x_{n+1} = h\beta_{-1} f(x_{n+1}, t_{n+1}) + \psi_n \qquad (5.128)$$

where:

$$\psi_n \triangleq -\sum_{j=0}^{k-1} \alpha_j x_{n-j} + \sum_{j=0}^{k-1} \beta_j f(x_{n-j}, t_{n-j}) \qquad (5.129)$$

is easily found from previously computed values. The nonlinear (5.128) can be solved by any of the standard methods, such as the fixed point method or Newton's method. Practical experience shows that, for non-stiff problems, fixed point is quite good, according to Lambert (1991), given a good initial candidate solution. Even better, one does not have to iterate the fixed point method until convergence. Instead, the iterations are cut short (a fixed number of iterations is applied) and the remaining error is accepted as simply part of the overall local error that is incurred when solving an ODE. However, for stiff problems (such as in circuit simulation), where implicit methods are required for stability reasons, this approach does not work well. Instead, the nonlinear (5.128) must be solved by Newton's method, and the iterations must be carried out until Newton convergence is achieved.

Predictor-Corrector Methods   When solving the implicit nonlinear (5.128) using either fixed point or Newton's method, what should be the initial candidate solution? The standard method is to first use an *explicit* LMS method to get a candidate value of x_{n+1}, and then use that as the initial solution for Newton's method or fixed point. As a result, a *pair* of LMS methods are used, an explicit one (called the *predictor*) and an implicit one (called the *corrector*). The use of predictor-corrector pairs is standard practice in all modern codes for solving ODEs. For non-stiff problems, a typical combination is to use an Adams-Bashforth predictor with an Adams-Moulton corrector. The result is called an Adams-Bashforth-Moulton (ABM) method. For stiff problems, the better approach is to extrapolate an interpolation polynomial as a predictor (see below) with a TR or BDF as the corrector.

It turns out that there is a significant advantage to be had by *requiring both the predictor and corrector to have the same order*. This greatly simplifies the LTE estimation, as we will see. It usually requires that the predictor have a step number, k, that is larger by 1 than that of the corrector. When the corrector iterations are not carried out to convergence, the accuracy and stability of the pair is dependent on both of them. One must analyze the pair, working together, to study their accuracy and stability properties; this is a significant complication. When the corrector is applied to convergence, as done in the stiff case, then the properties of the pair are identical to those of the corrector alone. No further detailed analysis of the pair is required. Thus, for our purposes, the only relevant part of predictor-corrector theory is the choice of predictor and the LTE estimation, which we cover below.

Although the use of predictor-corrector methods has been tested in the circuits research literature, it is not clear if any existing simulators use it. Instead, a simpler method that is often employed is to use the solution at the previous time-point as the initial candidate solution. The original SPICE program used only TR with no predictor because, according to Nagel (1975), the advantages of higher order methods did not outweigh their complexities. Nevertheless, we will give a brief coverage of how predictor-corrector methods *can* be applied to circuits.

Practical experience shows that the best predictor for stiff problems is to use an interpolation polynomial, rather than a regular LMS method, as follows. It can be shown, as we will see below, that an interpolation polynomial on $(p + 1)$ points, which itself is a valid LMS method, has order p. Thus, to get a predictor of order p, we use a degree $\leq p$ polynomial that interpolates (passes through) the most recent $(p + 1)$ previous values:

$$(t_n, x_n) \quad (t_{n-1}, x_{n-1}) \quad \cdots \quad (t_{n-p}, x_{n-p}) \tag{5.130}$$

and that polynomial is extrapolated to get an initial candidate value of x_{n+1}, which we denote by $x_{n+1}^{(0)}$. Thus, for the first-order BDF (BE), we interpolate the points (t_n, x_n) and (t_{n-1}, x_{n-1}) with a straight line and extrapolate that line to find $x_{n+1}^{(0)}$. For the second-order BDF (BDF2), or for TR, we interpolate (t_n, x_n), (t_{n-1}, x_{n-1}), and (t_{n-2}, x_{n-2}) with a parabola and extrapolate to get $x_{n+1}^{(0)}$. In general, there is a unique polynomial of degree no more than p that interpolates (passes through) the previous $(p + 1)$ points.

In the following, to complete our coverage of accuracy of LMS methods, we will discuss polynomial interpolation and LTE estimation.

5.3.7 Interpolation Polynomial

There are two general canonical forms for an interpolation polynomial, depending on whether we have *equidistant data* or not. When previous time-steps are all identical (same h) we say that we have equidistant data; all the numerical methods we have seen so far assume fixed time-step and equidistant data. Non-equidistant data arise in the context of variable time-step methods, and there are two approaches to deal with that:

1. Interpolate previous data values at a (new) equidistant time mesh, and restart the method from that point with new equidistant data.
2. Re-derive the LMS methods to use non-equidistant data.

We will return to this topic later on, but first, on the topic of interpolation, we give a theorem, from Dahlquist and Björck (2008), that provides the interpolation error.

Theorem 5.7. *Let* $x(t) : \mathbb{R} \to \mathbb{R}$ *and consider the discrete* $(p + 1)$ *time-points* $t_n > t_{n-1} > \cdots > t_{n-p}$ *which, in general, may not be equidistant. Let* $x^{(p+1)}(t)$ *be continuous in the smallest interval that contains the points* t *and* $t_n, \ldots, t_{n-p}$, *which we denote by* $\mathcal{J}(t, t_n, \ldots, t_{n-p})$. *Let* $P_n(t)$ *be the unique degree* $\leq p$ *polynomial that passes through (interpolates) the points:*

$$(t_n, x(t_n)) \quad (t_{n-1}, x(t_{n-1})) \quad \cdots \quad (t_{n-p}, x(t_{n-p})) \tag{5.131}$$

Then, there exists a point $\xi \in \mathcal{J}(t, t_n, \ldots, t_{n-p})$ such that:

$$x(t) - P_n(t) = \frac{x^{(p+1)}(\xi)}{(p+1)!} \prod_{j=0}^{p} (t - t_{n-j}) \tag{5.132}$$

When the points $t_n, \ldots, t_{n-p}$ are equidistant, the polynomial can be compactly expressed as the *Newton-Gregory backward interpolation formula*, which makes use of the *backward difference operator*. When the points $t_n, \ldots, t_{n-p}$ are not equidistant, the polynomial can be expressed as the *Newton divided difference interpolation formula*, which makes use of the *divided difference operator*. We describe these two possibilities, below.

Backward Difference Operator Let x_j, $j = 0, 1, 2, \ldots$, be a sequence in $\mathbb{R}^m$, then we define the *backward difference operator*, denoted ∇, as one that generates the sequence:

$$\nabla x_j = x_j - x_{j-1}, \quad j = 1, 2, \ldots \tag{5.133}$$

and higher powers are defined recursively, so that $\nabla^k x_j = \nabla^{k-1}(\nabla x_j)$. Because $\nabla(x_j - x_{j-1}) = \nabla x_j - \nabla x_{j-1}$, and:

$$\nabla^k x_j = \nabla^{k-2}[\nabla(\nabla x_j)] = \nabla^{k-2}[\nabla(x_j - x_{j-1})] \tag{5.134}$$

then:

$$\nabla^k x_j = \nabla^{k-1} x_j - \nabla^{k-1} x_{j-1} \tag{5.135}$$

Thus, for example:

$$\nabla^2 x_n = \nabla(x_n - x_{n-1}) = (x_n - x_{n-1}) - (x_{n-1} - x_{n-2}) \tag{5.136}$$

$$= x_n - 2x_{n-1} + x_{n-2} \tag{5.137}$$

and:

$$\nabla^3 x_n = \nabla(x_n - 2x_{n-1} + x_{n-2}) \tag{5.138}$$

$$= x_n - 3x_{n-1} + 3x_{n-2} - x_{n-3} \tag{5.139}$$

When x_j are the result of evaluating a continuous function $x(t)$ at the discrete time-points t_j, then we simply write $\nabla x(t_j) = x(t_j) - x(t_{j-1})$, etc.

Newton-Gregory Backward Interpolation Polynomial If we let $t = t_n + sh$, where $s \in \mathbb{R}$, $s = (t - t_n)/h$, then the interpolation polynomial of degree $\leq p$ that passes through the $(p + 1)$ points:

$$(t_n, x_n) \quad (t_{n-1}, x_{n-1}) \quad \cdots \quad (t_{n-p}, x_{n-p}) \tag{5.140}$$

where h is the separation between the time-points, can be expressed as:

$$P_n(t) = x_n + \sum_{k=1}^{p} \left[\frac{1}{k!} \prod_{j=0}^{k-1} (s+j) \right] \nabla^k x_n \qquad (5.141)$$

which is called the Newton-Gregory backward interpolation formula. For example, suppose we want to interpolate the points:

$$(t_n, x_n) \quad (t_{n-1}, x_{n-1}) \quad (t_{n-2}, x_{n-2}) \qquad (5.142)$$

then $p = 2$ and:

$$P_n(t) = x_n + s\nabla x_n + \frac{1}{2}s(s+1)\nabla^2 x_n \qquad (5.143)$$

When x_{n-j} are the result of evaluating at t_{n-j} a function $x(t) \in \mathbb{R}$, which is differentiable up to $(p+1)$, then (5.132) leads to the interpolation error:

$$x(t) - P_n(t) = h^{p+1} \frac{x^{(p+1)}(\xi)}{(p+1)!} \prod_{j=0}^{p} (s+j) \qquad (5.144)$$

where ξ is a point in the smallest interval that contains t and $t_n, \ldots, t_{n-p}$. When $x(t) \in \mathbb{R}^m$, then $x^{(p+1)}(\xi)$ is replaced by vector $\overline{x}^{(p+1)}(\xi)$ in which each component is evaluated at a different ξ_i, forming the vector ξ.

For use as a predictor, to be evaluated at t_{n+1}, we are interested only in the setting $s = 1$, so that the general Newton-Gregory backward interpolation formula reduces to:

$$x_{n+1}^{(0)} = x_n + \sum_{k=1}^{p} \nabla^k x_n \qquad (5.145)$$

and the case $p = 2$, for example, becomes:

$$x_{n+1}^{(0)} = x_n + \nabla x_n + \nabla^2 x_n \qquad (5.146)$$

$$= 3x_n - 3x_{n-1} + x_{n-2} \qquad (5.147)$$

which makes for a very simple predictor for the equidistant TR or BDF2.

Polynomial Predictor   The polynomial predictor is an explicit LMS method with no β_{n-j} coefficients and with:

$$\sum_{j=0}^{k-1} \alpha_j x_{n-j} = -P_n(t_{n+1}) = -x_n - \sum_{j=1}^{p} \nabla^j x_n \qquad (5.148)$$

What is the order of this explicit LMS method? Because it is the unique polynomial that interpolates the $(p+1)$ points, then it is clear that it must be *exact* if applied to $x(t) = t^q$, for $q = 0, 1, \ldots, p$. By this we mean that, under the localizing assumption, it would have a zero LTE and zero residual. Looking at the linear difference operator, it is easy to see that this gives:

$$C_0 = C_1 = \cdots = C_p \tag{5.149}$$

because the fact that $x^{(k)}(t_n) = 0$, for all $k > q$ leads to:

- When $q = 0$ and $x(t) = 1$, we have $0 = R_{n+1} = C_0$, so $C_0 = 0$.
- When $q = 1$ and $x(t) = t$, we have $0 = R_{n+1} = C_0 t_n + C_1 h$, so $C_1 = 0$.
- And so on, until $q = p$ and $C_p = 0$.

And, using the above error expression (5.144), we have for the $s = 1$ case:

$$x(t_{n+1}) - P_n(t_{n+1}) = h^{p+1}\overline{x}^{(p+1)}(\xi) \tag{5.150}$$

and one can easily show that $\overline{x}^{(p+1)}(\xi) = x^{(p+1)}(t_n) + 0(h)$, so that:

$$x(t_{n+1}) - P_n(t_{n+1}) = h^{p+1}x^{(p+1)}(t_n) + 0(h^{p+2}) \tag{5.151}$$

which, because $P_n(t_{n+1})$ interpolates (exactly) the previous points (the localizing assumption), means that $P_n(t_{n+1}) = \tilde{x}_{n+1}$, so that the LTE is:

$$\text{LTE} = R_{n+1} = h^{p+1}x^{(p+1)}(t_n) + 0(h^{p+2}) \tag{5.152}$$

in this (explicit) case being equal to the residual. As a result, this polynomial predictor is of order p and has an error constant $C_{p+1} = 1$.

Deriving the BDFs   We digress briefly to see another way in which the BDFs can be derived, which makes use of an interpolation polynomial over past data points. This alternate approach will be useful in our future study of DAEs. For a BDF of order $p = k$, consider the interpolation polynomial $P_{n+1}(t)$, of degree $\leq k$, that interpolates the $k + 1$ equidistant points:

$$(t_{n+1}, x_{n+1}), (t_n, x_n), \ldots, (t_{n-k+1}, x_{n-k+1}) \tag{5.153}$$

As we saw previously, the Newton-Gregory backward interpolation polynomial is:

$$P_{n+1}(t) = x_{n+1} + \sum_{i=1}^{k} \left[\frac{1}{i!} \prod_{j=0}^{i-1}(s + j) \right] \nabla^i x_{n+1} \tag{5.154}$$

where $t = t_{n+1} + sh$, h is the time-step, and $s \in \mathbb{R}$, so that:

$$x(t) \approx x_{n+1} + \sum_{i=1}^{k} \left[\frac{1}{i!} \prod_{j=0}^{i-1}(s + j) \right] \nabla^i x_{n+1} \tag{5.155}$$

If we differentiate both sides of the above, we can write:

$$x'(t) \approx \frac{1}{h} \sum_{i=1}^{k} \delta_i(s) \nabla^i x_{n+1} \qquad (5.156)$$

where:

$$\delta_i(s) = \frac{1}{i!} \frac{d}{ds} \prod_{j=0}^{i-1} (s+j) \qquad (5.157)$$

At t_{n+1}, with $s = 0$, it is easy to show that:

$$\delta_0(0) = 0, \quad \text{and} \quad \delta_i(0) = 1/i, \forall i = 1, 2, \dots \qquad (5.158)$$

so that:

$$x'(t_{n+1}) \approx \frac{1}{h} \sum_{i=1}^{k} \frac{1}{i} \nabla^i x_{n+1} \qquad (5.159)$$

and, because $x'(t) = f(x, t)$, this motivates the introduction of the BDF, in backward difference form (hence the name), as:

$$f(x_{n+1}, t_{n+1}) = \frac{1}{h} \sum_{i=1}^{k} \frac{1}{i} \nabla^i x_{n+1} \qquad (5.160)$$

which is equivalent to the formulas given earlier for the BDFs.

Divided Difference Operator   To deal with non-equidistant data, we need the concept of a *divided difference operator*, as follows. For a sequence $\{x_n\}$ corresponding to the time-points $\{t_n\}$, define the *zeroth divided difference* of $\{x_n\}$ relative to t_i, denoted $x[t_i]$, as simply the value of x corresponding to t_i:

$$x[t_i] = x_i \qquad (5.161)$$

The *first divided difference* of $\{x_n\}$ with respect to t_i and t_{i-1} is denoted $x[t_i, t_{i-1}]$ and is defined as:

$$x[t_i, t_{i-1}] = \frac{x[t_i] - x[t_{i-1}]}{t_i - t_{i-1}} \qquad (5.162)$$

The *second divided difference* of $\{x_n\}$ with respect to t_i, t_{i-1}, and t_{i-2} is denoted $x[t_i, t_{i-1}, t_{i-2}]$ and defined as:

$$x[t_i, t_{i-1}, t_{i-2}] = \frac{x[t_i, t_{i-1}] - x[t_{i-1}, t_{i-2}]}{t_i - t_{i-2}} \qquad (5.163)$$

In general, the k-th divided difference of $\{x_n\}$ relative to $t_i, t_{i-1}, \ldots, t_{i-k}$, is:

$$x[t_i, t_{i-1}, \ldots, t_{i-k+1}, t_{i-k}] = \frac{x[t_i, \ldots, t_{i-k+1}] - x[t_{i-1}, \ldots, t_{i-k}]}{t_i - t_{i-k}} \qquad (5.164)$$

Thus, for example, given three data points, the divided difference is:

$$x[t_n, t_{n-1}, t_{n-2}] = \frac{\frac{x_n - x_{n-1}}{t_n - t_{n-1}} - \frac{x_{n-1} - x_{n-2}}{t_{n-1} - t_{n-2}}}{t_n - t_{n-2}} \qquad (5.165)$$

and, if we denote $h_n = t_n - t_{n-1}$ and $h_{n-1} = t_{n-1} - t_{n-2}$, then:

$$x[t_n, t_{n-1}, t_{n-2}] = \frac{1}{h_{n-1} + h_n}\left(\frac{x_n - x_{n-1}}{h_n} - \frac{x_{n-1} - x_{n-2}}{h_{n-1}}\right) \qquad (5.166)$$

Newton Divided Difference Interpolation Formula   Consider the $(p+1)$ points:

$$(t_n, x_n) \quad (t_{n-1}, x_{n-1}) \quad \cdots \quad (t_{n-p}, x_{n-p}) \qquad (5.167)$$

where, in general, the time-points are not equidistant. Then, the degree $\leq p$ polynomial that interpolates these points can be expressed as the Newton divided difference interpolation polynomial:

$$P_n(t) = x_n + \sum_{k=1}^{p}\left[\prod_{j=0}^{k-1}(t - t_{n-j})\right] x[t_n, \ldots, t_{n-k}] \qquad (5.168)$$

For example, suppose we want to interpolate the points:

$$(t_n, x_n) \quad (t_{n-1}, x_{n-1}) \quad (t_{n-2}, x_{n-2}) \qquad (5.169)$$

then $p = 2$ and:

$$P_n(t) = x_n + (t - t_n)x[t_n, t_{n-1}] + (t - t_n)(t - t_{n-1})x[t_n, t_{n-1}, t_{n-2}] \qquad (5.170)$$

leading to:

$$P_n(t) = x_n + (t - t_n)\frac{(x_n - x_{n-1})}{t_n - t_{n-1}} + \frac{(t - t_n)(t - t_{n-1})}{t_n - t_{n-2}}\left(\frac{x_n - x_{n-1}}{t_n - t_{n-1}} - \frac{x_{n-1} - x_{n-2}}{t_{n-1} - t_{n-2}}\right)$$

For use as a predictor, we are interested only in the setting $t = t_{n+1}$, so that the general formula becomes:

$$x_{n+1}^{(0)} = x_n + \sum_{k=1}^{p}\left[\prod_{j=0}^{k-1}(t_{n+1} - t_{n-j})\right] x[t_n, \ldots, t_{n-k}] \qquad (5.171)$$

and the case $p = 2$ becomes:

$$x_{n+1}^{(0)} = x_n + (t_{n+1} - t_n) \frac{(x_n - x_{n-1})}{t_n - t_{n-1}}$$
$$+ \frac{(t_{n+1} - t_n)(t_{n+1} - t_{n-1})}{t_n - t_{n-2}} \left(\frac{x_n - x_{n-1}}{t_n - t_{n-1}} - \frac{x_{n-1} - x_{n-2}}{t_{n-1} - t_{n-2}} \right)$$

which makes for a simple predictor for the non-equidistant TR or BDF2.

5.3.8 Estimating the LTE

It is always better to vary the time-step during the process of solving an ODE. This is done by monitoring the LTE, as in the following framework:

1. A step is taken, and x_{n+1} is found.
2. The LTE is computed and compared to a prespecified threshold.
3. If the LTE is too large, the step is not accepted, the time-step h is reduced, and the step is re-attempted.
4. If the LTE is smaller than the threshold, the step is accepted. If the LTE is very small, the time-step is increased for future steps.

Since the LTE is the error relative to the *exact* solution, one may well wonder how exactly it is to be found in practice. In fact, there is no exact way of finding the LTE. Generally, we estimate the PLTE and use that as a proxy for the LTE. But, even that is not easy, and will require some approximation. We now examine several ways in which the PLTE can be estimated.

In the context of predictor-corrector methods, the standard method for estimating the PLTE uses *Milne's estimate*, after Milne (1949), which is as follows. Let $x_{n+1}^{(0)}$ be the initial candidate solution, obtained by using an order p *predictor* based on an extrapolated degree p interpolation polynomial. Then, based on (5.151), we can write:

$$x(t_{n+1}) - \tilde{x}_{n+1}^{(0)} = h^{p+1} x^{(p+1)}(t_n) + 0(h^{p+2}) \tag{5.172}$$

where $\tilde{x}_{n+1}^{(0)}$ is the value of $x_{n+1}^{(0)}$ under the localizing assumption. Let x_{n+1} be the final solution obtained by an order p *corrector*, with error constant C_{p+1}, using Newton's method, applied to convergence. Then, as we already know:

$$x(t_{n+1}) - \tilde{x}_{n+1} = C_{p+1} h^{p+1} x^{(p+1)}(t_n) + 0(h^{p+2}) \tag{5.173}$$

where, again, $\tilde{x}_{n+1}$ is the value of x_{n+1} under the localizing assumption. Subtracting the first equation from the second, we get:

$$(1 - C_{p+1}) h^{p+1} x^{(p+1)}(t_n) = (\tilde{x}_{n+1} - \tilde{x}_{n+1}^{(0)}) + 0(h^{p+2}) \tag{5.174}$$

from which:

$$C_{p+1}h^{p+1}x^{(p+1)}(t_n) = \frac{C_{p+1}}{1 - C_{p+1}}\left(\tilde{x}_{n+1} - \tilde{x}_{n+1}^{(0)}\right) + 0(h^{p+2}) \qquad (5.175)$$

where the left-hand side is simply the PLTE. Therefore, the PLTE is given by:

$$\text{PLTE} = \frac{C_{p+1}}{1 - C_{p+1}}\left(\tilde{x}_{n+1} - \tilde{x}_{n+1}^{(0)}\right) + 0(h^{p+2}) \qquad (5.176)$$

which, because we do not, in practice, have access to the values under the localizing assumption, motivates the practical estimate:

$$\text{PLTE} \approx \frac{C_{p+1}}{1 - C_{p+1}}\left(x_{n+1} - x_{n+1}^{(0)}\right) \qquad (5.177)$$

with the expectation that the difference between the two is negligible.

If not using a predictor-corrector pair, as would seem to be the case in most simulators, then alternate approaches are available, as follows.

Richardson Extrapolation   Richardson extrapolation is the ancestor of many *extrapolation methods* used in numerical analysis, whose purpose is to increase the accuracy. It is an old method, dating back to 1927 or earlier. A special application of it leads to a PLTE estimate similar to Milne's, and we summarize this here only for the case of 1-step LMS methods. Having found x_{n-1} and x_n, let $h = t_{n+1} - t_n$, with $t_n - t_{n-1} = \alpha h$, where $\alpha > 0$, and use a 1-step LMS method to find x_{n+1} in two ways:

1. Using a time-step h, find x_{n+1} from x_n, denote the solution by $x_{n+1}^{(1)}$, and under the localizing assumption denote the solution by $\tilde{x}_{n+1}^{(1)}$, so that:

$$x(t_{n+1}) - \tilde{x}_{n+1}^{(1)} = \text{LTE} = C_{p+1}h^{p+1}x^{(p+1)}(t_n) + 0(h^{p+2}) \qquad (5.178)$$

2. Using time-step $t_{n+1} - t_{n-1} = (1 + \alpha)h$, find x_{n+1} from x_{n-1}, and under the localizing assumption denote the solution by $\tilde{x}_{n+1}^{(2)}$, so that:

$$x(t_{n+1}) - \tilde{x}_{n+1}^{(2)} = \text{LTE} = C_{p+1}((1 + \alpha)h)^{p+1}x^{(p+1)}(t_{n-1}) + 0(h^{p+2}) \qquad (5.179)$$

By a Taylor series expansion of $x^{(p+1)}(t)$ around t_n, it is easy to see that $x^{(p+1)}(t_{n-1}) = x^{(p+1)}(t_n) + 0(h)$, so that we can write:

$$x(t_{n+1}) - \tilde{x}_{n+1}^{(2)} = C_{p+1}((1 + \alpha)h)^{p+1}x^{(p+1)}(t_n) + 0(h^{p+2}) \qquad (5.180)$$

Subtracting (5.178) from (5.180), we get:

$$\tilde{x}_{n+1}^{(1)} - \tilde{x}_{n+1}^{(2)} = ((1 + \alpha)^{p+1} - 1) \times \text{PLTE} + 0(h^{p+2}) \qquad (5.181)$$

from which:

$$\text{PLTE} = \frac{\tilde{x}_{n+1}^{(1)} - \tilde{x}_{n+1}^{(2)}}{(1+\alpha)^{p+1} - 1} + 0(h^{p+2}) \tag{5.182}$$

In practice, with no access to the values under the localizing assumption, this motivates the practical estimate:

$$\text{PLTE} \approx \frac{x_{n+1}^{(1)} - x_{n+1}^{(2)}}{(1+\alpha)^{p+1} - 1} \tag{5.183}$$

This approach is expensive because x_{n+1} is computed twice, but the cost is worthwhile in certain special cases, such as in the case where TR is being used with extrapolation, as we will see later on.

Estimating the PLTE When not using predictor-corrector methods, and if the expensive Richardson extrapolation is not an acceptable option, we can estimate the PLTE directly by first estimating $x^{(p+1)}(t_n)$, and then computing:

$$\text{PLTE} = C_{p+1}h^{p+1}x^{(p+1)}(t_n) \tag{5.184}$$

There are two ways of estimating $x^{(p+1)}(t_n)$, depending on whether we have equidistant data or not. Studying both cases is important because, even for 1-step methods, the estimation of $x^{(p+1)}(t_n)$ requires more than two time-points and therefore often involves non-equidistant data. If the data is equidistant, we have access to a method for estimating $x^{(p+1)}(t_n)$ based on *backward differences*. If the data is not equidistant, we have access to a method for estimating $x^{(p+1)}(t_n)$ based on *divided differences*. We will describe these, below.

Equidistant Data We saw earlier the notion of backward differences. For a sequence $x(t_j)$, where the time-points t_j are equidistant, it can be shown that:

$$h^{p+1}x^{(p+1)}(t_n) = \nabla^{p+1}x(t_n) + 0(h^{p+2}) \tag{5.185}$$

which motivates the estimate:

$$h^{p+1}x^{(p+1)}(t_n) \approx \nabla^{p+1}x(t_n) \tag{5.186}$$

Not having access to $x(t_n)$ values, we make the further approximation:

$$h^{p+1}x^{(p+1)}(t_n) \approx \nabla^{p+1}x_n \tag{5.187}$$

leading to the PLTE estimate:

$$\text{PLTE} \approx C_{p+1}\nabla^{p+1}x_n \tag{5.188}$$

For example, for TR with equidistant data:

$$\text{PLTE} \approx C_3\nabla^3 x_n = -\frac{1}{12}(x_n - 3x_{n-1} + 3x_{n-2} - x_{n-3}) \tag{5.189}$$

Non-Equidistant Data   We saw earlier the notion of divided differences for a sequence x_n corresponding to time-points t_n. One can also define the same in connection with a time function $x(t)$ that is evaluated at a sequence of time-points t_n, leading to a sequence $x(t_n)$. Thus, the *zeroth divided difference* of $x(t)$ relative to t_i is:

$$x[t_i] = x(t_i) \qquad (5.190)$$

The *first divided difference* of $x(t)$ with respect to t_i and t_{i-1} is:

$$x[t_i, t_{i-1}] = \frac{x[t_i] - x[t_{i-1}]}{t_i - t_{i-1}} \qquad (5.191)$$

and so on. In this context, the following result is useful:

Theorem 5.8. *If $x(t) : \mathbb{R} \to \mathbb{R}$ is k times differentiable, then there exists a $\xi \in [t_{i-k}, t_i]$ such that:*

$$x^{(k)}(\xi) = k! x[t_i, \ldots, t_{i-k}] \qquad (5.192)$$

If we apply this result on the interval $[t_{n-p}, t_{n+1}]$, which contains t_n, then:

$$x^{(p+1)}(\xi) = (p+1)! x[t_{n+1}, \ldots, t_{n-p}] \qquad (5.193)$$

When $x(t) \in \mathbb{R}^m$, then $x^{(p+1)}(\xi)$ is replaced by vector $\overline{x}^{(p+1)}(\xi)$ in which each component is evaluated at a different $\xi_i \in [t_{n-p}, t_{n+1}]$, and the ξ_i terms form a vector ξ. One can easily show that $x^{(p+1)}(t_n) = \overline{x}^{(p+1)}(\xi) + 0(t_{n+1} - t_{n-p})$, so that, with $h_{n+1} \triangleq t_{n+1} - t_n$, we have:

$$h_{n+1}^{p+1} x^{(p+1)}(t_n) = h_{n+1}^{p+1}(p+1)! x[t_{n+1}, \ldots, t_{n-p}] + 0\left((t_{n+1} - t_{n-p})^{p+2}\right) \qquad (5.194)$$

from which, for a one-step method of order p, for example:

$$\text{PLTE} \approx C_{p+1} h_{n+1}^{p+1}(p+1)! x[t_{n+1}, \ldots, t_{n-p}] \qquad (5.195)$$

This is the LTE estimate used in SPICE for TR and BE. For example, for a first-order method, with $p = 1$, we find:

$$\text{PLTE} \approx \frac{2C_2 h_{n+1}^2}{h_{n+1} + h_n} \left(\frac{x_{n+1} - x_n}{h_{n+1}} - \frac{x_n - x_{n-1}}{h_n} \right) \qquad (5.196)$$

where, notice, we have used the available x_{n-1}, x_n, and x_{n+1} in place of the unavailable $x(t_{n-1})$, $x(t_n)$, and $x(t_{n+1})$. When $p = 2$, we find:

$$\text{PLTE} \approx \frac{6C_3 h_{n+1}^3}{t_{n+1} - t_{n-2}} \left(\frac{\frac{x_{n+1}-x_n}{t_{n+1}-t_n} - \frac{x_n-x_{n-1}}{t_n-t_{n-1}}}{t_{n+1} - t_{n-1}} - \frac{\frac{x_n-x_{n-1}}{t_n-t_{n-1}} - \frac{x_{n-1}-x_{n-2}}{t_{n-1}-t_{n-2}}}{t_n - t_{n-2}} \right) \qquad (5.197)$$

In the equidistant case, this reduces to:

$$\text{PLTE} \approx C_3 \left(x_{n+1} - 3x_n + 3x_{n-1} - x_{n-2} \right) = C_3 \nabla^3 x_{n+1} \qquad (5.198)$$

which, for TR, becomes:

$$\text{PLTE} \approx -\frac{1}{12} \nabla^3 x_{n+1} \qquad (5.199)$$

5.4 STABILITY OF LMS METHODS

We saw earlier the notion of stability, in relation to the presence of perturbations in both the differential problem and the difference method. We saw that a *problem* is said to be *totally stable* or *well-posed* if its solution is not extremely sensitive to perturbations. Likewise, we saw that a *numerical method* is said to be *zero-stable* if, as $h \to 0$, its solution is not extremely sensitive to perturbations.

These notions were satisfying, in the sense that the notion of stability of the *numerical method* was analogous to the notion of stability of the *problem*. Keep in mind that, while the problem is a continuous-time differential equation, the numerical method is a discrete time difference equation. Effectively, numerical methods solve a difference equation instead of solving the differential equation. Our hope, and goal, is that the difference equation would be a good proxy for the differential equation. Indeed, this is what the property of *consistency* is all about. It is meant to ensure that, as $h \to 0$, the difference equation is actually solving the correct differential equation, and not some other.

However, even if a numerical method is zero-stable, there are things that can "go wrong" with it when h is non-zero. These issues are related only to the *numerical method* and not to the *problem* itself. The problem is always a $h = 0$ problem; it is in continuous-time. These issues arise because we are using a discrete-time difference equation, with a non-zero h, to solve a continuous-time differential system. Thus, these issues are somewhat artificial and their study will not be as satisfying and as well motivated as that of zero-stability.

One thing that *can* go wrong with the use of a numerical method is that local errors can accumulate over time. Naturally, one would expect *some* level of error accumulation. But when it becomes so excessive that the resulting solution is not reliable, the method becomes useless. Another thing that can go wrong is that the method can be too sensitive to the initial conditions. Here, too, one would expect the method to be sensitive, to *some* extent, to its initial conditions. However, extreme sensitivity can make the method, again, useless. A method that is inadequate due to such issues is said to be *unstable*.

While it is easy to give a qualitative description of what stability, and instability, mean, it is not as easy to describe it quantitatively and formally. Indeed, there are various ways of describing stability in quantitative terms, some of which are more suitable to some problem domains than to others. Broadly speaking there are two *theories* of stability, a classical linear stability theory, and a more recent nonlinear stability theory. The latter is not fully developed yet, and only some results are available. Those available results are not commonly used in computer codes, because it turns out that the linear theory is adequate in most cases. Thus, we will focus on linear stability theory.

5.4.1 Linear Stability Theory

Linear stability theory proceeds as follows. Consider a general homogeneous linear differential system with constant coefficients:

$$x' = Ax \qquad\qquad (5.200)$$

whose solution vector $x(t)$ is assumed to die down to 0 as $t \to \infty$, for any finite initial condition. This is called a *test system* or a *test equation*; it is a system for which the stability of a numerical method will be *tested*. For simplicity, and because it does not affect the results, it is assumed that the system matrix has distinct eigenvalues. Then, linear stability theory asks the following question: for this general class of problem, does the candidate numerical method *also* produce solutions that die down to 0 as $t \to \infty$? If *yes*, then that is a "good thing," and we call the method stable, in one sense or another. If *not*, then that's a "bad thing," and we call the method unstable. A method that is deemed to be stable in this way is found, most often, to be also good for solving more general problems in practice, including linear problems whose solutions do not necessarily die down to zero, linear problems with time-varying coefficients, and general nonlinear problems. Basically, such methods are found to not excessively accumulate errors and to not be excessively sensitive to initial conditions.

We may want to refer to a homogeneous linear problem whose solution dies down to zero over time as a "stable problem." One is tempted, in fact, to think of any systems that do not have this property as being impractical or non-physical. After all, if its solution does not die down to zero over time, then the problem would seem to be one that has infinite memory of its past, or that infinitely amplifies the values of its initial condition. Either way, it does not sound like a very useful or practical system. However, one should exercise caution with this line of thinking. In practice, if one encounters a homogeneous system whose solutions grow over time, the system may be actually designed to operate that way. For example, an oscillator requires this property, and a linearization of an oscillator around some operating point would produce a homogeneous linear system with non-decaying outputs. Or, it may be that the system is modeled in this way for only a limited time period in its operational cycle. So, in general, while one may want to call such systems "unstable," one should not assume that they are somehow badly modeled, non-physical, or ill-posed.[2] Given this notion of stability of homogeneous linear problems, then one can succinctly say

---

[2]Going back to the definition of well-posed (totally stable) problems given earlier, recall that that definition is with respect to a given finite time period of operation $[t_0, t_f]$. A system can have solutions that grow over time and still be well-posed (totally stable) over a finite time period.

that classical linear stability theory asks the following question: *when applied to a (homogeneous, linear) differential problem that is "stable," is my difference method "stable" as well?*. In other words: *does it produce decaying numerical solutions for stable linear homogeneous problems?*

That is *all* that the theory does! Significantly, the theory makes no mention of how the method may behave when the problem is unstable. That is important too; as mentioned above, a homogeneous problem whose solutions do not die down over time may well be encountered in practice. Thus, the definitions and results of linear stability theory should be used with the above caveats in mind. One should try to check, in practice, how a given numerical method would perform on systems that are slightly unstable.

We will now proceed with the detailed study of linear stability theory, by first studying the notions of a *test equation* and of *absolute stability*. We will then define the concept of a *stability region*, and give the two key properties of *A-stability* and *stiff stability*. There are other definitions of stability that can be given, as in Lambert (1991), but it will suffice for us to focus on these two.

5.4.2   The Test Equation

As our *test system*, we have introduced the arbitrary homogeneous linear differential system with constant coefficients:

$$x' = Ax \tag{5.201}$$

whose solution vector $x(t)$ is assumed to die down to 0, as $t \to \infty$, for any finite initial condition, and we have assumed that the eigenvalues λ_i of A are distinct. As we saw earlier, the exact *general solution* for such a system is given by:

$$x(t) = \sum_{i=1}^{m} c_i e^{\lambda_i t} q_i \tag{5.202}$$

where q_i is the eigenvector corresponding to λ_i, for all $i = 1, 2, \ldots, m$, and the c_i are constant coefficients. Because eigenvalues can be complex, $\lambda_i \in \mathbb{C}$, some of the individual terms in the summation in (5.202) may be *complex* time functions. However, the presence of complex conjugate λ_i, and the possibility of complex q_i, leads to an overall solution which is real. Because the system is assumed to have a solution that dies down over time for *any* (finite) initial condition, then it follows that:

$$\Re(\lambda_i) < 0, \quad \forall i \tag{5.203}$$

so that all eigenvalues are strictly in the left half-plane.

If we apply a k-step LMS method to the above test system, the resulting *difference system* is:

$$\sum_{j=-1}^{k-1} \alpha_j x_{n-j} = h \sum_{j=-1}^{k-1} \beta_j A x_{n-j} \tag{5.204}$$

or:

$$\sum_{j=-1}^{k-1} \left(\alpha_j I - h\beta_j A\right) x_{n-j} = 0 \tag{5.205}$$

where I is the $m \times m$ identity matrix. We are interested to learn what conditions must be imposed so that the numerical solution $\{x_n\}$ dies down to zero over time, in other words, to ensure that:

$$\lim_{n\to\infty} \|x_n\| = 0 \tag{5.206}$$

Because the λ_i are distinct, it is known that there exists a nonsingular matrix Q such that $\Lambda = Q^{-1}AQ$ is a diagonal matrix whose diagonal elements are $\lambda_1, \lambda_2, \ldots, \lambda_m$. We now define a sequence $\{y_n\}$ by $y_n = Q^{-1}x_n$, so that $x_n = Qy_n$, and we pre-multiply (5.205) by Q^{-1}, leading to:

$$\sum_{j=-1}^{k-1} \left(\alpha_j I - h\beta_j \Lambda\right) y_{n-j} = 0 \tag{5.207}$$

In general, Q can have complex entries, so that $\{y_n\}$ is a *complex* sequence. Because both I and Λ are diagonal matrices, this system is *decoupled*, and we can write it as:

$$\sum_{j=-1}^{k-1} \left(\alpha_j - h\beta_j \lambda_i\right) y_{n-j\,(i)} = 0, \quad \text{for } i = 1, 2, \ldots, m \tag{5.208}$$

so that it is a collection of m single-variable difference equations. Because $x_n = Qy_n$ and Q is nonsingular, then clearly (5.206) is satisfied if and only if:

$$\lim_{n\to\infty} \|y_n\| = 0 \tag{5.209}$$

which can be written as a condition to be met by each of the m difference equations, as:

$$\lim_{n\to\infty} |y_{n\,(i)}| = 0, \quad \text{for } i = 1, 2, \ldots, m \tag{5.210}$$

In other words, we can express the original stability question in relation to a single-variable as follows: if $\{z_n\}$ is a scalar, possibly *complex*, sequence, what conditions must be met by the difference equation:

$$\sum_{j=-1}^{k-1} \left(\alpha_j - h\beta_j \lambda\right) z_{n-j} = 0 \tag{5.211}$$

where $\lambda \in \mathbb{C}$, so that its solution $\{z_n\}$ satisfies:

$$\lim_{n\to\infty} |z_n| = 0 \tag{5.212}$$

It is enough to answer this question *once*, in connection with the z-system, because the m individual y-systems are exactly identical. Thus, effectively, we are concerned with the solution of a single-variable complex *test equation*:

$$x' = \lambda x, \quad \text{where } x, \lambda \in \mathbb{C}, \quad \text{with} \quad \Re(\lambda) < 0 \tag{5.213}$$

whose exact solution is $x(t) = ce^{\lambda t}$, and with the application of an LMS method to this test equation, leading to the difference equation:

$$\sum_{j=-1}^{k-1} \gamma_j x_{n-j} = 0 \tag{5.214}$$

where $\gamma_j \triangleq (\alpha_j - \beta_j h\lambda)$, whose solution $\{x_n\}$ is required to satisfy:

$$\lim_{n\to\infty} |x_n| = 0 \tag{5.215}$$

Equation (5.214) is a *homogeneous linear difference equation with constant coefficients*, because the α_j and β_j coefficients are independent of n. For example, for the case of FE, applied to the test equation $x' = \lambda x$, the resulting difference equation is:

$$x_{n+1} = x_n + h\lambda x_n \tag{5.216}$$

or:

$$x_{n+1} - (1 + h\lambda)x_n = 0 \tag{5.217}$$

For the case of BE, the difference equation is:

$$x_{n+1} = x_n + h\lambda x_{n+1} \tag{5.218}$$

or:

$$(1 - h\lambda)x_{n+1} - x_n = 0 \tag{5.219}$$

And, for TR, the difference equation is:

$$x_{n+1} = x_n + \frac{h\lambda}{2}(x_{n+1} + x_n) \tag{5.220}$$

or:

$$(1 - h\lambda/2)x_{n+1} - (1 + h\lambda/2)x_n = 0 \tag{5.221}$$

Such difference equations admit a *general solution* that depends on some arbitrary constants, whose values can in practice be found using initial conditions. The solutions are not hard to find, as we will now see.

Consider the value of the left-hand side of (5.214), upon substitution of the candidate solution $x_n = r_1^n$, for some non-zero $r_1 \in \mathbb{C}$:

$$\sum_{j=-1}^{k-1} \gamma_j r_1^{n-j} = r_1^{n-k+1} \left(\gamma_{-1} r_1^k + \gamma_0 r_1^{k-1} + \cdots + \gamma_{k-2} r_1 + \gamma_{k-1} \right) \qquad (5.222)$$

It is easy to see that a sequence of the form $x_n = r_1^n$ is a solution of (5.214), if r_1 is a root of the so-called *characteristic polynomial*:

$$\sum_{j=-1}^{k-1} \gamma_j r^{k-j-1} = \gamma_{-1} r^k + \gamma_0 r^{k-1} + \cdots + \gamma_{k-2} r + \gamma_{k-1} \qquad (5.223)$$

Indeed, if r_i is a root of the characteristic polynomial, with multiplicity μ_i, it can be shown that the following sequences are all valid solutions:

$$\{r_i^n\}, \quad \{n r_i^n\}, \quad \{n^2 r_i^n\}, \quad \ldots, \quad \{n^{\mu_i - 1} r_i^n\} \qquad (5.224)$$

The set of all such sequences, corresponding to every root r_i, forms a so-called *fundamental system* of *linearly independent solutions* of (5.214). It can be shown that the *general solution* of (5.214) can be written as a linear combination of the solutions that form this fundamental system.

Given the above, and in order for the *general solution* of the difference equation (5.214), resulting from application of the candidate numerical method to the test equation, to die down to zero over time, it is clear that we must require that $|r_i| < 1$, for all roots r_i of its characteristic polynomial (5.223).

5.4.3 Absolute Stability

The characteristic polynomial of the difference equation can be written in terms of the first and second characteristic polynomials of the LMS method, as:

$$\pi(r, \hat{h}) = \sum_{j=-1}^{k-1} (\alpha_j - \beta_j \hat{h}) r^{k-j-1} = \rho(r) - \hat{h}\sigma(r), \quad \text{where } \hat{h} \triangleq h\lambda \qquad (5.225)$$

and is referred to as the *stability polynomial* of the LMS method. Clearly, $|x_n| \to 0$ as $n \to \infty$ if all the roots of $\pi(r, \hat{h})$ are strictly inside the unit circle of the complex plane, which motivates the following definition.

Definition 5.11. (Absolute Stability) A LMS method is said to be *absolutely stable* for a given $\hat{h}$ if, for that $\hat{h}$, all the roots of $\pi(r, \hat{h})$ are strictly inside the unit circle. Otherwise, it is said to be *absolutely unstable* for that $\hat{h}$.

Where, in the complex plane, are the roots of $\pi(r, \hat{h})$? Note that any common roots of $\rho(r)$ and $\sigma(r)$ are also roots of $\pi(r, \hat{h})$, for any $\hat{h}$. Otherwise, *any* point

r_0 of the complex plane which is not a root of $\sigma(r)$ can be a root of $\pi(r, \hat{h})$ for *some* value of $\hat{h}$, namely for:

$$\hat{h}_0 = \frac{\rho(r_0)}{\sigma(r_0)} \tag{5.226}$$

In most cases, $\rho(r)$ and $\sigma(r)$ have no common roots, so that r_i is a root of $\pi(r, \hat{h})$ if and only if $\hat{h} = \rho(r_i)/\sigma(r_i)$. In fact, because the roots of a polynomial are continuous functions of its coefficients, the roots of $\pi(r, \hat{h})$ are *parameterized* by $\hat{h}$; they move in the plane as $\hat{h}$ is varied. For some values of $\hat{h}$, all the roots would be strictly inside the unit circle and the method would be absolutely stable. For some other values of $\hat{h}$, one or more roots would not be strictly inside the unit circle and the method would not be absolutely stable.

We are clearly interested in the values of $\hat{h}$ for which all the roots are strictly inside the unit circle, and the LMS method is absolutely stable.

Region of Absolute Stability The above discussion motivates the following definition.

Definition 5.12. (Region of Absolute Stability) A LMS method is said to have a *region of absolute stability* $\mathcal{R}_A$, in the complex plane, if it is absolutely stable for all $\hat{h} \in \mathcal{R}_A$.

The region of absolute stability is a characteristic of the LMS method, and we can plot it for a given method, as a fixed region of the complex plane. It is useful to visualize the relation between $\mathcal{R}_A$ and the unit circle as a mapping from the $\hat{h}$-plane to the r-plane, as shown in Fig. 5.1.

For a given specific system, $x' = Ax$, with known eigenvalues, we may be able to set the time-step h so as to ensure that $\hat{h} \in \mathcal{R}_A$. This raises the possibility that one may apply time-step control for stability, and not just for accuracy. This is not a "good thing," however, because we would prefer to be able to set the time-step based *only* on considerations of accuracy. If one must satisfy considerations of *both* accuracy and stability, one has much less flexibility in the

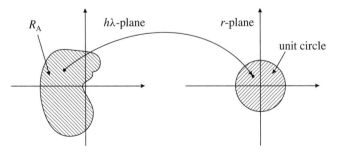

Figure 5.1: The region of absolute stability in the $h\lambda$-plane maps to the interior of the unit circle in the r-plane.

choice of time-step. In such cases, one is forced to take short time-steps much more often. We will return to this issue in our discussion of stiffness; for now, we are interested to identify the region of absolute stability.

In this regard, one can show (see Lambert (1991)) that the stability polynomial of any *convergent* LMS method must have a root r_1 that approaches 1 as $\hat{h} \to 0$. Furthermore, by examining the *rate* at which this root approaches 1, one arrives at the following conclusion, to quote from Lambert (1991), *"the region of absolute stability of any convergent LMS method cannot contain the positive real axis in the neighborhood of the origin."*

Finding the Region of Absolute Stability   One way of finding the region of absolute stability, called the *boundary locus method* is as follows. Because the roots of a polynomial are continuous functions of its coefficients, then the roots of the stability polynomial are continuous functions of the parameter $\hat{h}$. Therefore, any continuous path in the $h\lambda$-plane that starts inside $\mathcal{R}_A$ and ends outside it must pass through a point that corresponds to the stability polynomial having a root of magnitude 1. These points define a *boundary* of $\mathcal{R}_A$, consisting of certain points in the $h\lambda$-plane for which at least one root of $\pi(r, \hat{h})$ has magnitude 1.

The set of *all* points in the plane where at least one root has magnitude 1 can be discovered by setting:

$$\pi(e^{i\theta}, \hat{h}) = 0 \tag{5.227}$$

keeping in mind that this set may include points that are not necessarily boundary points of $\mathcal{R}_A$. This leads to a way to discover the boundary, based on:

$$\hat{h} = \frac{\rho(e^{i\theta})}{\sigma(e^{i\theta})} = \frac{\sum_{j=-1}^{k-1} \alpha_j e^{i\theta(k-j-1)}}{\sum_{j=-1}^{k-1} \beta_j e^{i\theta(k-j-1)}} \tag{5.228}$$

$$= \frac{\alpha_{-1} e^{ik\theta} + \alpha_0 e^{i(k-1)\theta} + \cdots + \alpha_{k-2} e^{i\theta} + \alpha_{k-1}}{\beta_{-1} e^{ik\theta} + \beta_0 e^{i(k-1)\theta} + \cdots + \beta_{k-2} e^{i\theta} + \beta_{k-1}} \tag{5.229}$$

By sweeping θ from 0 to 2π, we can discover the boundary. Once the boundary has been found, we can then test the regions on both sides of the boundary to identify the region of absolute stability, $\mathcal{R}_A$, as we will now demonstrate for a few test cases.

Stability of Forward Euler   For the case of forward Euler (FE), $x_{n+1} = x_n + hf_n$, or:

$$x_{n+1} - x_n = hf_n \tag{5.230}$$

we have $k = 1$ and $\alpha_{-1} = 1$, $\alpha_0 = -1$, $\beta_0 = 1$, so that:

$$\rho(z) = z - 1 \quad \text{and} \quad \sigma(z) = 1 \tag{5.231}$$

(which have no common roots) and the stability polynomial is:

$$\pi(r, \hat{h}) = \rho(r) - \hat{h}\sigma(r) = r - \hat{h} - 1 \tag{5.232}$$

with the single root:

$$r_1 = \hat{h} + 1 \tag{5.233}$$

and it is clear that, in order for r_1 to be inside the unit circle, we must keep $\hat{h}$ inside the circle of radius 1 centered at -1. Indeed, using the boundary locus method, the boundary of $\mathcal{R}_A$ is given by:

$$\hat{h} = \frac{\rho(e^{i\theta})}{\sigma(e^{i\theta})} = e^{i\theta} - 1 \tag{5.234}$$

which traces the circle of radius 1 centered at -1, as shown in Fig. 5.2, and we know that $\mathcal{R}_A$ is the *inside* of this circle because, for the convergent FE, recall, $\mathcal{R}_A$ cannot contain parts of the *positive* real axis near the origin.

What do we learn from the region of absolute stability FE? Because $\Re(\lambda) < 0$ and $h > 0$, then $\hat{h} = h\lambda$ is strictly in the left half-plane, and can always be "brought into" the region $\mathcal{R}_A$ by a small enough h. This is useful—FE can always be made stable by the use of a small enough time-step—but this is not a surprise because we already know that FE is zero-stable. However, the problem with FE is that the region $\mathcal{R}_A$ is quite limited and small. For large λ, a very small time-step may be required to achieve stability. In practice, one finds that a very small time-step is often required to maintain stability and control error accumulation when using FE. This renders the method somewhat useless; better methods are available.

Stability of Backward Euler For the case of backward Euler (BE), $x_{n+1} = x_n + hf_{n+1}$, or:

$$x_{n+1} - x_n = hf_{n+1} \tag{5.235}$$

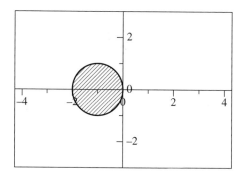

Figure 5.2: The region of absolute stability for FE.

we have $k = 1$ and $\alpha_{-1} = 1$, $\alpha_0 = -1$, $\beta_{-1} = 1$, so that:

$$\rho(z) = z - 1 \quad \text{and} \quad \sigma(z) = z \qquad (5.236)$$

(which have no common roots) and the stability polynomial is:

$$\pi(r, \hat{h}) = \rho(r) - \hat{h}\sigma(r) = r - \hat{h}r - 1 \qquad (5.237)$$

with the single root:

$$r_1 = \frac{1}{1 - \hat{h}} \qquad (5.238)$$

and it is clear that, in order for r_1 to be inside the unit circle, we must keep $\hat{h}$ outside the circle of radius 1 centered at 1. Indeed, using the boundary locus method, we have that the boundary of $\mathcal{R}_A$ is given by:

$$\hat{h} = \frac{\rho(e^{i\theta})}{\sigma(e^{i\theta})} = 1 - e^{-i\theta} \qquad (5.239)$$

which traces the circle of radius 1 centered at 1, as shown in Fig. 5.3, and we know that $\mathcal{R}_A$ is the *outside* of this circle because, for the convergent BE, recall, $\mathcal{R}_A$ cannot contain parts of the *positive* real axis near the origin.

What do we learn from the region of absolute stability BE? An immediate observation is that the region of absolute stability is much larger than that of FE. Generally, it is often the case that implicit methods (like BE) have larger regions of absolute stability than explicit methods (like FE). Practical experience is that methods with larger regions of absolute stability out-perform those with smaller regions. Furthermore, because $\Re(\hat{h}) < 0$ and $\mathcal{R}_A$ contains the whole left half-plane, it is clear that BE is absolutely stable for *any* value of time-step h. Thus, when using BE, time-step control can be based solely on considerations of accuracy, without worrying about stability.

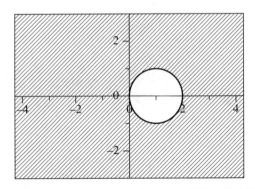

Figure 5.3: The region of absolute stability for BE.

However, notice that $\mathcal{R}_A$ contains most of the right half-plane as well. As a result, when applied to a problem with slightly positive $\Re(\lambda_i)$, BE may erroneously produce stable decaying responses. As discussed earlier, such problems may well arise in practice, so that one should exercise caution when using BE—it may be "too stable." It would seem that, ideally, we would want $\mathcal{R}_A$ to consist of the full left half-plane and no more.

Stability of the Trapezoidal Rule   For the case of the trapezoidal rule (TR):

$$x_{n+1} - x_n = h \left(\frac{1}{2} f_{n+1} + \frac{1}{2} f_n \right) \tag{5.240}$$

we have $k = 1$ and $\alpha_{-1} = 1$, $\alpha_0 = -1$, $\beta_{-1} = 1/2$, $\beta_0 = 1/2$, so that:

$$\rho(z) = z - 1 \quad \text{and} \quad \sigma(z) = \frac{z}{2} + \frac{1}{2} \tag{5.241}$$

(which have no common roots) and the stability polynomial is:

$$\pi(r, \hat{h}) = \rho(r) - \hat{h}\sigma(r) = r - \hat{h}\frac{r}{2} - \frac{\hat{h}}{2} - 1 \tag{5.242}$$

with the single root:

$$r_1 = \frac{1 + \hat{h}/2}{1 - \hat{h}/2} \tag{5.243}$$

for which it can be easily shown that $|r_1| < 1$ if and only if $\Re(\hat{h}) < 0$, and that $|r_1| = 1$ if and only if $\Re(\hat{h}) = 0$. Indeed, using the boundary locus method, we have that the boundary of $\mathcal{R}_A$ is given by:

$$\hat{h} = \frac{\rho(e^{i\theta})}{\sigma(e^{i\theta})} = 2 \left(\frac{e^{i\theta} - 1}{e^{i\theta} + 1} \right) \tag{5.244}$$

which traces the imaginary axis, with $\Re(\hat{h}) = 0$ and $\Im(\hat{h}) = 2\tan(\theta/2)$, as shown in Fig. 5.4, and we know that $\mathcal{R}_A$ is the *left* half-plane because, for the convergent TR, recall, $\mathcal{R}_A$ cannot contain parts of the *positive* real axis near the origin.

What do we learn from the region of absolute stability TR? As with BE, because $\mathcal{R}_A$ contains the whole left half-plane, TR is absolutely stable for any choice of time-step. Contrary to BE, however, and because $\mathcal{R}_A$ does not include any parts of the right half-plane, then TR does not suffer from the issue identified above with BE; it will not cause true growing responses to die down to zero; it is not "too stable." TR is not ideal, however, and there are issues to watch out for in practice, as we will see below in the study of stiffness.

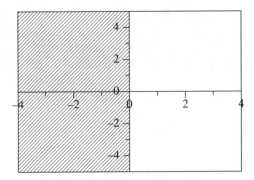

Figure 5.4: The region of absolute stability for TR.

5.4.4 Stiff Systems

Stiffness is a phenomena that is often exhibited by ODE systems from a wide range of applications, including circuit simulation. It does not have a precise mathematical definition, and thus is better termed a *phenomena* rather than a *property*. The visible "symptom" of stiffness is that one finds that very short time-steps are required to maintain accuracy and stability, and the solution takes a long time to complete. Stiffness usually arises in connection with numerical methods that have a finite region of absolute stability, such as FE, but is not restricted to such cases.

As described in Lambert (1991), a number of ways have been proposed to explain and define stiffness mathematically. For example, one approach is to say that a linear constant coefficient system with $\Re(\lambda_i) < 0$, $\forall i$, is *stiff* if the ratio of its largest $|\Re(\lambda_i)|$ to its smallest $|\Re(\lambda_i)|$ is very large. Another, related, method is to say that stiffness occurs when some components of the solution decay much more rapidly than others. A third, more general, view is that stiffness occurs when stability requirements, rather than accuracy requirements, dictate the time-step that should be used. However, counter-examples can be given to show that each of these characterizations misses the mark, in some way, for some stiff systems.

The following, slightly paraphrased from Lambert (1991), is perhaps the best way to characterize stiffness today: *if a numerical method with a finite $\mathcal{R}_A$, applied to a system with any initial conditions, is forced to use, in a certain time interval, a time-step which is exceedingly small in relation to the smoothness of the exact solution in that interval, then the system is said to be* stiff *in that interval.* This "definition" is not very helpful because it does not provide an *a priori* method to check whether a system is stiff or not; unfortunately, this is the state of the art today.

TR and the BDFs are probably the best modern methods for solving stiff systems, but one should keep in mind that there is no "silver bullet" to tackle stiffness in all cases. In the context of linear stability theory, one can identify certain traits that aid in combating stiffness. One key requirement, for example, is to ensure that there are no stability-imposed restrictions on the time-step. Thus, methods like FE, with a very limited region of absolute stability, are to be avoided. Instead, methods like BE and TR whose regions of absolute stability include the whole left half-plane, are quite attractive. This motivates the following definition.

Definition 5.13. (A-stability) A LMS method is said to be *A-stable* if its region of absolute stability includes the whole left half-plane, i.e., if $\{\hat{h}|\Re(\hat{h}) < 0\} \subset \mathcal{R}_A$.

Thus, both BE and TR are A-stable. At this point, it is useful to introduce the following key result, the second of Dahlquist's "barriers."

Theorem 5.9. *(The second Dahlquist barrier)*

1. An explicit LMS method cannot be A-stable.
2. The order of an A-stable LMS method cannot exceed 2.
3. The 2nd order A-stable LMS method with the smallest error constant is TR.

In other words, if we insist on A-stability, then TR is the most accurate LMS method! Thus, A-stability is viewed as being a bit too restrictive and some alternate definitions of stability have been proposed. We will consider only one of these, called stiff stability.

5.4.5 Stiff Stability

One way to slacken the requirements of A-stability is by appealing to the types of problems typically encountered in practice and to the properties of band-limited signals. This was done by Gear (1971) and (as later modified by others) leads to the following definition.

Definition 5.14. (Stiff Stability) A numerical method is said to be *stiffly stable* if its region of absolute stability includes $\mathcal{R}_1$ and $\mathcal{R}_2$, where, with positive real a and c:

$$\mathcal{R}_1 = \{\hat{h}|\Re(\hat{h}) < -a\} \tag{5.245}$$

$$\mathcal{R}_2 = \{\hat{h}| - a \le \Re(\hat{h}) < 0, |\Im(\hat{h})| \le c\} \tag{5.246}$$

In Fig. 5.5, we show the type of region $\mathcal{R}_1 \cup \mathcal{R}_2$ that must be included in $\mathcal{R}_A$ in order for an LMS method to be stiffly stable. It is clear that an A-stable method is also stiffly stable, but not vice versa. Both TR and BE are stiffly stable, but other methods can be stiffly stable as well, such as all the BDFs, as we will now see.

Stability of the BDFs It is easy to verify that BDF2 is A-stable and stiffly stable, with the region of absolute stability shown in Fig. 5.6. Thus, both the first order BDF (BE) and the 2nd order BDF (BDF2) are A-stable and stiffly stable.

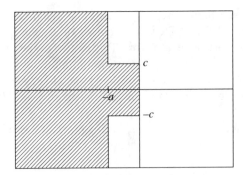

Figure 5.5: The shaded region must be part of the region of absolute stability for any stiffly stable system.

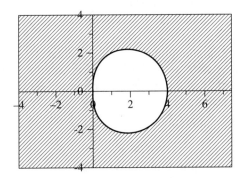

Figure 5.6: The region of absolute stability for BDF2.

BDF2 has an error constant of $-2/9$ which, as we would expect, is larger (in magnitude) than that of TR, which is $-1/12$. The remaining BDFs, of orders 3, 4, 5, and 6, are stiffly stable as well, but not A-stable, as we see from their regions of absolute stability, shown in Fig. 5.7, Fig. 5.8, Fig. 5.9, and Fig. 5.10, and their error constants are given in Table 5.1.

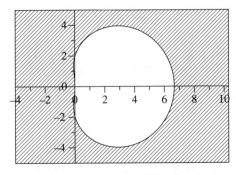

Figure 5.7: The region of absolute stability for the 3rd order BDF.

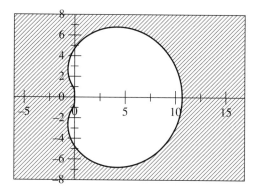

Figure 5.8: The region of absolute stability for the 4th order BDF.

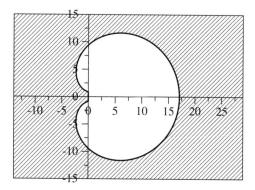

Figure 5.9: The region of absolute stability for the 5th order BDF.

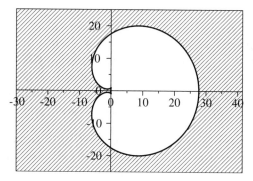

Figure 5.10: The region of absolute stability for the 6th order BDF.

Table 5.1: Error constants for the BDFs.

$k = p$:	1	2	3	4	5	6
C_{p+1}:	$-\dfrac{1}{2}$	$-\dfrac{2}{9}$	$-\dfrac{3}{22}$	$-\dfrac{12}{125}$	$-\dfrac{10}{137}$	$-\dfrac{20}{343}$

5.4.6 Remarks

This is a good point to summarize our findings and make a few remarks. Firstly, with regard to stiff systems:

1. It can be shown that no explicit LMS method or explicit Runge-Kutta method can be A-stable or stiffly stable. Thus, one must use *implicit* methods to tackle stiff systems.

2. When using implicit methods on stiff nonlinear problems, it is found in practice that one must use Newton's method (not a fixed point method) and one must iterate *to convergence*.

More generally, the following summary points are worth reiterating:

1. The great advantage of methods that are A-stable, like TR, BE, and BDF2, is that their stability imposes no restrictions on time-step. This is because their region of absolute stability includes the whole left half-plane. This allows one to efficiently handle stiff systems using these methods. With these methods, stable (negative real part) eigenvalues of the differential system lead to stable (within the unit circle) roots of the difference system. In other words, an A-stable method produces a stable difference system, for any given stable differential test system.

2. TR is the prototypical A-stable system—its region of absolute stability consists exactly of the left half-plane.

3. Some methods, like BE and BDF2, have bigger regions of absolute stability than TR, but they run the risk of being "too stable," as we saw earlier. For some systems which have eigenvalues with small positive real parts, the response from such methods would be erroneously decaying down to zero.

4. Non-A-stable but stiffly stable methods, like the BDFs of orders 3–6, have regions of absolute stability that do not include the full left half-plane. Nevertheless, these methods are very good at handling practical stiff systems.

In circuit simulation, the most popular methods are BE, TR, and BDF2. Higher order BDFs are difficult to implement, especially with regard to variable time-step approaches; it is not clear if they are worth the effort. TR is probably the *most* commonly used method, because:

1. It is a one-step method, thus easier to implement than the two-step BDF2, but it has the same order as the BDF2.

2. It is a second-order method, thus more accurate than BE.

3. It does not suffer from the possibility of being "too stable" as is the case with BE and BDF2.

But, as we commented earlier, TR is not perfect, and it does suffer from a weakness, commonly referred to as *trapezoidal ringing*, which we now explain.

5.5 TRAPEZOIDAL RINGING

Consider the standard test equation:

$$x' = \lambda x, \quad \text{with} \quad x, \lambda \in \mathbb{C}, \quad \text{and} \quad \Re(\lambda) < 0 \qquad (5.247)$$

whose exact solution is $x(t) = ce^{\lambda t}$, where $c \in \mathbb{C}$. Given that $t_{n+1} = t_0 + h(n + 1)$, we can write the exact solution at t_{n+1} as:

$$x(t_{n+1}) = x(t_0) \left(e^{h\lambda}\right)^{n+1} \qquad (5.248)$$

Suppose we use TR, $x_{n+1} - x_n = \frac{h}{2}(f_{n+1} + f_n)$, to solve the above system. Then, substituting $f_n = f(x_n, t_n) = \lambda x_n$, and similarly for f_{n+1}, into the TR expression, gives:

$$x_{n+1} = \left(\frac{1 + h\lambda/2}{1 - h\lambda/2}\right) x_n = x_0 \left(\frac{1 + h\lambda/2}{1 - h\lambda/2}\right)^{n+1} \qquad (5.249)$$

Thus, the term $(1 + h\lambda/2)/(1 - h\lambda/2)$ is supposed to approximate the term $e^{h\lambda}$ of the exact solution and, for simplicity, we focus on the nth terms and compare the two series:

$$x_n = x_0 \left(\frac{1 + h\lambda/2}{1 - h\lambda/2}\right)^n \qquad (5.250)$$

and:

$$x(t_n) = x(t_0) \left(e^{h\lambda}\right)^n \qquad (5.251)$$

We recognize the term $(1 + h\lambda/2)/(1 - h\lambda/2)$ as the single root, r_1, of the stability polynomial of TR:

$$r_1 = \left(\frac{1 + h\lambda/2}{1 - h\lambda/2}\right) \qquad (5.252)$$

which, we already know, has $|r_1| < 1$ because $\Re(\lambda) < 0$. Also, because $|e^z| = e^{\Re(z)}$, we also know that $|e^{h\lambda}| < 1$. Thus, both r_1^n and $e^{h\lambda n}$ go to zero as $n \to \infty$, which we already expect for the exact solution (because $\Re(\lambda) < 0$) and for TR (because it is A-stable).

However, consider what happens in the case when $|h\Re(\lambda)|$ is large. A large-magnitude, negative $\Re(\lambda)$ would correspond to a fast transient in the true solution (due to $|e^z| = e^{\Re(z)}$), such as what may be observed in digital circuits when a signal makes a logic transition. A large h is desirable and, of course, is the whole point of using an A-stable method like TR. As $|h\lambda| \to \infty$, it is easy to see that r_1 approaches -1 on the real axis, maintaining $|r_1| < 1$, while $e^{h\lambda}$ approaches 0. As a result, as n increases, we have that the term $e^{h\lambda n}$ (and the true solution, therefore) goes to zero very quickly, while the term r_1^n (and the numerical solution, therefore) goes to zero very slowly, and with alternating signs. In practice, this leads to a slowly damped oscillating error, that is commonly referred to as *trapezoidal ringing*, such as shown in Fig. 5.11.

To overcome this problem, one must either reduce the time-step or else resort to another method, such as BDF2, which does not suffer from ringing but is more complex to implement. Other alternatives involve trying to "fix" the ringing problem, such as by using TR in conjunction with *smoothing*, as we now discuss.

5.5.1 Smoothing

Smoothing is a technique, similar to the application of a filter to a noisy signal, that aims to reduce ringing, such as observed with TR, by "smoothing out" the waveform, as follows. Suppose we are using TR and we have already computed x_{n+1} from x_n, and suppose that t_{n-1}, t_n, and t_{n+1} are equidistant, so that $t_n - t_{n-1} = t_{n+1} - t_n$. Then, we can recompute another, hopefully better, value of x_n as follows:

$$\hat{x}_n = \frac{1}{4}\left(x_{n-1} + 2x_n + x_{n+1}\right) \tag{5.253}$$

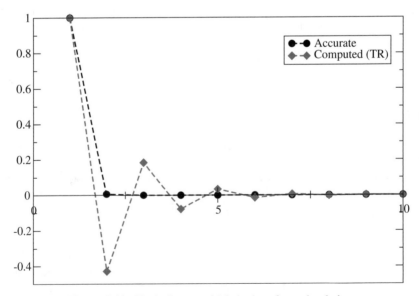

Figure 5.11: Typical trapezoidal ringing, from simulation.

and we adopt this weighted average, of x_n and its two neighbors, as the new value of x_n:

$$x_n = \hat{x}_n \tag{5.254}$$

and then re-compute x_{n+1} using standard TR before marching on forward.

There is no perfect or best recipe for when or how often to apply smoothing. Typically, it should be applied right after any "event" that is expected to generate fast transients, or whenever one suspects that ringing may occur. Smoothing is obviously expensive because of the cost of computing x_{n+1} twice, so it is not always worth the effort. Furthermore, practical experience shows that smoothing by itself does not guarantee significant improvement in the quality of the solution. However if, in addition, we also apply *extrapolation*, whose purpose is to improve the accuracy, then smoothing becomes worthwhile, as reported in Lindberg (1971), both in terms of computational cost and quality of solution. In the next section, we will describe extrapolation and its application to the TR. When extrapolation is being used anyway, then LTE estimation can be done using the technique we saw earlier, based on Richardson extrapolation. Thus, fully-featured implementations of TR employ smoothing, extrapolation, and LTE estimation by means of Richardson extrapolation.

What if $t_n - t_{n-1} \neq t_{n+1} - t_n$? There are no good alternatives to standard smoothing in this case. Perhaps one can explore alternative smoothing techniques that use unequal weights for x_{n-1} and x_{n+1}, or other possibilities. It is not clear if this has been attempted or tested in the field.

5.5.2 Extrapolation

Extrapolation is a scheme by which accuracy can be improved by combining the numerical solutions at different time-step settings. We first solve the system over $[t_0, t]$ using a *fixed* time-step h_1, starting from some initial condition x_0. We then re-solve the system, over the same time-span and starting with the same initial condition, but using a different *fixed* time-step $h_2 < h_1$. The goal, then, is to extrapolate those two results to hopefully come closer to the $h = 0$ (i.e., the exact) solution, which is our ultimate goal. Extrapolation can be repeatedly applied, producing iterative refinement of the solution, but we will describe only a single application of it.

For a given fixed time-step h over $[t_0, t]$, suppose that the error between the computed and exact solutions at time t, is given by the expansion:

$$E(h) \triangleq \sum_{j=1}^{\infty} a_j h^{\gamma_j}, \quad \text{where} \quad \gamma_1 < \gamma_2 < \cdots \tag{5.255}$$

and where the coefficients a_j do not depend on h; this is a typical situation, but notice that $E(h)$ is not an LTE, but rather a *global* truncation error (GTE). Now, suppose that two solutions are found over $[t_0, t]$, corresponding to two different (fixed) time-step values h_1 and $h_2 < h_1$, and let $\rho \triangleq (h_2/h_1) < 1$. If we denote the two computed solutions at t by $x^{(1)}(t)$ and $x^{(2)}(t)$, then:

$$x(t) = x^{(1)}(t) + \sum_{j=1}^{\infty} a_j h_1^{\gamma_j} \tag{5.256}$$

$$x(t) = x^{(2)}(t) + \sum_{j=1}^{\infty} a_j h_2^{\gamma_j} \tag{5.257}$$

where the coefficients a_j and exponents γ_j are the same in both expressions. Proceeding as in Ralston and Rabinowitz (2001), if we multiply (5.256) by $h_2^{\gamma_1}$ and (5.257) by $h_1^{\gamma_1}$, then subtract the two results and solve for $x(t)$, we find that the terms with $h_i^{\gamma_1}$ cancel out, and we have:

$$x(t) = \left(\frac{x^{(2)}(t) - \rho^{\gamma_1} x^{(1)}(t)}{1 - \rho^{\gamma_1}} \right) + \sum_{j=2}^{\infty} \left(\frac{\rho^{\gamma_j} - \rho^{\gamma_1}}{1 - \rho^{\gamma_1}} \right) a_j h_1^{\gamma_j} \tag{5.258}$$

Thus, we can compute an improved solution at t as:

$$\hat{x}(t) = \left(\frac{x^{(2)}(t) - \rho^{\gamma_1} x^{(1)}(t)}{1 - \rho^{\gamma_1}} \right) \tag{5.259}$$

for which:

$$x(t) - \hat{x}(t) = \sum_{j=2}^{\infty} b_j h_1^{\gamma_j} \tag{5.260}$$

where:

$$b_j \triangleq \left(\frac{\rho^{\gamma_j} - \rho^{\gamma_1}}{1 - \rho^{\gamma_1}} \right) a_j \tag{5.261}$$

Crucially, the order of the global error has been improved from γ_1 to γ_2. Note, we do not need to know the values of a_j or b_j in order to use extrapolation; we merely find $\hat{x}(t)$ and reap the benefits of the increased order. We do need to know γ_1, but that is known for any given LMS method.[3]

Extrapolation is applicable to many numerical problems, such as for finding derivatives, computing definite integrals, as well as solving ODEs. But, it is obviously not cheap, because we have to re-solve the system over the same time-span and because we have to use a fixed time-step across that time-span. So, it is not clear that it is always worth the effort. However, because of a special property of TR, which we describe next, extrapolation becomes especially compelling in that case.

---

[3]In general, it can be shown that the global truncation error (GTE) of an LMS method of order p is $O(h^p)$, in contrast to the LTE which is $O(h^{p+1})$, so that $\gamma_1 = p$. For example, TR has an LTE which is $O(h^3)$ but a GTE which is only $O(h^2)$, so that an order of 1 is "lost" due to error accumulation, and $\gamma_1 = 2$.

Even Powers Expansion The trapezoidal rule is called a *symmetric* rule because, if the sign of h is reversed, then the roles of x_n and x_{n+1} are reversed. Such rules are also called *reflexive* by some authors. Because of this, and under certain general conditions having to do with differentiability of the system function, $f(x, t)$, then it can be shown, as in Stetter (1965), that the global truncation error of the trapezoidal rule has an expansion in terms of only even powers of h (so that $\gamma_1 = 2$, $\gamma_2 = 4$, $\gamma_3 = 6$, etc.):

$$E(h) = \sum_{j=1}^{\infty} c_j h^{2j} \tag{5.262}$$

where, as before, the c_j coefficients are independent of h (this expansion is *asymptotic* in the sense that is only accurate as $h \to 0$). The consequences of this, when applying extrapolation to TR, are immediately clear and compelling: whereas, in general, we would expect extrapolation to improve the order by an increment of 1, in this case it *doubles* the order, from 2 to 4. This makes it worthwhile to apply extrapolation to TR. Experimental evidence, according to Lindberg (1971), shows that using extrapolation *and* smoothing in an implementation of TR gives good accuracy with reduced ringing.

Applicability to simulation: One difficulty in using extrapolation for circuit simulation is the fact that the time-steps h_1 and h_2 must be *fixed* over $[t_0, t]$. One can envision allowing both h_1 and h_2 to be relatively large, but even then, the use of a fixed time-step would seem to be expensive for simulation. It is not clear if this has been attempted or used in circuit simulation, even while "bending the rules" somewhat by relaxing the fixed time-step requirement. It may make sense, for example, to apply extrapolation over a limited, short, time-span, instead of over the whole time period from the beginning of simulation time.

5.6 VARIABLE TIME-STEP METHODS

General purpose ODE codes always include the ability to vary the time-step, because that provides a considerable speed advantage. This is also the case for circuit simulation, especially given the widely varying time-scales employed in digital signaling. Generally speaking, a variable time-step approach requires three components:

1. A way to estimate the local error (PLTE).
2. A strategy for deciding, based on the estimated error, whether to increase or decrease the time-step, and by how much.
3. A way to implement the time-step change, by accordingly updating the numerical method being used.

Regarding the first item, we have already discussed PLTE estimation at length, and we will briefly return to this later on for circuit-specific methods. Regarding the second item, a typical approach is to compare the error estimate against a threshold, in order to decide whether a change of time-step is required. The threshold typically includes both a relative and an absolute error term, which are user-specified. As for deciding by how much to change the time-step, commonly used methods are mostly heuristic, and there are no theoretical results for what constitutes a "best" strategy. A typical strategy is to double and halve the time-step when needed, while restricting how large and how small the time-step is allowed to become. We will return to time-step control later on in a circuit-specific context. As for the third item, this is a rich and detailed area of study, which we now summarize briefly.

5.6.1 Implementing a Change of Time-Step

It should be clear that one-step methods, like FE, BE, and TR, present no difficulties at all with regard to implementing a change of time-step. One simply sets a new value of h for the next step, and moves on. Difficulties arise only in connection with $k > 1$ multistep methods, such as the BDFs of orders higher than 1, as well as many other LMS methods. The previous data, x_n, x_{n-1}, x_{n-2}, etc., are separated by the previous value of time-step, which may be different from the new, intended, $(x_{n+1} - x_n)$. This, in fact, is the main disadvantage of LMS methods compared to Runge-Kutta methods, but there are many ways to deal with this. The available solutions are in two categories:

1. Interpolation methods: use an interpolation polynomial to obtain the required back-data at a mesh with the required (new) time-step value. As a result, the method is restarted as a fresh fixed time-step method, using the new value of the time-step. The interpolation polynomial can be stored and updated in various ways; some highly efficient techniques have been developed for this. For circuit simulation, and because the higher-order BDFs are rarely used, one probably does not need to use the most advanced methods.

2. Variable-coefficient methods: these methods *re-derive* the whole LMS theory based on an LMS template that has non-equidistant data. Here, too, there is a large body of work, and the techniques are quite involved; we will only give highlights of such methods.

In the following, we will summarize both of these categories.

5.6.2 Interpolation Methods

One way to implement a time-step change is to use an interpolation polynomial to interpolate the back-data at the missing time-points. For the BDFs, this requires interpolation of the x_{n-j} values only, and no interpolation of the f_{n-j} values is required. Suppose we have just solved for x_n, so that we have the back-data:

$$(t_n, x_n), (t_{n-1}, x_{n-1}), (t_{n-2}, x_{n-2}), \ldots \tag{5.263}$$

with $h = (t_n - t_{n-1}) = (t_{n-1} - t_{n-2}) = \cdots$. Suppose we now want to implement a change of time-step, using a new value of time-step αh, where $\alpha > 0$, to find the next solution point. Effectively, we are starting a new time mesh which we denote $t_{n+1}^{(\alpha)}$, $t_{n+2}^{(\alpha)}$, etc., where $t_n^{(\alpha)} = t_n$ and $\alpha h = t_{n+1}^{(\alpha)} - t_n^{(\alpha)} = t_{n+2}^{(\alpha)} - t_{n+1}^{(\alpha)} = \cdots$. To make the description more concrete, suppose we are using BDF2:

$$x_{n+1} - \frac{4}{3}x_n + \frac{1}{3}x_{n-1} = \frac{2}{3}hf_{n+1} \tag{5.264}$$

Finding the next solution point at $t_{n+1}^{(\alpha)}$, which we will denote by $x_{n+1}^{(\alpha)}$, using BDF2 requires a value of the solution at $t_n^\alpha = t_n$, which we have, as $x_n^{(\alpha)} = x_n$, and a value of the solution at $t_{n-1}^{(\alpha)} \triangleq t_n - \alpha h$, i.e., $x_{n-1}^{(\alpha)}$, which we do *not* have! To be able to proceed, we need a value of $x_{n-1}^{(\alpha)}$, and we can find it by interpolating the previous data $x_n, x_{n-1}, x_{n-2}, \ldots$ using an interpolation polynomial. For higher order BDFs, other back-data values have to be estimated as well. But, how many back-data points x_{n-j} should we use for interpolation? In other words, what should be the degree of the interpolation polynomial?

Recall, the interpolation error for a polynomial over $k+1$ equidistant data points is $O(h^{k+1})$, which is the same order as the LTE of the k-step BDF. Thus, it is sufficient in practice to interpolate the previous $k+1$ points, when using a k-step BDF method. For BDF2, this means that we interpolate the previous three points x_n, x_{n-1}, and x_{n-2}, using a quadratic equation (a parabola).

In general, as we saw earlier, the Newton-Gregory backward interpolation polynomial over $k+1$ equidistant points, $x_n, x_{n-1}, \ldots, x_{n-k}$, is:

$$P_n(t) = x_n + \sum_{i=1}^{k} \left[\frac{1}{i!} \prod_{j=0}^{i-1}(s+j) \right] \nabla^i x_n \tag{5.265}$$

where $t = t_n + sh$, h is the time-step, and $s \in \mathbb{R}$. For high-order methods, maintaining and updating this polynomial over time becomes a significant overhead that can be expensive if not done well. There are highly developed methods for doing this efficiently, as described in Lambert (1991). However, for circuit simulation, where BDF2 is often the highest order method used, requiring a simple parabola, these advanced methods are "overkill." Instead, for the BDF2 case, we can simply write:

$$P_n(t) = x_n + s\nabla x_n + \frac{s(s+1)}{2}\nabla^2 x_n \tag{5.266}$$

$$= x_n + s(x_n - x_{n-1}) + \frac{s(s+1)}{2}\left(x_n - 2x_{n-1} + x_{n-2} \right) \tag{5.267}$$

so that we can interpolate for the value at $t = t_{n-1}^{(\alpha)}$ as:

$$x_{n-1}^{(\alpha)} = \frac{1}{2}(2 - 3\alpha + \alpha^2)x_n + \alpha(2 - \alpha)x_{n-1} - \frac{\alpha}{2}(1 - \alpha)x_{n-2} \qquad (5.268)$$

With the required back-data thus complete and available, we simply move forward in time using the standard fixed time-step BDF2.

5.6.3 Variable-Coefficient Methods

The study of variable-coefficient methods is too detailed and involved for our purposes, and beyond the scope of our study. Suffice it to say that, one can re-derive the LMS methods, their predictors, their truncation errors, etc., for the case of non-equidistant data. Note that one-step LMS methods require no modification to their formulas, but may require new predictors and new Milne's LTE estimates.

For the BDFs, the derivation is not that hard; we simply use an interpolation formula for the non-equidistant data case to derive the formulas, and proceed as usual. For BDF2, the re-derived LMS method is as follows:

$$x_{n+1} - \left[\frac{(h_{n+1} + h_n)^2}{h_n(2h_{n+1} + h_n)}\right]x_n + \left[\frac{h_{n+1}^2}{h_n(2h_{n+1} + h_n)}\right]x_{n-1} = \left[\frac{h_{n+1}(h_{n+1} + h_n)}{2h_{n+1} + h_n}\right]f_{n+1}$$

where $h_{n+1} \triangleq (t_{n+1} - t_n)$ and $h_n \triangleq (t_n - t_{n-1})$, which reduces to the regular BDF2 when $h_{n+1} = h_n = h$, and the PLTE is this:

$$\text{PLTE} = -\left[\frac{h_{n+1}^2(h_{n+1} + h_n)^2}{6(2h_{n+1} + h_n)}\right]x^{(3)}(t_n) \qquad (5.269)$$

which reduces to the regular result when $h_{n+1} = h_n = h$, of $-(2/9)h^3 x^{(3)}(t_n)$. For TR, being a one-step method, its formula is unchanged, and we have:

$$\text{PLTE} = -\frac{1}{12}h_{n+1}^3 x^{(3)}(t_n) \qquad (5.270)$$

A predictor can be used based on extrapolation of an interpolation polynomial, as we saw earlier, which for TR and BDF2 is this:

$$x_{n+1}^{(0)} = \left[\frac{h_{n+1}(h_{n+1} + h_n)}{h_{n-1}(h_n + h_{n-1})}\right]x_{n-2} - \left[\frac{h_{n+1}(h_{n+1} + h_n + h_{n-1})}{h_n h_{n-1}}\right]x_{n-1}$$
$$+ \left[\frac{(h_{n+1} + h_n)(h_{n+1} + h_n + h_{n-1})}{h_n(h_n + h_{n-1})}\right]x_n$$

For TR, this leads to the Milne's PLTE estimate:

$$\text{PLTE} \approx \frac{-h_{n+1}^2}{h_{n+1}^2 + 2(h_{n+1} + h_n)(h_{n+1} + h_n + h_{n-1})}\left(x_{n+1} - x_{n+1}^{(0)}\right) \qquad (5.271)$$

These results are really all that is needed for circuit simulation from the theory of variable-coefficient methods.

5.6.4   Variable Step Variable Order (VSVO) Methods

Finally, it should be mentioned that general purpose ODE codes vary not only the time-step, but also the order, so they are called *variable step variable order* (VSVO) methods. For example, one approach is to monitor the LTE in several formulas at once, and pick the one with the lowest LTE to proceed forward. These methods have been tested in circuit simulation, and it was found that, for integrated circuits simulation at least, the order most often selected is two. Thus, the earliest simulators, including SPICE, were based on the use of (only) second-order methods like TR and BDF2. During circuit simulation, one mostly uses TR, and it is easy to switch-over to BE or BDF2 if ringing becomes a significant problem. For reference, Brayton et al. (1972) describe a VSVO implementation of circuit simulation based on the BDFs 1–6.

5.7   APPLICATION TO CIRCUIT SIMULATION

In applying the above ODE solution methods to circuit simulation, one finds that efficiency gains are possible by making use of (another variety of) *companion models*. Similar to the case of nonlinear resistive circuits, companion models are useful so as to quickly construct the numerical problem to be solved upon a simple inspection of the circuit. Companion models are built for dynamic elements, for different numerical solution methods. Inspecting the circuit, one then (conceptually) replaces every dynamic element by its companion model. This effectively transforms a (possibly nonlinear) *differential* problem into a (possibly nonlinear) *algebraic* problem. The solution can then be found by repeatedly applying a linear or nonlinear solver, like GE or Newton's method, at every time-point. Such techniques are implemented in all modern circuit simulators.

Before we present this standard solution approach using companion models, however, we will start with a brief overview of relevant solution methods for DAEs. Recall that, in general, circuit equations are DAEs, not ODEs. Recall also that the solution methods for DAEs are not as fully developed, and that many ODE methods are used in practice to solve DAEs, but that this is possible with some, not all DAEs, and that the guarantees of accuracy and/or stability of these ODE methods no longer necessarily hold when they are applied to DAEs. Indeed, as pointed out earlier, there are DAEs for which even the best known ODE methods lead to unstable behavior. One common way in which ODE methods are applied to DAEs is by means of so-called *direct discretization*, which we will describe below. Direct discretization works well in practice for many low-index (index 1 or 2) DAEs; it is the method of choice for circuit simulation. After that, we will address issues of companion models, element stamps, error estimation, and time-step control.

The material in this section is based on a number of sources, including Ascher and Petzold (1998), McCalla (1988), Ogrodzki (1994), Pillage et al. (1995), and Kundert (1995).

5.7.1 From DAEs to Algebraic Equations

Recall that, in the most general case, the MNA system is a DAE of the general form:

$$\mathfrak{F}\left(x, x', t\right) = 0 \tag{5.272}$$

where $\mathfrak{F} : (\mathbb{R}^m \times \mathbb{R}^m \times \mathbb{R}) \to \mathbb{R}^m$. As we did with ODEs, suppose we have solved for $x_i \approx x(t_i)$ for $i \leq n$ and we now need to find $x_{n+1} \approx x(t_{n+1})$, by means of some numerical method. We can write the system equation at time t_{n+1} as:

$$\mathfrak{F}\left(x(t_{n+1}), x'(t_{n+1}), t_{n+1}\right) = 0 \tag{5.273}$$

and the essence of direct discretization is to then remove the dependence of this equation on $x'(t_{n+1})$. Once this is done, the equation becomes an *algebraic equation*; it is no longer *differential*. This is done in a way that is specific to the numerical method being used, and is quite different for implicit versus explicit LMS methods, as we will see.

BE Direct Discretization   In order to see the way forward, it is instructive to recall how we arrived at some simple numerical methods for ODEs, such as BE. Given an ODE $x' = f(x, t)$, the justification for the BE method, in the context of ODEs, can be re-stated as the following argument. We first write the ODE at t_{n+1}, and use $x(t_{n+1}) \approx x_{n+1}$, so that:

$$x'(t_{n+1}) = f(x(t_{n+1}), t_{n+1}) \approx f(x_{n+1}, t_{n+1}) \triangleq f_{n+1} \tag{5.274}$$

We then note that, from a Taylor series expansion, we have:

$$x(t_n) \approx x(t_{n+1}) - hx'(t_{n+1}) \tag{5.275}$$

which, combined with $x(t_n) \approx x_n$ and $x(t_{n+1}) \approx x_{n+1}$, leads to:

$$x'(t_{n+1}) \approx \frac{x_{n+1} - x_n}{h} \tag{5.276}$$

Then, equating the two expressions for $x'(t_{n+1})$ from (5.274) and (5.276) provides the motivation for the BE method:

$$x_{n+1} = x_n + hf_{n+1} \tag{5.277}$$

This round-about way of re-stating the motivation for BE may seem pointless, but its value will immediately become apparent as we now apply it to motivate a BE-inspired solution method for DAEs. Indeed, we can follow the same above process for a general DAE (5.272), by first writing it at t_{n+1} as (5.273), then using $x(t_{n+1}) \approx x_{n+1}$, to write:

$$\mathfrak{F}\left(x_{n+1}, x'(t_{n+1}), t_{n+1}\right) \approx 0 \qquad (5.278)$$

and then combining this with (5.276), leading to the BE-inspired numerical method:

$$\mathfrak{F}\left(x_{n+1}, \frac{x_{n+1} - x_n}{h}, t_{n+1}\right) = 0 \qquad (5.279)$$

This is a valid numerical solution method because it provides an (implicit) algebraic equation for finding x_{n+1}, given x_n. This approach for solving a DAE is called a *direct discretization* approach; in this case, it is based on BE. Effectively, a *differential equation* (5.273) (the DAE) has been transformed into an *algebraic equation* (5.279). Whether this equation is easily solvable or not depends on the DAE. Note that all that is needed from "the past," in this case, is the previous data value, x_n.

BDF Direct Discretization   We can extend the BE-inspired approach to the rest of the BDF family, as follows. At time t_{n+1}, we use (5.278) and recall the approximation, introduced earlier (5.159), as part of the alternate derivation of the BDFs, based on polynomial interpolation:

$$x'(t_{n+1}) \approx \frac{1}{h} \sum_{i=1}^{k} \frac{1}{i} \nabla^i x_{n+1} = \frac{1}{h\beta_{-1}} \sum_{j=-1}^{k-1} \alpha_j x_{n-j} \qquad (5.280)$$

Combining (5.280) with (5.278), leads to the BDF-inspired numerical method:

$$\mathfrak{F}\left(x_{n+1}, \frac{1}{\beta_{-1} h} \sum_{j=-1}^{k-1} \alpha_j x_{n-j}, t_{n+1}\right) = 0 \qquad (5.281)$$

Here, too, we have a viable (direct discretization) approach for stepping forward in time that depends on previous x_i data only. In contrast, we will see that the case of TR leads to an approach which requires additional information about the past.

TR Direct Discretization   In the case of TR, we write the familiar DAE approximation at time t_{n+1}, (5.278), and recall the result, derived earlier (5.61) from a Taylor series expansion in connection with TR, namely:

$$x(t_{n+1}) - x(t_n) = \frac{h}{2}\left[x'(t_{n+1}) + x'(t_n)\right] + 0(h^3) \qquad (5.282)$$

which we now combine with $x(t_n) \approx x_n$ and $x(t_{n+1}) \approx x_{n+1}$, leading to:

$$x'(t_{n+1}) \approx \frac{2}{h} (x_{n+1} - x_n) - x'(t_n) \tag{5.283}$$

We then combine this with (5.278), leading to the TR-inspired numerical method:

$$\mathfrak{F}\left(x_{n+1}, \frac{2}{h} (x_{n+1} - x_n) - x'(t_n), t_{n+1}\right) = 0 \tag{5.284}$$

A distinguishing feature of this method is that it requires not only the past data x_n, but also the past value of the derivative $x'(t_n)$. In general, this would not be a viable approach, because of the need to compute the past values of the derivatives. Derivative estimation is feasible, but not easy, and certainly requires a dedicated numerical method for accurate estimation, such as a BDF. However, we will see that for circuit simulation, because of the special form of the element equations, this issue is easy to deal with and does not present any complications.

Discretization in Circuit Equations   To see how the above discretization techniques can be carried out in circuit simulation, the first step is to recall the constitutive equations of the MNA system:

1. KCL:
$$Ai = 0 \tag{5.285}$$

   where i is a vector of all the branch currents.
2. KVL:
$$A^T v = u \tag{5.286}$$

   where u is a vector of all branch voltages and v the nodal voltages vector.
3. The element equations, be they resistive or dynamic. For the resistive elements, their equations can be collected in matrix form as:

$$E_r \begin{bmatrix} i \\ u \end{bmatrix} = g(x) \tag{5.287}$$

   where x is the MNA vector of variables, E_r is a diagonal matrix whose elements are either 0 or 1, and $g(x)$ is a vector function representing all the (possibly nonlinear) resistive elements. For the dynamic elements, their equations can be collected in matrix form as:

$$E_d \begin{bmatrix} i \\ u \end{bmatrix} = D(x)x'(t) \tag{5.288}$$

   where E_d is another diagonal matrix whose elements are either 0 or 1, and $D(x)$ is a matrix function representing all the (possibly nonlinear) dynamic elements. Note, this is not to be confused with the $D(x)$ matrix that we employed earlier in (5.14).

Altogether, these four equations form the system of equations (5.272) that has been the subject of our direct discretization approach. We will refer to the four equations (5.285), (5.286), (5.287), and (5.288) as the *unraveled* form of the circuit equations. Instead of applying direct discretization to the DAE (5.272), the key idea is to apply it to each of the four constitutive equations separately. Of these four equations, it is only the last one (5.288) that has a time dependence, and is impacted by discretization. Therefore, discretization is applied directly to only this last set of equations (5.288), which are the dynamic element equations.

This approach leads directly to the concept of a *companion model* for dynamic elements, as we will see below in connection with different numerical methods. In every case, we will focus on a single capacitor and a single inductor with only self capacitance/inductance. The method can be easily extended to the case of elements with mutual capacitance/inductance.

5.7.2 FE Discretization

Consider first the use of the simplest numerical method, forward Euler (FE). To simplify the notation, we will use $i(t)$ to denote the current in a dynamic element and $u(t)$ to denote its branch voltage. For FE, we start by writing the dynamic element equations at t_n:

$$i(t_n) = C(u(t_n))u'(t_n) \quad \text{and} \quad u(t_n) = L(i(t_n))i'(t_n) \qquad (5.289)$$

then, using $i(t_n) \approx i_n$ and $u(t_n) \approx u_n$, we have:

$$i_n \approx C(u_n)u'(t_n) \quad \text{and} \quad u_n \approx L(i_n)i'(t_n) \qquad (5.290)$$

then, using the familiar result from the Taylor series expansion, we write:

$$i_n \approx C(u_n)\left(\frac{u_{n+1} - u_n}{h}\right) \quad \text{and} \quad u_n \approx L(i_n)\left(\frac{i_{n+1} - i_n}{h}\right) \qquad (5.291)$$

leading to the FE-inspired direct discretization scheme:

$$u_{n+1} = u_n + \left(\frac{h}{C(u_n)}\right)i_n \quad \text{and} \quad i_{n+1} = i_n + \left(\frac{h}{L(i_n)}\right)u_n \qquad (5.292)$$

Because the previous data, i_n and u_n, are *available*, then the voltages across all capacitors and the currents in all inductors, at t_{n+1}, are *known*. No solution of the overall system of equations is required in order to reveal the values of capacitor voltages or inductor currents at t_{n+1}. This curious result is a special feature of explicit methods like FE; such methods have an interesting *predictive* quality, which is also a source of their weakness.

We can plug all these values in the other equations (of the unraveled form), which we would then solve for the remaining variables at t_{n+1}. This can be efficiently achieved by replacing all capacitors and inductors in the netlist by ideal voltage and current sources, and solving the resulting circuit. Each capacitor is replaced by a voltage source of value u_{n+1}, from (5.292), and each inductor is replaced by a current source of value i_{n+1}, from (5.292). Once the whole circuit is solved at t_{n+1}, using possibly Newton's method and MNA, we would then repeat the process for the next time-point, etc.

Companion Models The above *replacements* for dynamic elements at t_{n+1} are called the *companion models* of these elements, and they depend on the LMS method being used. For the case of FE, as we have seen, the companion models for the dynamic elements are as shown in Fig. 5.12. Once all dynamic elements have been replaced by their companion models, the rest of the circuit is resistive, possibly nonlinear, and can be solved, typically using Newton's method, in which each iteration requires solution of the familiar linear system:

$$J(x^{(k)})x^{(k+1)} = s^{(k)} \tag{5.293}$$

The unraveled form of the equations is never explicitly constructed. Because the FE companion models consist of only sources, their *values* affect only the RHS vector, $s^{(k)}$, and *not* the Jacobian (i.e., the MNA) matrix $J(x^{(k)})$. Thus, the Jacobian is independent of the time-step h and of the values of the dynamic elements in the network. It does not change over time! This is a great advantage of the FE case; among other things, it means that, for a *linear* dynamic circuit, the Jacobian needs to be built and factored only once!

Note that, from one time-point to the next, the values of the companion models may change, but the *topology* of the netlist remains fixed. Both i_{n+1} and u_{n+1} for each dynamic element must be stored and used at the next time-point, as i_n and u_n, in order to update the FE companion models. Thus, once the MNA system is built, after the first time-step, it never has to be rebuilt again; only the element stamp values are updated over time.

Element Stamps From the above, it is easy to see that the FE element stamps are as follows. For a capacitor, we denote by $V_{n+1}^{(k)}$ the value of the voltage source

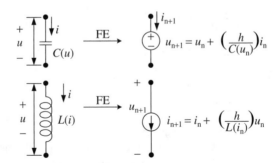

Figure 5.12: Companion models for the case of FE.

that replaces the capacitor at time t_{n+1}, for finding $x^{(k+1)}$, i.e.:

$$V_{n+1}^{(k)} \triangleq u_n + \left(\frac{h}{C(u_n)}\right) i_n \tag{5.294}$$

so that the (already) linearized companion model of the capacitor, in the k-th Newton iteration at t_{n+1}, has the element equation:

$$u = V_{n+1}^{(k)} \tag{5.295}$$

From this, we see that a capacitor has the FE MNA stamp shown in Table 5.2, where the row and column labels in the table refer to the corresponding rows and columns of the MNA system equation, as usual. Similarly, for an inductor, denoting:

$$I_{n+1}^{(k)} \triangleq i_n + \left(\frac{h}{L(i_n)}\right) u_n \tag{5.296}$$

the FE element stamp, representing the element equation $i = I_{n+1}^{(k)}$ is shown in Table 5.3.

5.7.3   BE Discretization

Consider now the case of the implicit backward Euler (BE). In contrast to FE, we write the dynamic element equations at t_{n+1}:

$$i(t_{n+1}) = C(u(t_{n+1}))u'(t_{n+1}) \quad \text{and} \quad u(t_{n+1}) = L(i(t_{n+1}))i'(t_{n+1}) \tag{5.297}$$

Table 5.2: Element stamp for a capacitor, based on FE discretization and linearization.

	$v+$	$v-$	i	RHS
$v+$			1	
$v-$			-1	
i	1	-1		$V_{n+1}^{(k)}$

Table 5.3: Element stamp for an inductor, based on FE discretization and linearization.

	RHS
$v+$	$-I_{n+1}^{(k)}$
$v-$	$I_{n+1}^{(k)}$

then, using $i(t_{n+1}) \approx i_{n+1}$ and $u(t_{n+1}) \approx u_{n+1}$, we have:

$$i_{n+1} \approx C(u_{n+1})u'(t_{n+1}) \quad \text{and} \quad u_{n+1} \approx L(i_{n+1})i'(t_{n+1}) \tag{5.298}$$

then, using the familiar result from the Taylor series expansion, we write:

$$i_{n+1} \approx C(u_{n+1})\left(\frac{u_{n+1} - u_n}{h}\right) \quad \text{and} \quad u_{n+1} \approx L(i_{n+1})\left(\frac{i_{n+1} - i_n}{h}\right) \tag{5.299}$$

For a capacitor, this leads to the BE-inspired direct discretization:

$$i_{n+1} = \left(\frac{C(u_{n+1})}{h}\right)u_{n+1} - \left(\frac{C(u_{n+1})}{h}\right)u_n \tag{5.300}$$

or, equivalently:

$$u_{n+1} = \left(\frac{h}{C(u_{n+1})}\right)i_{n+1} + u_n \tag{5.301}$$

For an inductor, we also get the BE-inspired direct discretization:

$$u_{n+1} = \left(\frac{L(i_{n+1})}{h}\right)i_{n+1} - \left(\frac{L(i_{n+1})}{h}\right)i_n \tag{5.302}$$

or, equivalently:

$$i_{n+1} = \left(\frac{h}{L(i_{n+1})}\right)u_{n+1} + i_n \tag{5.303}$$

In contrast to the FE case, we do not here have immediate solutions for the i_{n+1} and u_{n+1} values. Instead, we have (algebraic) relations between i_{n+1} and u_{n+1}, that include, among other parameters, past data values u_n and i_n.

For linear C and L, the above discretization schemes are simplified as follows. For a linear capacitor C, we have:

$$i_{n+1} = \left(\frac{C}{h}\right)u_{n+1} - \left(\frac{C}{h}\right)u_n \tag{5.304}$$

or, equivalently:

$$u_{n+1} = \left(\frac{h}{C}\right)i_{n+1} + u_n \tag{5.305}$$

For a linear inductor L, we have:

$$u_{n+1} = \left(\frac{L}{h}\right)i_{n+1} - \left(\frac{L}{h}\right)i_n \tag{5.306}$$

or, equivalently:

$$i_{n+1} = \left(\frac{h}{L}\right) u_{n+1} + i_n \tag{5.307}$$

To solve for the t_{n+1} responses, we can replace all the differential equations of the dynamic elements (in the unraveled form) by these algebraic equations. As in the case of FE, this can be more efficiently achieved by:

1. Replacing every dynamic element in the netlist by a *companion model* that embodies the above terminal characteristics between its i_{n+1} and u_{n+1}.
2. Constructing the MNA system for the resulting netlist using element stamps and solving it, typically using Newton's method.

This is repeated for every time-step, knowing that the topology of the netlist remains fixed, but only the element values may change.

Companion Models for Linear Elements   We first consider the BE companion models in the linear case. It is easy to see that, for the BE case, and for a *linear C* and L, the companion models can be given as the *linear* circuits in Fig. 5.13. Note, we have deliberately chosen the form of the equations that allows a companion model based on a current source, instead of a voltage source. The advantages of this for MNA are obvious; whenever possible, we use this Norton form of the companion model instead of the Thévenin form. In the BE case, the companion model values affect both the RHS vector and the system matrix, i.e., the Jacobian. However, if the dynamic elements are linear, and if a fixed time-step is used, then their contributions to the Jacobian do not change over time.

Once the netlist has been solved at t_{n+1}, we then need to compute u_{n+1} for every capacitor and i_{n+1} for every inductor. These are required at the next time-point, in order to update the model parameter values that go into (5.304) and (5.307), namely u_n and i_n. It is trivial to find u_{n+1} for every capacitor, as simply the difference of its two node voltages. As for i_{n+1} in every inductor, we can compute it using (5.307), provided we first compute u_{n+1} across it, which, as for capacitors, is very easy to do. This requires knowledge of i_n, which should have

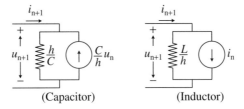

(Capacitor)   (Inductor)

Figure 5.13: Companion models for linear L and C, for the case of BE.

been stored at the previous time-point, and therefore at all previous time-points, going back to the beginning of simulation time, t_0, where i_0 is required! This generates a requirement that, as part of the DC Analysis run at t_0, one should find and store the current for every inductor. We will briefly digress to discuss this further.

A Note on Initialization Recall that the simpler methods of performing DC Analysis involve stripping the circuit of its dynamic elements, by disabling all the L and C elements. Disabling dynamic elements means that every capacitor is replaced by an open circuit, and so the voltage across that open circuit can be easily found from the DC Analysis results. It also means that every inductor is replaced by a short circuit, typically represented by a 0 V voltage source, whose current is also easily available from the MNA solution vector. More complex methods, such as pseudo-transient, maintain and add to the dynamic elements in the circuit. However, pseudo-transient also provides an initial state for the whole circuit, including capacitor voltages and inductor currents: they are all set to zero. Thus, in all cases, DC Analysis can provide the initial "state" of all dynamic elements: currents in inductors and voltages across capacitors.

Element Stamps for Linear Elements The above BE companion models are already linear, so that no further linearization is required in the Newton loop. The corresponding element stamps are as follows. For a linear capacitor, denote:

$$G_{n+1}^{(k)} = \frac{C}{h} \quad \text{and} \quad I_{n+1}^{(k)} = -\frac{C}{h}u_n \tag{5.308}$$

so that the (already) linearized element equation is:

$$i = G_{n+1}^{(k)}u + I_{n+1}^{(k)} \tag{5.309}$$

and the resulting MNA element stamp is as shown in Table 5.4. For a linear inductor, denote:

$$G_{n+1}^{(k)} = \frac{L}{h} \quad \text{and} \quad I_{n+1}^{(k)} = i_n \tag{5.310}$$

so that the (already) linearized element equation is:

$$i = G_{n+1}^{(k)}u + I_{n+1}^{(k)} \tag{5.311}$$

and the resulting MNA element stamp is shown in Table 5.5.

Table 5.4: Element stamp for a linear capacitor, based on BE discretization and linearization.

	$v+$	$v-$	RHS
$v+$	$G_{n+1}^{(k)}$	$-G_{n+1}^{(k)}$	$-I_{n+1}^{(k)}$
$v-$	$-G_{n+1}^{(k)}$	$G_{n+1}^{(k)}$	$I_{n+1}^{(k)}$

Table 5.5: Element stamp for a linear inductor, based on BE discretization and linearization.

	$v+$	$v-$	RHS
$v+$	$G_{n+1}^{(k)}$	$-G_{n+1}^{(k)}$	$-I_{n+1}^{(k)}$
$v-$	$-G_{n+1}^{(k)}$	$G_{n+1}^{(k)}$	$I_{n+1}^{(k)}$

Companion Models for Nonlinear Elements   For a *nonlinear* capacitor, we use (5.300) instead of (5.301) so as to get a valid (explicit, in this case voltage-controlled) nonlinear element form:

$$i_{n+1} = \frac{C(u_{n+1})}{h}(u_{n+1} - u_n) \triangleq g_C(u_{n+1}) \qquad (5.312)$$

Notice that, if we were to use (5.301), we would get an element equation in the unacceptable *implicit* form $u_{n+1} = g(u_{n+1}, i_{n+1})$. For a *nonlinear* inductor, we use (5.302) instead of (5.303) so as to get a valid (explicit, in this case current-controlled) nonlinear element form:

$$u_{n+1} = \frac{L(i_{n+1})}{h}(i_{n+1} - i_n) \triangleq g_L(i_{n+1}) \qquad (5.313)$$

The companion models, in this case nonlinear, can be as shown in Fig. 5.14. For a capacitor, finding u_{n+1} to plug into (5.312), as u_n at the next time-point, is just as simple as in the linear capacitor case (voltage difference). For an inductor, as we will see below, it is also easy to compute i_{n+1}, to plug into (5.313) as i_n at the next time-point.

Element Stamps for Nonlinear Elements   In the nonlinear case, the BE element stamps can be found as follows. For a nonlinear capacitor $C(u)$, the BE companion model has the element equation $i = g_C(u)$ which, linearized in the

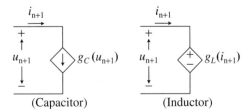

(Capacitor)   (Inductor)

Figure 5.14: Companion models for nonlinear L and C, for the case of BE.

Newton loop at t_{n+1}, becomes:

$$i = g'_C \left(u_{n+1}^{(k)} \right) \left(u - u_{n+1}^{(k)} \right) + g_C \left(u_{n+1}^{(k)} \right) \tag{5.314}$$

$$= g'_C \left(u_{n+1}^{(k)} \right) u + \left[g_C \left(u_{n+1}^{(k)} \right) - g'_C \left(u_{n+1}^{(k)} \right) u_{n+1}^{(k)} \right] \tag{5.315}$$

$$\triangleq G_{n+1}^{(k)} u + I_{n+1}^{(k)} \tag{5.316}$$

which represents a familiar Norton type model, with the element stamp shown in Table 5.6. Recall, the terms of this element stamp depend, not only on $u_{n+1}^{(k)}$, but also on u_n at the previous time-point, which is available.

For a nonlinear inductor $L(i)$, the BE companion model has the element equation $u = g_L(i)$ which, linearized in the Newton loop at t_{n+1}, becomes:

$$u = g'_L \left(i_{n+1}^{(k)} \right) \left(i - i_{n+1}^{(k)} \right) + g_L \left(i_{n+1}^{(k)} \right) \tag{5.317}$$

$$= g'_L \left(i_{n+1}^{(k)} \right) i + \left[g_L \left(i_{n+1}^{(k)} \right) - g'_L \left(i_{n+1}^{(k)} \right) i_{n+1}^{(k)} \right] \tag{5.318}$$

$$\triangleq R_{n+1}^{(k)} i + V_{n+1}^{(k)} \tag{5.319}$$

which is a Thévenin model, but can also be written in the Norton form:

$$i = \left(\frac{1}{R_{n+1}^{(k)}} \right) u - \left(\frac{V_{n+1}^{(k)}}{R_{n+1}^{(k)}} \right) \tag{5.320}$$

$$\triangleq G_{n+1}^{(k)} u + I_{n+1}^{(k)} \tag{5.321}$$

for which the element stamp is the same form as for the nonlinear capacitor, but the values of $G_{n+1}^{(k)}$ and $I_{n+1}^{(k)}$ are, of course, computed differently.

The current $i_{n+1}^{(k+1)}$ is needed, to be used as $i_{n+1}^{(k)}$ in the next Newton iteration, as it is required to update the affine approximation and element stamp. One can easily find this using (5.321), based on $u_{n+1}^{(k+1)}$ (easily computable from the MNA vector), $i_{n+1}^{(k)}$, and i_n (both available from previous work). It is prudent, when using this "current update" during the Newton loop, to ensure that the current value has converged before loop termination. Thus, one should check the step-sizes of the current vector representing all the inductor currents, which may not

Table 5.6: Element stamp for a nonlinear capacitor, based on BE discretization and linearization.

	$v+$	$v-$	RHS
$v+$	$G_{n+1}^{(k)}$	$-G_{n+1}^{(k)}$	$-I_{n+1}^{(k)}$
$v-$	$-G_{n+1}^{(k)}$	$G_{n+1}^{(k)}$	$I_{n+1}^{(k)}$

be part of the MNA vector. Here, too, we rely on having the inductor currents at t_0, the beginning of the simulation time, as a by-product of the DC Analysis run.

5.7.4 TR Discretization

The flow of the argument should be clear by now:

1. For a given LMS method, apply discretization to the individual dynamic element equations to discover relations between i_{n+1} and u_{n+1}.
2. Construct companion models that embody these same relations as their terminal characteristics.
3. The network is solved at every time-point by replacing all dynamic elements by their companion models and solving the resulting (possibly nonlinear) resistive circuit.

The topology of the network, and the structure of the MNA matrix, do not change over time; only the element stamp values may change. It is easy to extend the BE case to the other BDFs, notably to the case of BDF2 which is used in some commercial simulators. This is left as an exercise for the reader. But the application of direct discretization to the case of TR offers an interesting "twist," as we will now see.

Consider the case of the trapezoidal rule (TR), an implicit method. As in the case of BE, we write the dynamic element equations at t_{n+1}:

$$i(t_{n+1}) = C(u(t_{n+1}))u'(t_{n+1}) \quad \text{and} \quad u(t_{n+1}) = L(i(t_{n+1}))i'(t_{n+1}) \qquad (5.322)$$

then, using $i(t_{n+1}) \approx i_{n+1}$ and $u(t_{n+1}) \approx u_{n+1}$, we have:

$$i_{n+1} \approx C(u_{n+1})u'(t_{n+1}) \quad \text{and} \quad u_{n+1} \approx L(i_{n+1})i'(t_{n+1}) \qquad (5.323)$$

then, using the familiar TR result from the Taylor series expansion, namely:

$$x'(t_{n+1}) \approx \frac{2}{h}(x_{n+1} - x_n) - x'(t_n) \qquad (5.324)$$

we get, for a capacitor:

$$i_{n+1} \approx C(u_{n+1})\left(\frac{2}{h}(u_{n+1} - u_n) - u'(t_n)\right) \qquad (5.325)$$

and, for an inductor:

$$u_{n+1} \approx L(i_{n+1})\left(\frac{2}{h}(i_{n+1} - i_n) - i'(t_n)\right) \qquad (5.326)$$

The presence of the derivatives $u'(t_n)$ and $i'(t_n)$ may seem problematic, but we can easily eliminate them by making use of:

$$i_n \approx C(u_n)u'(t_n) \quad \text{and} \quad u_n \approx L(i_n)i'(t_n) \tag{5.327}$$

which give:

$$u'(t_n) \approx \frac{i_n}{C(u_n)} \quad \text{and} \quad i'(t_n) \approx \frac{u_n}{L(i_n)} \tag{5.328}$$

both of which are computable from previous data. Replacing $u'(t_n)$ and $i'(t_n)$ by these values, leads to, for a capacitor:

$$i_{n+1} \approx C(u_{n+1})\left(\frac{2}{h}(u_{n+1} - u_n) - \frac{i_n}{C(u_n)}\right) \tag{5.329}$$

and, for an inductor:

$$u_{n+1} \approx L(i_{n+1})\left(\frac{2}{h}(i_{n+1} - i_n) - \frac{u_n}{L(i_n)}\right) \tag{5.330}$$

For a capacitor, this leads to the TR-inspired direct discretization:

$$i_{n+1} = C(u_{n+1})\left(\frac{2}{h}(u_{n+1} - u_n) - \frac{i_n}{C(u_n)}\right) \tag{5.331}$$

or:

$$u_{n+1} = \left(\frac{h}{2C(u_{n+1})}\right)i_{n+1} + \left[u_n + \left(\frac{h}{2C(u_n)}\right)i_n\right] \tag{5.332}$$

For an inductor, we get the TR-inspired direct discretization:

$$u_{n+1} = L(i_{n+1})\left(\frac{2}{h}(i_{n+1} - i_n) - \frac{u_n}{L(i_n)}\right) \tag{5.333}$$

or:

$$i_{n+1} = \left(\frac{h}{2L(i_{n+1})}\right)u_{n+1} + \left[i_n + \left(\frac{h}{2L(i_n)}\right)u_n\right] \tag{5.334}$$

For linear C and L, these relations are simplified as:

$$i_{n+1} = \left(\frac{2C}{h}\right)u_{n+1} - \left[i_n + \left(\frac{2C}{h}\right)u_n\right] \tag{5.335}$$

and:

$$i_{n+1} = \left(\frac{h}{2L}\right)u_{n+1} + \left[i_n + \left(\frac{h}{2L}\right)u_n\right] \tag{5.336}$$

where we have selected the forms that lead to a companion model that includes a current source, instead of a voltage source.

As in the case of BE, this being another implicit method, we do not have immediate solutions for the i_{n+1} and u_{n+1} values. Instead, we have (algebraic) relations between i_{n+1} and u_{n+1} that include, among other parameters, past data values u_n and i_n. To solve for the t_{n+1} responses, we replace all the differential equations of the dynamic elements (in the unraveled form) by these algebraic equations. And, as in the BE case, we can more efficiently do this by replacing the dynamic elements by their TR companion models, which we now consider.

Companion Models for Linear Elements   We first consider the TR companion models in the linear case. It is easy to see that, for the TR case, and for a *linear* C and L, the companion models can be given as the *linear* circuits in Fig. 5.15, where, for a capacitor, we have:

$$G_{eq} = \frac{2C}{h} \quad \text{and} \quad I_{eq} = i_n + \left(\frac{2C}{h} \right) u_n \qquad (5.337)$$

and, for an inductor:

$$G_{eq} = \frac{h}{2L} \quad \text{and} \quad I_{eq} = i_n + \left(\frac{h}{2L} \right) u_n \qquad (5.338)$$

and where, again, we have selected the forms of the companion models that are based on a current source, instead of a voltage source.

Once the netlist has been solved at t_{n+1}, we need to compute u_{n+1} and i_{n+1} for every dynamic element. These are required at the next time-point, to update the model parameter values that go into (5.335) and (5.336), namely u_n and i_n. It is trivial to find u_{n+1} for every L and C, as simply the difference of its two node voltages, and i_{n+1} can then be found using (5.335) and (5.336). As in the case of BE, we also rely on having all inductor currents available at the start of simulation time, as a result of DC Analysis.

Finally, construction of the MNA element stamps is very similar to the BE case, and is left as an exercise for the reader.

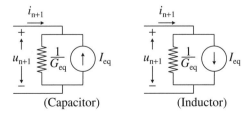

(Capacitor)   (Inductor)

Figure 5.15: TR companion models for linear elements.

Companion Models for Nonlinear Elements   In the nonlinear case, as was the case with BE, we must use the explicit forms of the element equations, which are, for a capacitor:

$$i_{n+1} = C(u_{n+1}) \left(\frac{2}{h}(u_{n+1} - u_n) - \frac{i_n}{C(u_n)} \right) \triangleq g_C(u_{n+1}) \tag{5.339}$$

and, for an inductor:

$$u_{n+1} = L(i_{n+1}) \left(\frac{2}{h}(i_{n+1} - i_n) - \frac{u_n}{L(i_n)} \right) \triangleq g_L(i_{n+1}) \tag{5.340}$$

The companion models, in this case nonlinear, can be as shown in Fig. 5.16. These models are linearized during the Newton loop and can both be put into a (linear) companion model in the standard Norton form:

$$i = G_{n+1}^{(k)}u + I_{n+1}^{(k)} \tag{5.341}$$

for which construction of the MNA element stamps is very similar to the BE case, and is left as an exercise for the reader.

In order to be able to update the element stamps throughout the Newton loop, the same comments apply as in the BE case. One must have in-hand u_n and i_n, from the previous time-point, relying on having available u_0 and i_0 at time t_0 from the DC Analysis run. One must also have in-hand $u_{n+1}^{(k)}$ and $i_{n+1}^{(k)}$, both of which should be available from the previous Newton iteration. From the results of this Newton iteration, one can compute $u_{n+1}^{(k+1)}$ from the MNA vector and $i_{n+1}^{(k+1)}$ using (5.341). Here, too, it is prudent to check the step sizes $\|i_{n+1}^{(k+1)} - i_{n+1}^{(k)}\|$ before declaring convergence of the Newton loop.

Example   Consider the circuit shown in Fig. 5.17 where, provided $u > -1/2$, the capacitance is:

$$C(u) = \frac{10^{-12}}{\sqrt{1 + 2u}} \tag{5.342}$$

which is similar to the characteristics of pn-junction depletion capacitance.

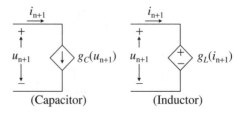

(Capacitor)   (Inductor)

Figure 5.16: TR companion models for nonlinear elements.

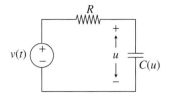

Figure 5.17: A nonlinear RC circuit.

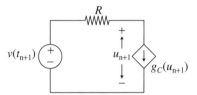

Figure 5.18: The circuit of Fig. 5.17, using companion models to find the solution at time t_{n+1}.

Using the strategy of direct discretization, we assume that the circuit has been solved up to time t_n, and we consider the circuit at t_{n+1}, as shown in Fig. 5.18, where, suppose we are using BE as the solution method, so that:

$$g_C(u_{n+1}) = \frac{10^{-12}}{\sqrt{1 + 2u_{n+1}}} \left(\frac{u_{n+1} - u_n}{h} \right) \qquad (5.343)$$

and:

$$g'_C(u_{n+1}) = \frac{10^{-12}/(2h)}{\sqrt{1 + 2u_{n+1}}} \left(\frac{2 + 3u_{n+1} + u_n}{1 + 2u_{n+1}} \right) \qquad (5.344)$$

Inside the Newton loop, the above nonlinear circuit is linearized, using the companion model introduced earlier, leading to the circuit shown in Fig. 5.19, where, as we saw earlier:

$$G^{(k)}_{n+1} = g'_C \left(u^{(k)}_{n+1} \right) \quad \text{and} \quad I^{(k)}_{n+1} = \left[g_C \left(u^{(k)}_{n+1} \right) - g'_C \left(u^{(k)}_{n+1} \right) u^{(k)}_{n+1} \right] \qquad (5.345)$$

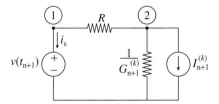

Figure 5.19: The circuit of Fig. 5.18, after replacement of the nonlinear controlled source by its linearized companion model.

so that:

$$G_{n+1}^{(k)} = \frac{10^{-12}/(2h)}{\sqrt{1+2u_{n+1}^{(k)}}} \left(\frac{2 + 3u_{n+1}^{(k)} + u_n}{1 + 2u_{n+1}^{(k)}} \right) \tag{5.346}$$

and:

$$I_{n+1}^{(k)} = \frac{10^{-12}}{\sqrt{1+2u_{n+1}^{(k)}}} \left(\frac{u_{n+1}^{(k)} - u_n}{h} \right) - \frac{10^{-12}/(2h)}{\sqrt{1+2u_{n+1}^{(k)}}} \left(\frac{2 + 3u_{n+1}^{(k)} + u_n}{1 + 2u_{n+1}^{(k)}} \right) u_{n+1}^{(k)} \tag{5.347}$$

With this, the MNA system $J_{n+1}^{(k)} x_{n+1}^{(k+1)} = s_{n+1}^{(k)}$ is then built using element stamps:

$$\begin{bmatrix} 1/R & -1/R & 1 \\ -1/R & (G_{n+1}^{(k)} + 1/R) & 0 \\ 1 & 0 & 0 \end{bmatrix} \begin{bmatrix} v_1 \\ v_2 \\ i_s \end{bmatrix} = \begin{bmatrix} 0 \\ -I_{n+1}^{(k)} \\ v(t_{n+1}) \end{bmatrix} \tag{5.348}$$

The overall solution flow, for this example and also in general, is as shown in Fig. 5.20.

5.7.5   Charge-Based and Flux-Based Models

Time discretization results in numerical errors, because we cannot take infinitesimally small time-steps. This, we know, and have come to expect by now. However, in the case of nonlinear dynamic elements, the type of companion models developed above can give *severe* numerical errors. The reasons for this are somewhat subtle and will be explained below. We will also see that there is an alternative; we will develop better companion models for nonlinear dynamic elements that do not have such problems.

This problem is often observed in the simulation of circuits with nonlinear capacitors, whose response is strongly dependent on capacitor *charges*. Examples include DRAMs and switched-capacitor filters. In such circuits, the problem is manifested as a *charge non-conservation* problem. Consider a capacitor whose voltage starts out as $v(t_0) = V$, varies arbitrarily over the

```
Input: Initial condition x_0 at t_0.
    for (n = 0; t_n ≤ T; n = n + 1) do
    t_{n+1} = t_n + h
    k = 0; x_{n+1}^(0) = x_n
    while (not converged) do {Newton loop}
        Update element stamps and the Jacobian.
        Solve J_{n+1}^(k) x_{n+1}^(k+1) = s_{n+1}^(k) for x_{n+1}^(k+1)
        k = k + 1
    x_{n+1} = x_{n+1}^(k)
```

Figure 5.20: Overall solution flow.

time interval $[t_0, t_1]$, then again becomes $v(t_1) = V$. If the initial charge on the capacitor, at time t_0, was $q(t_0) = Q$, then the final charge, at t_1, should also be $q(t_1) = Q$, because $v(t_1) = v(t_0)$. If, as a result of numerical integration of currents, it is not, then we say that charge has not been conserved. A tell-tale sign of charge non-conservation is finding a (simulated) DC current flowing in a capacitor.

This has spurred the development of charge-based models for nonlinear capacitors, and for semiconductor devices with internal nonlinear capacitance. There is ample empirical evidence that these charge-based models are much better at conserving charge during circuit simulation. Likewise, there is a flux conservation issue with nonlinear inductors, and a similar development of flux-conserving models has been done for them. In the following, we will examine the roots of this problem, for capacitors, and see how the improved class of charge-based models can be developed.

Consider the case of a nonlinear capacitor using BE discretization. In the preceding, we developed the BE companion model for a nonlinear capacitor, with voltage u and current i, starting from the element equation:

$$i(t) = C(u)u'(t) \tag{5.349}$$

which we wrote at t_{n+1}, then used the finite difference approximation of the derivative, from the Taylor series expansion, to motivate the BE discretization:

$$i_{n+1} = C(u_{n+1})\left(\frac{u_{n+1} - u_n}{h}\right) = \left(\frac{C(u_{n+1})}{h}\right)u_{n+1} - \left(\frac{C(u_{n+1})}{h}\right)u_n \tag{5.350}$$

a function which we earlier denoted as $i_{n+1} = g_C(u_{n+1})$.

We could have proceeded differently, as follows. Let $q(t)$ be the charge on the capacitor. Because it is a function of the branch voltage u, we will also write it as $q(u)$, and we know that:

$$C(u) \triangleq \frac{dq}{du} \tag{5.351}$$

Capacitor current is the rate of change of charge, so that:

$$i(t) = \frac{dq}{dt} \tag{5.352}$$

which we can write at t_{n+1} and, using the finite difference approximation for the derivative, we can write:

$$i(t_{n+1}) \approx \frac{q(t_{n+1}) - q(t_n)}{h} \tag{5.353}$$

and we can use this, along with $i(t_{n+1}) \approx i_{n+1}$, $q(t_n) \approx q(u_n)$, and $q(t_{n+1}) \approx q(u_n)$, to motivate the BE discretization:

$$i_{n+1} = \frac{q(u_{n+1}) - q(u_n)}{h} = \frac{q(u_{n+1})}{h} - \frac{q(u_n)}{h} \qquad (5.354)$$

a function which we denote as $i_{n+1} = g_Q(u_{n+1})$.

Now, let us consider how these two discretization schemes (5.350) and (5.354), lead to linearized models in the Newton loop. For our original (capacitance-based) model, $i_{n+1} = g_C(u_{n+1})$, we have:

$$g_C' \left(u_{n+1}^{(k)} \right) = \left(\frac{C'(u_{n+1}^{(k)})}{h} \right) u_{n+1}^{(k)} + \frac{C(u_{n+1}^{(k)})}{h} - \left(\frac{C'(u_{n+1}^{(k)})}{h} \right) u_n \qquad (5.355)$$

which gives the (linearized) companion model $i = G_{n+1}^{(k)} u + I_{n+1}^{(k)}$, where:

$$G_{n+1}^{(k)} = \frac{C'(u_{n+1}^{(k)})}{h} \left(u_{n+1}^{(k)} - u_n \right) + \frac{C(u_{n+1}^{(k)})}{h} \qquad (5.356)$$

$$I_{n+1}^{(k)} = -\frac{C'(u_{n+1}^{(k)})}{h} \left(u_{n+1}^{(k)} - u_n \right) u_{n+1}^{(k)} - \left(\frac{C(u_{n+1}^{(k)})}{h} \right) u_n \qquad (5.357)$$

For the new (charge-based) model, $i_{n+1} = g_Q(u_{n+1})$, we have:

$$g_Q' \left(u_{n+1}^{(k)} \right) = \frac{C(u_{n+1}^{(k)})}{h} \qquad (5.358)$$

which gives the (linearized) companion model $i = G_{n+1}^{(k)} u + I_{n+1}^{(k)}$, where:

$$G_{n+1}^{(k)} = \frac{C(u_{n+1}^{(k)})}{h} \qquad (5.359)$$

$$I_{n+1}^{(k)} = \frac{q(u_{n+1}^{(k)})}{h} - \frac{q(u_n)}{h} - \left(\frac{C(u_{n+1}^{(k)})}{h} \right) u_{n+1}^{(k)} \qquad (5.360)$$

The two models are quite different, in both $G_{n+1}^{(k)}$, which contributes to the Jacobian, and $I_{n+1}^{(k)}$, which contributes to the RHS vector. Note that the form of the model is the same, and the companion model circuit diagram is the same, but the values of the elements are different. The difference in the Jacobian may be inconsequential, but the difference in the RHS vector can be a serious problem. Recall, the exact value of the Jacobian is not important, as long as the Newton loop converges—think of the Newton-chord method, for example. In the case of a *linear* capacitor, with $C'(u) = 0$, the two models become exactly the same, and there is no problem with charge conservation.

Both models are approximate, and both will have numerical errors, but the capacitance-based model gives additional error, at least due to the following reason. Notice that, in (5.350), it is as if the capacitance $C(u_{n+1})$ is assumed fixed across the interval $[u_n, u_{n+1}]$, and $C(u_n)$ is not part of the model. In fact, the rate of change of capacitance with respect to voltage is not part of the model at all. The model would give the same value whether $C(u_n) \ll C(u_{n+1})$ or $C(u_n) \gg C(u_{n+1})$, for example. The charge-based model does not make this sort of assumption; it "monitors" the charge at both ends of the interval $[u_n, u_{n+1}]$.

In practice, charge-based models give much better performance, especially in terms of charge conservation. Charge-based models do not *perfectly* conserve charge; some small charge error is inevitable due to truncation, roundoff, and error accumulation. But they give charge errors that are much smaller and quite acceptable, compared with capacitance-based models. We will now look at one example that shows errors resulting from the use of capacitance-based models.

Example   Consider the circuit shown in Fig. 5.21, consisting of a constant current source and a nonlinear capacitor, with the following settings:

$$i(t) = I > 0 \quad \text{(fixed current source)}$$

$$q(v) = v^2/2 \quad \text{(nonlinear capacitor)}$$

$$h = t_1 - t_0 \quad \text{(one time-step)}$$

$$v(t_0) = \sqrt{2Ih}$$

and we are interested in $v(t_1)$, the capacitor voltage at time t_1. Notice that the capacitance is $C(v) = v$ and that the *exact* voltage solution is governed by the differential equation:

$$v\frac{dv}{dt} + I = 0 \tag{5.361}$$

with $v(t_0) = \sqrt{2Ih}$, from which it is easy to find that:

$$v(t_1)^2 = v(t_0)^2 - 2Ih = 0 \tag{5.362}$$

so that $v(t_1) = 0$ is the exact voltage at time t_1.

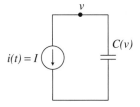

Figure 5.21: A simple test circuit.

We will now solve this circuit with BE discretization, in two ways, using a capacitance-based model and a charge-based model. We assume that $v_0 = v(t_0)$ is exact, and we seek a $v_1 \approx v(t_1)$. Using the capacitance-based BE companion model, writing KCL at t_1 gives:

$$-I = g_C(v_1) = \left(\frac{C(v_1)}{h}\right) v_1 - \left(\frac{C(v_1)}{h}\right) v_0 \tag{5.363}$$

leading to:

$$v_1^2 - v_0 v_1 + v_0^2/2 = 0 \tag{5.364}$$

which is a quadratic equation with no (real) solution! Newton's method, applied to this equation, would not converge. Alternatively, it may give $v_0^2/4$ as solution, in case $v_0^2/4$ is smaller than the error tolerance. Either way, the situation is problematic.

Using the charge-based BE companion model, writing KCL at t_1 gives:

$$-I = g_Q(v_1) = \frac{q(v_1)}{h} - \frac{q(v_0)}{h} \tag{5.365}$$

and this equation is actually *exact* because of the constant current in this case, so we should expect zero truncation error. Indeed, using $q(v_i) = v_i^2/2$, we get:

$$-I = \frac{v_1^2}{2h} - \frac{v_0^2}{2h} = \frac{v_1^2}{2h} - I \tag{5.366}$$

so that the solution is $v_1 = 0$, and Newton's method is actually guaranteed to converge on this problem, because of the ease of solving $f(x) = x^2/2h$.

Again, one indication of a problem, and a cause for concern, is the fact that (5.363) depends only on $C(v_1)$ and not on $C(v_0)$. It gives the same solution as if we had a constant fixed $C \triangleq C(v_1)$. One might think that the use of TR, instead of BE, would "fix" this problem because TR "monitors" the derivative at both ends of the interval. However, this not true because, using a capacitance-based model, TR leads to the following equation, to be solved for v_1:

$$4v_1^2 - 5v_0 v_1 + v_0^2 = 0 \tag{5.367}$$

whose solutions are v_0 and $v_0/4$, both of which are incorrect.

Sensitivity and Stability   Ogrodzki (1994) shows another test case which illustrates that, in the presence of nonlinear capacitance, and especially if we have strong local nonlinearity, we get heightened sensitivity of numerical integration errors to time-step sizes when capacitance-based models are used, and that these problems disappear when charge-based models are used. Building on early work by Calahan (1972), Ogrodzki (1994) also argues that capacitance-based models

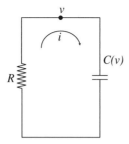

Figure 5.22: An RC circuit with a nonlinear capacitor.

can lead to error accumulation and amplification over time. We make further illustration of this effect with the following example.

Consider the circuit shown in Fig. 5.22, which makes use of a nonlinear capacitor, assumed to have been initially charged, whose exact solution obeys:

$$v = -Ri = -R\frac{dq}{dt} = -RC(v)\frac{dv}{dt} \tag{5.368}$$

Using the shorthand $v_n \triangleq v(t_n)$ and $v_{n+1} \triangleq v(t_{n+1})$, and considering the solution from t_n to $t_{n+1} = t_n + h$, we find:

$$\int_{v_n}^{v_{n+1}} \frac{C(v)}{v} dv = -\frac{h}{R} \tag{5.369}$$

We are interested in the effect of error in v_n on the solution v_{n+1}; the sensitivity of v_{n+1} to v_n is captured by the derivative:

$$\frac{dv_{n+1}}{dv_n} \tag{5.370}$$

Taking the derivative of both sides of (5.369) with respect to v_n, leads to:

$$\frac{dv_{n+1}}{dv_n} = \left[\left(\frac{v_n}{v_{n+1}}\right)\left(\frac{C_{n+1}}{C_n}\right)\right]^{-1} \tag{5.371}$$

where we have introduced $C_n \triangleq C(v_n)$ and $C_{n+1} \triangleq C(v_{n+1})$, to simplify the notation.

We now explore the response in the case of BE discretization with charge-based models, which gives:

$$v_{n+1} = -Ri_{n+1} = -\frac{R}{h}q(v_{n+1}) + \frac{R}{h}q(v_n) \tag{5.372}$$

from which:

$$\frac{dv_{n+1}}{dv_n} = -\frac{RC_{n+1}}{h}\frac{dv_{n+1}}{dv_n} + \frac{RC_n}{h} \tag{5.373}$$

which leads to:

$$\frac{dv_{n+1}}{dv_n} = \left[\left(\frac{h}{RC_n}\right) + \left(\frac{C_{n+1}}{C_n}\right)\right]^{-1} \tag{5.374}$$

which is not *exactly* the same sensitivity as in the exact solution, nor would we expect it to be, but it does not show signs of trouble.

Using capacitance-based models, we start with:

$$v_{n+1} = -Ri_{n+1} = -\frac{RC_{n+1}}{h}v_{n+1} + \frac{RC_{n+1}}{h}v_n \tag{5.375}$$

and, using the notation $C'_{n+1} \triangleq \frac{d}{dv}C(v)|_{v_{n+1}}$, we eventually arrive at:

$$\frac{dv_{n+1}}{dv_n} = \left[1 + \left(\frac{h}{RC_{n+1}}\right) + \left(\frac{C'_{n+1}}{C_{n+1}}\right)(v_{n+1} - v_n)\right]^{-1} \tag{5.376}$$

$$\approx \left[1 + \left(\frac{h}{RC_{n+1}}\right) + \left(\frac{C_{n+1} - C_n}{C_{n+1}}\right)\right]^{-1} \tag{5.377}$$

and here we see signs of trouble if/when the nonlinearity of $C(v)$ is strong. In that case, it is possible for the third term on the right-hand side to become "negative enough," causing the overall sensitivity dv_{n+1}/dv_n to increase tremendously. In such cases, error is amplified, and instability may be observed, unless a very small time-step is used.

Truncation Error We make yet one more exploration of the difference between capacitance-based models and charge-based models, by looking at the Taylor series. Exactly how much error do we incur when we truncate the Taylor series?

Recall that BE discretization starts with a Taylor series:

$$x(t_n) = x(t_{n+1}) - hx'(t_{n+1}) + \frac{h^2}{2}x''(t_{n+1}) + 0(h^3) \tag{5.378}$$

from which:

$$x'(t_{n+1}) = \frac{x(t_{n+1}) - x(t_n)}{h} + \frac{h}{2}x''(t_{n+1}) + 0(h^2) \tag{5.379}$$

For a capacitance-based model $i = C(u)u'$, we have:

$$u'(t_{n+1}) = \frac{u(t_{n+1}) - u(t_n)}{h} + \frac{h}{2}u''(t_{n+1}) + 0(h^2) \tag{5.380}$$

from which, using the shorthand $u''_{n+1} \triangleq u''(t_{n+1})$, we get:

$$i(t_{n+1}) = \frac{C_{n+1}}{h}u(t_{n+1}) - \frac{C_{n+1}}{h}u(t_n) + \frac{h}{2}C_{n+1}u''_{n+1} + 0(h^2) \tag{5.381}$$

in which the leading error term $(\frac{h}{2})C_{n+1}u''_{n+1}$ is insensitive to C_n and $C'(u)$. For a charge-based model $i = q'$, we have:

$$q'(t_{n+1}) = \frac{q(t_{n+1}) - q(t_n)}{h} + \frac{h}{2}q''(t_{n+1}) + 0(h^2) \qquad (5.382)$$

from which, using the shorthand $u'_{n+1} \triangleq u'(t_{n+1})$, we get:

$$i_{n+1} = \frac{q(u_{n+1})}{h} - \frac{q(u_n)}{h} + \frac{h}{2}\frac{d}{dt}\left(\frac{dq}{du}\bigg|_{u_{n+1}} u'_{n+1}\right) + 0(h^2) \qquad (5.383)$$

$$= \frac{q(u_{n+1})}{h} - \frac{q(u_n)}{h} + \frac{h}{2}\left(C'_{n+1}u'_{n+1} + C_{n+1}u''_{n+1}\right) + 0(h^2) \qquad (5.384)$$

The difference between the two leading error terms is:

$$\Delta = \frac{h}{2}C'_{n+1}u'_{n+1} \qquad (5.385)$$

which can be significant if h is not very small and/or if C'_{n+1} is large. Again, we see a problem in case of strong nonlinearity of capacitance with respect to voltage.

RHS Charge Terms   As we saw above, the charge-based companion model for a nonlinear capacitor contributes charge terms $q(u_j)/h$ to the MNA RHS vector:

$$i = \left(\frac{C(u^{(k)}_{n+1})}{h}\right)u + \left[\frac{q(u^{(k)}_{n+1})}{h} - \frac{q(u_n)}{h} - \left(\frac{C(u^{(k)}_{n+1})}{h}\right)u^{(k)}_{n+1}\right] \qquad (5.386)$$

With capacitance-based models, these charge terms are absent, and the RHS vector depends only on capacitance. In order for both models to produce the same results, the simulation when using capacitance-based models must implicitly compute charges somehow. Given only $C(v) = \frac{dq}{dv}$, this can only be done (implicitly) using numerical integration of currents. This makes capacitance-based models much more susceptible to numerical errors, compared with simply evaluating the charge equations. As we saw above, evaluation of $i^{(k)}_{n+1}$ for the capacitor incurs error in the leading error term, relative to the charge-based formulation, of $hC'_{n+1}u'_{n+1}/2$. Thus, using explicit charge terms in the RHS ensures that charge is the same for the same voltage, and helps avoid charge non-conservation.

Explicit Methods The above development of a charge-based model for BE can be applied to other implicit formulas as well, including TR and the BDFs. This leads to new element stamps that represent the charge-based models of the nonlinear dynamic elements. However, explicit integration (such as by using FE) for charge-based models requires an invertible $q(u)$, for the following reason. Suppose we want to use FE for a nonlinear capacitor, then, starting with $i(t_n) = q'(t_n)$, we write:

$$i_n = \frac{q(u_{n+1})}{h} - \frac{q(u_n)}{h} \tag{5.387}$$

from which we would need to compute:

$$u_{n+1} = q^{-1}\left(hi_n + q(u_n)\right) \tag{5.388}$$

which may not be well-defined. Thus, in general, using explicit methods with charge-based models is not possible.

Convergence Criteria As we saw above in the case of a nonlinear capacitor, charge-based models introduce capacitance terms in the Jacobian and charge terms in the RHS vector. Otherwise, using charge-based models for the nonlinear elements does not alter the rest of the MNA formulation; the solution vector contains no charge variables, for example. However, for better charge conservation, it is reported that, as part of the convergence check in the Newton loop, one should also check that $\|u^{(k+1)} - u^{(k)}\|$ has become (close enough to) zero, where u is a vector of all the capacitor branch voltages.

Furthermore, it is possible to incorporate a charge-based local error check, as follows. Define the two error metrics, in capacitor branch voltages and charges:

$$\tau_u = |u_n - u(t_n)| \qquad \tau_q = |q(u_n) - q(u(t_n))| \tag{5.389}$$

then, write the Taylor series expansion:

$$q(u(t_n)) = q(u_n) + (u(t_n) - u_n)\, q'(u_n) + \cdots \tag{5.390}$$

so that:

$$\tau_q \approx |C(u_n)|\tau_u \tag{5.391}$$

Thus, once τ_u has been estimated, such as by using an LTE estimate, we can use the above to estimate τ_q and check that against its own threshold, in order to determine if the time-step size is acceptable. Some simulators use an internal charge error tolerance CHGTOL for this purpose. According to Yang et al. (1983), it is advisable that the tolerances in voltage, charge, and current be set "appropriately," i.e., so they are commensurate with each other. Likewise, for use during the Newton loop, it is possible to express the step sizes in charge in terms of step sizes in voltage:

$$s_q \approx |C(u^{(k)})|s_u \tag{5.392}$$

so that it is possible to also check convergence in the charge step sizes.

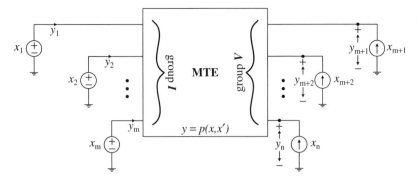

Figure 5.23: A general template for multiterminal elements.

The above comments also apply to the case of a nonlinear inductor, so we may check current step-sizes, and possibly the LTE and step-sizes of flux.

5.7.6 Multiterminal Elements

As with resistive multiterminal elements (MTE), dynamic MTEs can be easily handled, provided some restrictions are made on their specification. We assume that the terminals of any MTE are partitioned up-front into two groups, a group $\mathcal{I}$ and a group $\mathcal{V}$, as shown in Fig. 5.23, such that, if we let vector x consist of the voltage signals at group $\mathcal{I}$ terminals and the current signals into group $\mathcal{V}$ terminals, and let vector y consist of the current signals into group $\mathcal{I}$ terminals and the voltage signals at group $\mathcal{V}$ terminals, then the MTE terminal characteristics can be expressed in the form:

$$y = p(x, x') \tag{5.393}$$

where $p(\cdot)$ is some general, possibly nonlinear, vector function. Notice that this restriction has some subtle implications:

1. If the MTE contains any internal capacitors, then they must be connected (only) to terminal nodes in the group $\mathcal{I}$.
2. If the MTE contains any internal inductors, then their currents must be equal to currents of terminals in the group $\mathcal{V}$.

More generally, if the MTE has any internal charges or fluxes, then they must depend only on terminal voltages and currents that are in x vector. This is a practical requirement, since any variables that carry circuit state must be "remembered," and the MNA vector is a convenient place for them.

Most MTEs of interest, e.g., semiconductor devices, can be modeled so that all their terminals are in group $\mathcal{I}$, and they are specified using $i = p(v, v')$. With this, it is clear that any terminal of an MTE may be viewed as connected to

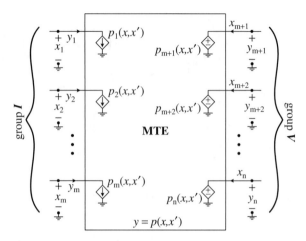

Figure 5.24: An equivalent circuit for a general multiterminal element.

a *two terminal*, possibly nonlinear, DCCS or DCVS, $y_i = p_i(x, x')$, as shown in Fig. 5.24. Thus, MTEs fit quite easily in the familiar MNA framework. In general, the model $y = p(x, x')$ may be either given explicitly or captured by a nonlinear dynamic equivalent circuit. The model may also come with certain additional features, such as different modes of operation, different levels of accuracy/complexity, etc.

In practice, most MTEs, such as MOSFETs, do not use the general form (5.393), but can be represented in the slightly simpler form:

$$y = g(x) + D(x)x'(t) \qquad (5.394)$$

in which one part, $g(x)$, is the familiar nonlinear resistive term we saw earlier, and the other, $D(x)x'(t)$, is the new dynamic term. Note, this $D(x)$ matrix is not to be confused with the $D(x)$ notation that we employed earlier in (5.14) and in (5.288). In our previous study of pseudo-transient and of companion models, we have focused on this less-general form. One can view this model as being *decoupled*, consisting of two models: a DC model and an AC or transient model. Most devices are specified in this way, by means of two models.

Discretization is applied to x', in the usual way, in order to build companion models for MTEs; linearization then leads to element stamps for MNA. The *model evaluation routine* in a circuit simulator is one that returns the MNA element stamp and RHS vector contributions, for the given operating point.

MTE Companion Models   Generation of companion models for MTEs involves a few complications, some more subtle than others, but none of which are insurmountable. If we focus on the less-general form (5.394) and, specifically, on its transient component:

$$y = D(x)x'(t) \qquad (5.395)$$

then each component of the model may involve several mutual capacitances or inductances, as we saw earlier:

$$i(t) = \sum_j C_j(x) x'_{(j)}(t) \qquad \text{or} \qquad v(t) = \sum_j L_j(x) x'_{(j)}(t) \qquad (5.396)$$

where we have employed the subscript notation introduced earlier, on page 210, using the more familiar v instead of u to represent branch voltages, and where as usual:

$$C_j(x) \triangleq \frac{\partial q}{\partial x_{(j)}} \qquad \text{and} \qquad L_j(x) \triangleq \frac{\partial \phi}{\partial x_{(j)}} \qquad (5.397)$$

Allowing for nonlinear capacitance/inductance, and in the interest of charge and flux conservation, we now express these relations differently, as:

$$i(t) = \sum_j q'_j(x) \qquad \text{or} \qquad v(t) = \sum_j \phi'_j(x) \qquad (5.398)$$

where each of the $q_j(x)$ or $\phi_j(x)$ terms can be thought of as representing a separate element, which can be a function of several voltages or currents. For example, given an element with element equation $i = q'(v)$, where v is a vector of node voltages, we construct the companion model by BE direct discretization:

$$i(t_{n+1}) = \frac{d}{dt} q(v(t)) \Big|_{t_{n+1}} \approx \frac{q(v_{n+1})}{h} - \frac{q(v_n)}{h} \triangleq g_Q(v_{n+1}) \qquad (5.399)$$

which we then linearize in the usual way as a nonlinear element with multiple controlling variables, which makes use of the several partial derivatives:

$$\frac{\partial g_Q(v)}{\partial v_{(j)}} \Big|_{v_{n+1}} = \frac{1}{h} \frac{\partial q(v)}{\partial v_{(j)}} \Big|_{v_{n+1}} = \frac{C_j(v_{n+1})}{h} \qquad (5.400)$$

which requires that the model provide means to compute $q(v)$ and $\partial q/\partial v_{(j)}$. The process then proceeds as usual. However, some further subtle details must be noted, in the interest of charge conservation, as we will now describe for the case of the MOSFET.

MOSFET Model   In a MOSFET, the internal capacitances depend on several voltages. In fact, there is more to it than this, as we will see that accurate modeling of the MOSFET requires the notion of a *non-reciprocal capacitance*. An early capacitance model for MOSFETs was given in Meyer (1971), and is called the Meyer model. In addition to the diffusion junction capacitors C_{JSB} and C_{JDB}, this model also includes:

$$C_{gs} \triangleq \frac{\partial Q_g}{\partial v_s}, \qquad C_{gd} \triangleq \frac{\partial Q_g}{\partial v_d}, \qquad \text{and} \qquad C_{gb} \triangleq \frac{\partial Q_g}{\partial v_b} \qquad (5.401)$$

where Q_g is the total gate charge and v_s, v_d, and v_b are the source, drain, and bulk node voltages, respectively. With nonlinear equations provided for the above capacitances, this model is quite sophisticated and works well in most cases.

However, it was found to lead to charge non-conservation for certain circuits whose response depends more on charge than on capacitance. The model leads to a situation where, as pointed out in Ward and Dutton (1978), *"the charge stored in a node is not equal to the integrated net current flowing into the node."* The problems were traced to several issues with this model, notably:

1. The absence of source-to-bulk and drain-to-bulk capacitors that represent the (rate of change of) space charge in the channel.
2. The fact that the Meyer capacitances are reciprocal, i.e., that the model assumes that, for example, $C_{gb} \triangleq \partial Q_g / \partial v_b = \partial Q_b / \partial v_g \triangleq C_{bg}$.

Ward and Dutton (1978) improved this model by including the missing capacitances and allowing the capacitors to be non-reciprocal, so that:

$$C_{gb} \triangleq \frac{\partial Q_g}{\partial v_b} \neq \frac{\partial Q_b}{\partial v_g} \triangleq C_{bg} \tag{5.402}$$

As a result, C_{gb} and C_{bg} are not really physical "capacitors," but are merely two measures of "capacitance," related to different space charges. Yang et al. (1983) later improved this model by incorporating charge terms in the RHS vector. The result is a charge conserving nonlinear capacitance model for the MOSFET that is in use in many modern simulators. Further details are available in Yang et al. (1983) and in Ruehli (1986).

This model, focusing only on its transient component, is based on the recognition of the presence of several *space charges*: Q_g is the total gate charge, Q_b is the total bulk (substrate, body) charge, and $(Q_s + Q_d)$ is the total channel charge (partitioned into source-side and drain-side portions). The model also includes junction and overlap capacitance, but we leave them out here, because they are not as problematic for simulation.

For any region of space with charge Q, the net current flowing into that region is given by dQ/dt. For the MOSFET, we write:

$$i_g = \frac{dQ_g}{dt}, \qquad i_b = \frac{dQ_b}{dt}, \qquad i_d = \frac{dQ_d}{dt}, \qquad \text{and} \quad i_s = \frac{dQ_s}{dt} \tag{5.403}$$

which, again, is only the transient part of the total current, and it also excludes the junction and overlap capacitor currents. The above current equations are then discretized and linearized in the usual way, leading to the MOSFET companion model and element stamp. To explore further, we can write:

$$i_g = \frac{\partial Q_g}{\partial v_{gb}} v'_{gb} + \frac{\partial Q_g}{\partial v_{gd}} v'_{gd} + \frac{\partial Q_g}{\partial v_{gs}} v'_{gs} \tag{5.404}$$

$$i_b = \frac{\partial Q_b}{\partial v_{bg}} v'_{bg} + \frac{\partial Q_b}{\partial v_{bd}} v'_{bd} + \frac{\partial Q_b}{\partial v_{bs}} v'_{bs} \tag{5.405}$$

$$i_d = \frac{\partial Q_d}{\partial v_{dg}} v'_{dg} + \frac{\partial Q_d}{\partial v_{db}} v'_{db} + \frac{\partial Q_d}{\partial v_{ds}} v'_{ds} \tag{5.406}$$

$$i_s = \frac{\partial Q_s}{\partial v_{sg}} v'_{sg} + \frac{\partial Q_s}{\partial v_{sb}} v'_{sb} + \frac{\partial Q_s}{\partial v_{sd}} v'_{sd} \tag{5.407}$$

leading to a matrix of 12 nonlinear non-reciprocal capacitances:

$$C_{ij} \triangleq \frac{\partial Q_i}{\partial v_{ij}} \tag{5.408}$$

For charge conservation, we require:

$$Q_g + Q_b + Q_d + Q_s = 0 \tag{5.409}$$

which, by taking derivatives, also leads to the expected KCL:

$$i_g + i_b + i_d + i_s = 0 \tag{5.410}$$

Some conclusions follow from this. Notably, KCL leads to:

$$\sum_{j \neq i} C_{ij} = \sum_{j \neq i} C_{ji}, \qquad \forall i \in \{g, b, d, s\} \tag{5.411}$$

so that the capacitances are not all independent. In fact, it can be shown that three of the twelve capacitances can be computed from the other nine. To ensure conservation of charge, these nine capacitances are computed from the provided model equations, then the three others are computed from them.

Thus, modern MOSFET models typically include equations to compute the following 3×3 capacitance matrix:

$$\begin{bmatrix} C_{gb} & C_{gd} & C_{gs} \\ C_{bg} & C_{bd} & C_{bs} \\ C_{dg} & C_{db} & C_{ds} \end{bmatrix} \tag{5.412}$$

Any approximation of the model capacitances, by omitting or simplifying some of them as in the Meyer model, can lead to charge non-conservation. As well, for better numerical results, it has been shown that charge terms should appear in the RHS vector, as we saw with 2-terminal nonlinear capacitors. This makes the charge computation more robust to numerical errors.

MTE with Internal Nodes Finally, it remains to consider the case of an MTE with internal nodes. For resistive MTEs, as we saw earlier, it is possible to eliminate the internal nodes, so that they do not show up as variables in the MNA vector. In the case of a dynamic MTE, one cannot do this if any internal nodes are terminals of any capacitors or if any internal branches are inductive. This is the reason behind our earlier requirement that capacitor voltages must be terminal voltages and inductor currents must be terminal currents. If, say, a MOSFET contains series resistance, then any internal nodes that are are terminals of C_{ij} capacitances must be made explicit new terminal nodes and must appear in the MNA vector. Other internal nodes do not carry state information and they may be eliminated in the same way as we did earlier in the resistive case.

5.7.7   Time-Step Control

Practical experience confirms the intuitive notion that small (large) time-steps must (can) be taken when the solution is changing quickly (slowly). Thus, it is imperative, for efficiency reasons, to use a variable time-step approach in simulation, typically based on an LTE metric. By the use of such schemes, it is hoped that the resulting LTE would be approximately constant, at some acceptable level, for all the time-steps taken during the simulation.

In one scheme for using LTE to manage the time-step, once the Newton loop has converged for a value of x_{n+1}, we check the PLTE at t_{n+1}, typically by checking whether:

$$\text{PLTE} \leq \epsilon \triangleq \epsilon_{rel}|x_{n+1}| + \epsilon_{abs} \tag{5.413}$$

where ϵ_{rel} and ϵ_{abs} are relative and absolute error tolerances, respectively. Instead of the normalizing vector $|x_{n+1}|$, some authors use a vector whose i-th entry is the largest value observed so far in $x_{k(i)}, \forall k \leq n+1$. Some others, instead, check whether:

$$\text{PLTE} \leq \epsilon \triangleq \epsilon_{rel}|x'(t_{n+1})| + \epsilon_{abs} \tag{5.414}$$

with some suitable approximation scheme to find $x'(t_{n+1})$, so that the LTE check is effectively *relaxed* when $x(t)$ is changing rapidly.

If the error meets the test criterion then we accept the solution x_{n+1}, otherwise we reject it, we backtrack to x_n, and we use a new value of time-step to re-attempt the step from t_n to t_{n+1}. If the error meets the test criterion by a large margin, then the solution x_{n+1} is (obviously) accepted and the time-step is reduced subsequently. We will, shortly, discuss how the new time-step value is chosen.

In some simulator implementations, only the LTEs in the voltage entries of the MNA vector are checked, and not in the current entries. The rationale given for this is that the current variables are "artificial," introduced only to be able to complete the MNA formulation and, therefore, do not merit an LTE check. However, this view is not universal and, in general, modern simulators may check the LTE in all MNA variables, as well as in charge and flux, as we saw earlier.

Choice of Time-Step   It is hard to predict up-front exactly what the time-step should be. In a *fixed time-step* regime, or for a *one-step* method, we have access to the simple result:

$$\text{PLTE}(t_{n+1}) = C_{p+1}h^{p+1}x^{(p+1)}(t_n) \tag{5.415}$$

from which, if the maximum tolerable error is a vector $\epsilon > 0$, then the time-step should obviously be:

$$h \leq \min_{\forall i} \left(\frac{\epsilon_{(i)}}{C_{p+1} x_{(i)}^{(p+1)}(t_n)} \right)^{\frac{1}{p+1}} \tag{5.416}$$

and this may be used to provide some guidance for setting the time-step. This result does not hold for the variable time-step case, except when using a one-step method, like BE or TR. In fact, for one-step methods, this result can be further extended, as in Ogrodzki (1994), as follows. If we, heuristically, assume that $x^{(p+1)}(t_n) \approx x^{(p+1)}(t_{n-1})$, then, $\forall i$:

$$\frac{\text{PLTE}_{(i)}(t_{n+1})}{\text{PLTE}_{(i)}(t_n)} \approx \left(\frac{h_{n+1}}{h_n} \right)^{p+1} \tag{5.417}$$

so that, if $\epsilon_n = \text{PLTE}(t_n)$ is the *computed* PLTE at t_n and ϵ is the *desired* PLTE at t_{n+1}, then we should take:

$$h_{n+1} \approx h_n \min_{\forall i} \left(\frac{\epsilon_{(i)}}{\epsilon_{n(i)}} \right)^{\frac{1}{p+1}} \tag{5.418}$$

as the next time-step. By the same logic, if we have backtracked from t_{n+1} because of too large a PLTE, then we should choose a smaller/corrected h_{n+1} based on:

$$h_{n+1}^{new} \approx h_{n+1} \min_{\forall i} \left(\frac{\epsilon_{(i)}}{\epsilon_{n+1(i)}} \right)^{\frac{1}{p+1}} \tag{5.419}$$

where ϵ is the *desired* PLTE and ϵ_{n+1} is the *computed* PLTE, both at t_{n+1}. It is not clear how effective these heuristic methods are in practice. Much simpler strategies are often employed, for example, by simply doubling or halving the time-step.

Alternatives   The use of LTE for time-step control can be expensive, due to the need to estimate the LTE and the wasted effort whenever a step is rejected. In addition, practical experience shows that LTE-based schemes can be overly conservative, leading to exceedingly small time-steps. For example, using a divided differences approach has been found to give LTE estimates that can be 10 times larger than the true LTE. Thus, alternatives have been sought, and one of them is as follows. It has been observed that slow (fast) signals that allow (require) a large (small) time-step turn out to also require a small (large) number of Newton iterations to converge. This suggests that one can make the choice of the next time-step value based on how many iterations were required to converge at the present time-point. For example, one scheme can be as follows:

1. If x_n is found to converge in less than, say, five Newton iterations, then double the time-step to find x_{n+1}, i.e., set $t_{n+1} - t_n = 2(t_n - t_{n-1})$.
2. Otherwise, if x_n is found to converge in more than, say, ten Newton iterations, then halve the time-step to find x_{n+1}, i.e., set $t_{n+1} - t_n = (t_n - t_{n-1})/2$.
3. Otherwise, keep the time-step fixed at its present value.

These are obviously heuristic schemes and there is no way to determine their effectiveness *a priori*; one must simply test them out in the field.

As well, the choice of time-step is guided by the input "breakpoints." Input signal sources are typically piece-wise linear (PWL) waveforms, and the points where two linear segments meet are called *breakpoints*. The time-step is always reduced in order to to make sure that breakpoint times coincide with solution time-points. In addition, the choice of the first time-step taken beyond a breakpoint is guided by the slope of that linear segment of the input waveform. For example, it may be set to 1/10 of the time-span of that segment. Finally, the time-step may be reduced at the end of the simulation time so that a solution time-point coincides with the end of simulation time, t_f.

5.7.8 Enhancements

It is reported, in McCalla (1988), that the total time spent performing Newton iterations breaks down as follows:

- About 80% of it is spent linearizing and evaluating nonlinear elements, i.e., generating the element stamps for the nonlinear elements, and
- The remaining 20% is spent on solving the linear equations.

Thus, there is great interest in reducing the computational effort expended on linearizing and evaluating the nonlinear elements. It is expensive to have to do this in every Newton iteration. One successful scheme for reducing this cost, from McCalla (1988), is as follows. At time t_{n+1}, treat any nonlinear dynamic elements as *linear* dynamic elements, whose values are computed based on x_n at t_n, so that:

$$C(u_{n+1}) \triangleq C(u_n) \quad \text{and} \quad L(i_{n+1}) \triangleq L(i_n) \tag{5.420}$$

If the LTE is in check, then the error introduced by this is usually acceptable. Of course, other (resistive) nonlinear elements must still be linearized and evaluated in every Newton iteration.

If one carries out this scheme to its extreme, then one can also apply this to the other (resistive) nonlinear elements, as well. This would mean that the Jacobian would remain fixed throughout the Newton iterations at a given time-point. Effectively, this would be a Newton-Chord method. Note, while the Jacobian may be fixed, updates to the RHS vector due to the nonlinear elements would still be required.

5.7.9 Overall Flow

Putting together all that we have seen so far, we can now give a basic "bare bones" simulation flow, including DC Analysis and Transient Analysis, with time-step control, as shown in Fig. 5.25.

Input: Initial time t_0 and final time T.
{Perform DC analysis at t_0:}
Disable all dynamic elements.
Choose an initial candidate solution at t_0.
while ($x(t_0)$ has not converged and not timed-out) **do** {Newton loop}
    For every (resistive) element in the network, evaluate its
    linearized element stamp and construct the MNA system.
    Solve the resulting MNA system using LU factorization.
endwhile
if (Newton loop has timed-out) **then**
    Abort! {Unable to find initial DC solution}
endif

{Perform Transient Analysis:}
Reinstate all dynamic elements.
Set $n = 0$, and choose an initial time-step h_1.
while ($t_n < T$) **do** {Time discretization loop}
    $t_{n+1} = t_n + h_{n+1}$
    Use $x(t_n)$ as the initial candidate solution at t_{n+1}.
    while ($x(t_{n+1})$ has not converged and not timed-out) **do** {Newton loop}
        For every element in the network, evaluate its discretized,
        linearized element stamp and construct the MNA system.
        Solve the resulting MNA system using LU factorization.
    endwhile
    if (Newton loop has timed-out) **then** {Need to reduce the time-step}
        $h_{n+1} = h_{n+1}/2$
    else
        Compute the LTE at t_{n+1}.
        if (LTE is too large) **then** {Need to reduce the time-step}
            $h_{n+1} = h_{n+1}/2$
        else {Accept the solution $x(t_{n+1})$ and move forward in time}
            $n = n + 1$
            if (LTE is too small) **then** {We can increase the time-step}
                $h_{n+1} = 2h_n$
            else {Keep the same time-step}
                $h_{n+1} = h_n$
            endif
        endif
    endif
    if (h_{n+1} is too small) **then**
        Abort! {Time-step too small}
    endif
endwhile

Figure 5.25: Overall simulation flow.

Notes   Additional reading is available in the following sources. In Lambert (1991), see chapters 1–6. In Ascher and Petzold (1998), see chapters 1–5 and 9–10. In Burden and Faires (2005), see chapters 3–5. In McCalla (1988), see chapter 5. In Ogrodzki (1994), see chapter 5. In Pillage et al. (1995), see chapters 4 and 10. Finally, much useful information is available in the classic texts Ruehli (1986) and Ruehli (1987).

The problem of transient analysis continues to be an interesting topic of research. While it is clear that the classical solutions are quite mature, it is also clear that they do not offer iron-clad guarantees of accuracy and stability. Besides the above general references on the topic, the research literature includes some noteworthy publications, including Lindberg (1971); Brayton et al. (1972); Van Bokhoven (1975); Ward and Dutton (1978); Yang et al. (1983); Hosea and Shampine (1996); Tischendorf (1996).

Problems

5.1. Show that FE, BE, and TR are consistent, zero-stable, and convergent.

5.2. Check whether the following LMS methods are convergent (after Lambert (1991)):

$$x_{n+1} + x_n - 2x_{n-1} = \frac{h}{4}\left[f(x_{n+1}, t_{n+1}) + 8f(x_n, t_n) + 3f(x_{n-1}, t_{n-1})\right]$$

$$x_{n+1} - x_n = \frac{h}{3}\left[3f(x_n, t_n) - 2f(x_{n-1}, t_{n-1})\right]$$

$$x_{n+1} + \frac{1}{4}x_n - \frac{1}{2}x_{n-1} - \frac{3}{4}x_{n-2} = \frac{h}{8}\left[19f(x_n, t_n) + 5f(x_{n-2}, t_{n-2})\right]$$

5.3. Write a program to solve the following IVP using FE, BE, and TR:

$$x'(t) = -\left(\frac{4}{3}\right)x(t), \qquad x(0) = 1$$

Generate and plot the solutions over the interval $0 \le t \le 10$, using three fixed time-step settings of $h = 0.5$, $h = 1.0$, and $h = 2.0$.

5.4. Write a program, using FE, BE, and TR to solve the following IVP:

$$x'(t) = \begin{bmatrix} 0.1 & -2 \\ 8 & 0.1 \end{bmatrix} x(t), \qquad x(0) = \begin{bmatrix} 0 \\ -2 \end{bmatrix}$$

whose exact solution is:

$$x(t) = \begin{bmatrix} e^{t/10}\sin(4t) \\ -2e^{t/10}\cos(4t) \end{bmatrix}$$

Use a time-step value of $h = 0.05$ and find the solutions over the interval $[0, 4]$.

5.5. Derive the expression for BDF2 in three ways:

(a) In the equidistant data case, using the polynomial basis $\{1, t, t^2, \ldots\}$.

(b) In the equidistant data case, using the Newton-Gregory backward interpolation polynomial.

(c) In the non-equidistant data case, using the Newton divided difference interpolation polynomial.

5.6. Show that, in the non-equidistant case, the general formula for a BDF method of order $p = k$ is given by:

$$f(x_{n+1}, t_{n+1}) = \sum_{i=1}^{k} \prod_{j=1}^{i-1} \left(\sum_{l=0}^{j-1} h_{n+1-l} \right) x[t_{n+1}, \ldots, t_{n+1-i}]$$

5.7. Consider the equidistant points $t_n, t_{n-1}, \ldots, t_{n-p}$, where the time-step is h, and let $x(t)$ be arbitrarily differentiable. If $\xi \in [t_{n-p}, t_n]$, and $\tau = h^{p+1} x^{(p+1)}(\xi) + 0(h^{p+2})$, show that $\tau = h^{p+1} x^{(p+1)}(t_n) + 0(h^{p+2})$.

5.8. Consider the non-equidistant data points $t_{n+1}, t_n, \ldots, t_{n-p}$, where $h_j \triangleq t_j - t_{j-1}$, and let $x(t)$ be arbitrarily differentiable. For any $\xi \in [t_{n-p}, t_{n+1}]$, and if $\tau = h_{n+1}^{p+1} x^{(p+1)}(t_n)$, show that $\tau = h_{n+1}^{p+1} x^{(p+1)}(\xi) + 0((t_{n+1} - t_{n-p})^{p+2})$.

5.9. For the case of TR with non-equidistant data, with $h_j \triangleq t_j - t_{j-1}$, derive a Milne's estimate for the LTE, in two steps:

(a) Show that the interpolation polynomial, extrapolated to provide an initial candidate solution $x_{n+1}^{(0)}$ at t_{n+1}, provides:

$$x_{n+1}^{(0)} = \left[\frac{h_{n+1}(h_{n+1} + h_n)}{h_{n-1}(h_n + h_{n-1})} \right] x_{n-2} - \left[\frac{h_{n+1}(h_{n+1} + h_n + h_{n-1})}{h_n h_{n-1}} \right] x_{n-1}$$
$$+ \left[\frac{(h_{n+1} + h_n)(h_{n+1} + h_n + h_{n-1})}{h_n(h_n + h_{n-1})} \right] x_n$$

(b) If such an extrapolation is used, followed by the application of Newton's method until convergence to solve the implicit TR for the solution x_{n+1}, show that the resulting Milne's estimate for the LTE would be this:

$$\text{LTE} \approx \frac{-h_{n+1}^2}{h_{n+1}^2 + 2(h_{n+1} + h_n)(h_{n+1} + h_n + h_{n-1})} \left(x_{n+1} - x_{n+1}^{(0)} \right)$$

5.10. For the single root of the stability polynomial for the trapezoidal rule, given by:

$$r_1 = \frac{1 + \hat{h}/2}{1 - \hat{h}/2}$$

show that:

(a) $|r_1| < 1$ if and only if $\Re(\hat{h}) < 0$.
(b) $|r_1| = 1$ if and only if $\Re(\hat{h}) = 0$.

5.11. Using the boundary locus method, and with $\theta \in [0, 2\pi]$, the boundary of the region of absolute stability of the trapezoidal rule is given by:

$$\hat{h} = 2\left(\frac{e^{i\theta} - 1}{e^{i\theta} + 1}\right)$$

Prove that $\Re(\hat{h}) = 0$ and $\Im(\hat{h}) = 2\tan(\theta/2)$.

5.12. Write a program to generate the region of absolute stability for the 5th order BDF, using the boundary locus method.

5.13. (Computer Project) Based on the nonlinear solver that was developed previously in problem 4.8, write a C or C++ implementation of a time-domain circuit simulator, based on the trapezoidal rule. Your implementation should be general, in the sense that it should accept any circuit description consisting of any combination of linear resistors, independent voltage and current sources, diodes, BJTs, MOSFETs, and linear capacitors and inductors. As in problem 4.8, you should use the simple DC models for the semiconductor devices given in chapter 4.

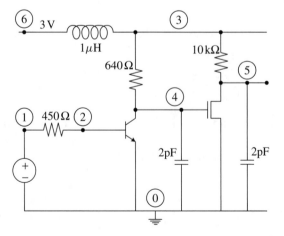

Figure 5.26: A test circuit for time-domain simulation.

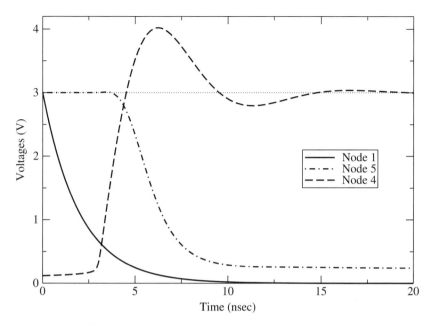

Figure 5.27: The solution for the circuit in Fig. 5.26.

Use your code to perform a time-domain simulation of the circuit in Fig. 5.26, based on $v_1(t) = 3e^{-t/\tau}$ Volts, where $\tau = 2$ nsec, over the interval $[0, 20\,\text{nsec}]$, and using the following parameters. For the MOSFET, $V_t = 0.6\,\text{V}$, $\lambda = 0.01/\text{V}$, and $\beta = 0.5\,\text{mA/V}^2$. For the BJT, $\alpha_F = 0.99$, $\alpha_R = 0.02$, $I_{es} = 2 \times 10^{-14}\,\text{A}$, $I_{cs} = 99 \times 10^{-14}\,\text{A}$, and $V_{Tc} = V_{Te} = 26\,\text{mV}$. For the Newton stopping criteria, use a relative tolerance of 0.1% and an absolute tolerance of 1 mV (for voltages) and $1\,\mu\text{A}$ (for currents). The overall flow of your solution should be as shown in Fig. 5.25, based on a minimum allowable time-step of 1psec. For the time-step control scheme in Fig. 5.25, you should use the following thresholds for the PLTE vector. To check if the PLTE is too large, use (5.413) based on a relative tolerance of 0.1% and an absolute tolerance of 1 mV (for voltages) and $1\,\mu\text{A}$ (for currents). To check if the PLTE is too small, use a relative tolerance of 0.001% and an absolute tolerance of $10\,\mu\text{V}$ (for voltages) and 10 nA (for currents). Generate a plot of the voltage waveforms at nodes 1, 4 and 5. The correct solution is shown in Fig. 5.27.

GLOSSARY

ABM: Adams-Bashforth-Moulton
AC: alternating current
AMD: average minimum degree
BBD: bordered block diagonal
BDF: backward differentiation formula
BDF2: second order BDF
BE: backward Euler
BJT: bipolar junction transistor
BS: backward substitution
BTF: block-triangular form
CCCS: current-controlled current source
CCS: controlled current source
CCVS: current-controlled voltage source
COLAMD: column average minimum degree
COLMMD: column multiple minimum degree
CVS: controlled voltage source
DAE: differential-algebraic equation
DC: direct current
DCCS: dynamic controlled current source
DCVS: dynamic controlled voltage source
DRAM: dynamic random access memory
ERO: elementary row operation
FE: forward elimination
FE: forward Euler
FET: field-effect transistor
FS: forward substitution
GE: Gaussian elimination

GEPP: GE with partial pivoting

GE/LU: shorthand, refers to both LU factorization and GE when implemented as in Gauss's method for LU factorization.

GJ: Gauss-Jacobi

GS: Gauss-Seidel

GTE: global truncation error

IEEE: institute of electrical and electronics engineers

IVP: initial value problem

KCL: Kirchoff's current law

KLU: "Clark Kent" LU factorization

KVL: Kirchoff's voltage law

LMS: linear multistep

LTE: local truncation error

MD: minimum degree

MMD: multiple minimum degree

MNA: modified nodal analysis

MOSFET: metal-oxide-semiconductor field-effect transistor

MTE: multiterminal element

NA: nodal analysis

ODE: ordinary differential equation

PLTE: principal local truncation error

PWL: piece-wise linear

RHS: right-hand side

SPD: symmetric positive definite

STA: sparse tableau analysis

SVD: singular value decomposition

TR: trapezoidal rule

VCCS: voltage-controlled current source

VCVS: voltage-controlled voltage source

VSVO: variable step variable order

P. Amestoy, T. A. Davis, and I. S. Duff. An approximate minimum degree ordering algorithm. *SIAM Journal on Matrix Analysis and Applications*, 17(4):886–905, October 1996.

U. M. Ascher and L. R. Petzold. *Computer Methods for Ordinary Differential Equations and Differential-Algebraic Equations*. SIAM, Philadelphia, PA, 1998.

R. G. Bartle. *The Elements of Real Analysis*. John Wiley & Sons, Ltd., New York, NY, second edition, 1976.

R. D. Berry. An optimal ordering of electronic circuit equations for a sparse matrix solution. *IEEE Transactions on Circuit Theory*, 18(1):40–50, January 1971.

R. K. Brayton, F. G. Gustavson, and G. D. Hachtel. A new efficient algorithm for solving differential-algebraic systems using implicit backward differentiation formulas. *Proceedings of the IEEE*, 60(1):98–108, January 1972.

R. L. Burden and J. D. Faires. *Numerical Analysis*. Thomson Books/Cole, Belmont, CA, eighth edition, 2005.

J. C. Butcher. Numerical methods for ordinary differential equations in the 20th century. *Journal of Computational and Applied Mathematics*, 125(1-2):1–29, December 2000.

D. A. Calahan. *Computer-Aided Network Design*. McGraw-Hill, Inc., New York, NY, revised edition, 1972.

L. O. Chua, C. A. Desoer, and E. S. Kuh. *Linear and Nonlinear Circuits*. McGraw-Hill Book Company, Inc., New York, NY, 1987.

L. O. Chua and P-M. Lin. *Computer-Aided Analysis of Electronic Circuits: Algorithms and Computational Techniques*. Prentice-Hall Inc., Englewood Cliffs, NJ, 1975.

G. Dahlquist and Å. Björck. *Numerical Methods in Scientific Computing*, volume One. SIAM, Philadelphia, PA, 2008.

T. A. Davis. *Direct Methods for Sparse Linear Systems*. SIAM, Philadelphia, PA, 2006.

T. A. Davis, J. R. Gilbert, S. I. Larimore, and E. G. Ng. A column approximate minimum degree ordering algorithm. *ACM Transactions on Mathematical Software*, 30(3):353–376, September 2004.

T. A. Davis and K. Stanley. Sparse LU factorization of circuit simulation matrices. In *SIAM Conference on Parallel Processing for Scientific Computing*, San Francisco, CA, February 2004.

J. E. Dennis, Jr. and R. B. Schnabel. *Numerical Methods for Unconstrained Optimization and Nonlinear Equations*. SIAM, Philadelphia, PA, 1996.

I. S. Duff, A. M. Erisman, and J. K. Reid. *Direct Methods for Sparse Matrices*. Oxford University Press, New York, NY, 1986.

D. Estévez Schwarz and C. Tischendorf. Structural analysis of electric circuits and consequences for MNA. *International Journal of Circuit Theory and Applications*, 28(2):131–162, March 2000.

C. W. Gear. *Numerical initial value problems in ordinary differential equations*. Prentice-Hall Inc., Englewood Cliffs, NJ, 1971.

A. George and J. Liu. The evolution of the minimum degree ordering algorithm. *SIAM Review*, 31(1):1–19, March 1989.

A. George and E. Ng. An implementation of Gaussian elimination with partial pivoting for sparse systems. *SIAM Journal on Scientific and Statistical Computing*, 6(2):390–409, April 1985.

J. R. Gilbert, C. Moler, and R. Schreiber. Sparse matrices in MATLAB: design and implementation. *SIAM Journal on Matrix Analysis and Applications*, 13(1):333–356, January 1992.

G. H. Golub and C. F. Van Loan. *Matrix Computations*. The Johns Hopkins University Press, Baltimore, MD, second edition, 1989.

G. D. Hachtel, R. K. Brayton, and F. G. Gustavson. The sparse tableau approach to network analysis and design. *IEEE Transactions on Circuit Theory*, 18(1):101–113, January 1971.

N. J. Higham. *Accuracy and Stability of Numerical Algorithms*. SIAM, Philadelphia, PA, second edition, 2002.

C.-W. Ho, A. E. Ruehli, and P. A. Brennan. The modified nodal approach to network analysis. *IEEE Transactions on Circuits and Systems*, 22(6):504–509, June 1975.

C. W. Ho, D. A. Zein, A. E. Ruehli, and P. A. Brennan. An algorithm for DC solutions in an experimental general purpose interactive circuit design program. *IEEE Transactions on Circuits and Systems*, 24(8):416–422, August 1977.

R. A. Horn and C. R. Johnson. *Matrix Analysis*. Cambridge University Press, New York, NY, 1985.

M. E. Hosea and L. F. Shampine. Analysis and implementation of TR-BDF2. *Applied Numerical Mathematics*, 20(1-2):21–37, February 1996.

T. E. Idleman, F. S. Jenkins, W. J. McCalla, and D. O. Pederson. SLIC—A simulator for linear integrated circuits. *IEEE Journal of Solid-State Circuits*, 6(4):188–203, August 1971.

C. T. Kelley. *Iterative Methods for Linear and Nonlinear Equations*. SIAM, Philadelphia, PA, 1995.

K. S. Kundert. *The Designer's Guide to SPICE and Spectre*. Kluwer Academic Publishers, Norwell, MA, 1995.

J. D. Lambert. *Numerical Methods for Ordinary Differential Systems: The Initial Value Problem*. John Wiley & Sons, Ltd., Chichester, UK, 1991.

B. Lindberg. On smoothing and extrapolation for the trapezoidal rule. *BIT Numerical Mathematics*, 11(1):29–52, March 1971.

H. M. Markowitz. The elimination form of the inverse and its application to linear programming. *Management Science*, 3(3):255–269, April 1957.

W. J. McCalla. *Fundamentals of Computer-Aided Circuit Simulation*. Kluwer Academic Publishers, Norwell, MA, 1988.

J. E. Meyer. MOS models and circuit simulation. *RCA Review*, 32(1):42–63, March 1971.

W. E. Milne. A note on the numerical integration of differential equations. *Journal of Research of the National Bureau of Standards*, 43(6):537–542, December 1949.

J.-M. Muller. *Elementary Functions—Algorithms and Implementation*. Birkhäuser, Boston, MA, second edition, 2006.

L. W. Nagel. *SPICE2: A Computer Program to Simulate Semiconductor Circuits*. PhD thesis, University of California, Berkeley, 1975. Memorandum No. ERL-M520.

L. W. Nagel and D. O. Pederson. Simulation program with integrated circuit emphasis. In *Proceedings of the Sixteenth Midwest Symposium on Circuit Theory*, Waterloo, Canada, April 12, 1973.

L. W. Nagel and R. A. Rohrer. Computer analysis of nonlinear circuits, excluding radiation (CANCER). *IEEE Journal of Solid-State Circuits*, 6(4):166–182, August 1971.

T. Nishi and L. O. Chua. Topological criteria for nonlinear resistive circuits containing controlled sources to have a unique solution. *IEEE Transactions on Circuits and Systems*, 31(8):722–741, August 1984.

J. Ogrodzki. *Circuit Simulation Methods and Algorithms*. CRC Press, Boca Raton, FL, 1994.

D. O. Pederson. A historical review of circuit simulation. *IEEE Transactions on Circuits and Systems*, 31(1):103–111, January 1984.

L. T. Pillage, R. A. Rohrer, and C. Visweswaraiah. *Electronic Circuit and System Simulation Methods*. McGraw-Hill Book Company, Inc., New York, NY, 1995.

W. H. Press, S. A. Teukolsky, W. T. Vetterling, and B. P. Flannery. *Numerical Recipes—The Art of Scientific Computing*. Cambridge University Press, New York, NY, third edition, 2007.

A. Ralston and P. Rabinowitz. *A First Course in Numerical Analysis*. Dover Publications, Inc., Mineola, NY, second edition, 2001.

G. Reißig. Extension of the normal tree method. *International Journal of Circuit Theory and Applications*, 27(2):241–265, March 1999.

A. E. Ruehli, editor. *Circuit Analysis, Simulation and Design—Part 1*. North-Holland, Amsterdam, 1986. Part 1 of 2, published as Volume 3 of *Advances in CAD for VLSI*.

A. E. Ruehli, editor. *Circuit Analysis, Simulation and Design—Part 2*. North-Holland, Amsterdam, 1987. Part 2 of 2, published as Volume 3 of *Advances in CAD for VLSI*.

Y. Saad. *Iterative Methods for Sparse Linear Systems*. SIAM, Philadelphia, PA, second edition, 2003.

M. Sipics. Sparse matrix algorithm drives SPICE performance gains. *SIAM News*, 40(4), May 2007. Available online at http://siam.org/news/news.php?id=1121.

H. J. Stetter. Asymptotic expansions for the error of discretization algorithms for nonlinear functional equations. *Numerische Mathematik*, 7(1):18–31, February 1965.

W. F. Tinney and J. W. Walker. Direct solution of sparse network equations by optimally ordered triangular factorization. *Proceedings of the IEEE*, 55(11):1801–1809, November 1967.

C. Tischendorf. *Solution of Index-2 Differential Algebraic Equations and its Application in Circuit Simulation*. PhD thesis, Humboldt University, Berlin, 1996.

W. M. G. Van Bokhoven. Linear implicit differentiation formulas of variable step and order. *IEEE Transactions on Circuits and Systems*, 22(2):109–115, February 1975.

J. Vlach and K. Singhal. *Computer Methods for Circuit Analysis and Design*. Van Nostrand Reinhold Co., New York, NY, second edition, 1994.

A. Vladimirescu. *The SPICE Book*. John Wiley & Sons, Ltd., New York, NY, 1994.

D. E. Ward and R. W. Dutton. A charge-oriented model for MOS transistor capacitances. *IEEE Journal of Solid-State Circuits*, 13(5):703–708, October 1978.

J. H. Wilkinson. *Rounding Errors in Algebraic Processes*. Prentice-Hall, Inc., Englewood Cliffs, NJ, 1963. Reprinted by Dover, New York, NY, 1994.

J. H. Wilkinson. *The Algebraic Eigenvalue Problem*. Oxford University Press, New York, NY, 1965. Reprinted, 1988.

P. Yang, B. D. Epler, and P. K. Chatterjee. An investigation of the charge conservation problem for MOSFET circuit simulation. *IEEE Journal of Solid-State Circuits*, 18(1):128–138, February 1983.

■■■■ INDEX

Circuit Simulation, by Farid N. Najm
Copyright © 2010 John Wiley & Sons, Inc.

311